T0297763

CAMBRIDGE LIBRARY COLLECTION

Books of enduring scholarly value

Life Sciences

Until the nineteenth century, the various subjects now known as the life sciences were regarded either as arcane studies which had little impact on ordinary daily life, or as a genteel hobby for the leisured classes. The increasing academic rigour and systematisation brought to the study of botany, zoology and other disciplines, and their adoption in university curricula, are reflected in the books reissued in this series.

The Natural History of Birds

Georges-Louis Leclerc, Comte de Buffon (1707–88), was a French mathematician who was considered one of the leading naturalists of the Enlightenment. An acquaintance of Voltaire and other intellectuals, he work as Keeper at the Jardin du Roi from 1739, and this inspired him to research and publish a vast encyclopaedia and survey of natural history, the ground-breaking *Histoire Naturelle*, which he published in forty-four volumes between 1749 and 1804. These volumes, first published between 1770 and 1783 and translated into English in 1793, contain Buffon's survey and descriptions of birds from the *Histoire Naturelle*. Based on recorded observations of birds both in France and in other countries, these volumes provide detailed descriptions of various bird species, their habitats and behaviours and were the first publications to present a comprehensive account of eighteenth-century ornithology. Volume 6 covers parrots, parakeets and other foreign birds.

The Natural History of Birds

From the French of the Count de Buffon

VOLUME 6

COMTE DE BUFFON
WILLIAM SMELLIE

CAMBRIDGE UNIVERSITY PRESS

Cambridge, New York, Melbourne, Madrid, Cape Town, Singapore,
São Paolo, Delhi, Dubai, Tokyo, Mexico City

Published in the United States of America by Cambridge University Press, New York

www.cambridge.org
Information on this title: www.cambridge.org/9781108023030

© in this compilation Cambridge University Press 2010

This edition first published 1793
This digitally printed version 2010

ISBN 978-1-108-02303-0 Paperback

THE

NATURAL HISTORY

OF

BIRDS.

FROM THE FRENCH OF THE

COUNT DE BUFFON.

ILLUSTRATED WITH ENGRAVINGS;

AND A

PREFACE, NOTES, AND ADDITIONS,

BY THE TRANSLATOR.

IN NINE VOLUMES.

VOL. VI.

LONDON:

PRINTED FOR A. STRAHAN, AND T. CADELL IN THE STRAND;
AND J. MURRAY, N⁰. 32, FLEET-STREET.

MDCCXCIII.

CONTENTS

OF THE

SIXTH VOLUME.

———————

CONTENTS.

The

CONTENTS.

A 3   *The*

CONTENTS.

The

CONTENTS.

The

CONTENTS.

The

CONTENTS.

CONTENTS.

5

The

CONTENTS.

VARIETIES

CONTENTS.

CONTENTS.

THE

FLY-BIRDS OF THE NATURAL SIZE.

THE

NATURAL HISTORY

OF

B I R D S.

***

The F L Y - B I R D.

L'Oifeau * *Moûche*, Buff.

OF all anim:ted beings, the Fly-bird is the
moft elegant in its form, and the moft
brilliant in its colours. The precious ftones
and metals polifhed by our art cannot be compared to this jewel of nature. Her miniature

* In Spanifh *Tomineios:* in Peruvian *Quinti* or *Quindé*, which
name obtains alfo in Paraguay : in Mexican *Huitzitzil* or *Hoitzitzil*,
Ourïffïa (fun-beam) : in Brazilian *Guianumbi*, which is generic.
It is alfo called *Viцililin* and *Guachichil* (flower-fucker) in Mexico.
Briffon terms it *Mellifuga* or honey-fucker; Linnæus *Trochilus*, or
little-top. In Englifh it is ufually known by the name of *humming-bird*. Mr. Pennant denominates it *honey-fucker*.

[The Mexican appellations of *Huitzitzil* and *Vicililin*, fignify *re-
generated*; which alludes to a notion entertained by the Indians
that in autumn this bird ftuck its bill into the trunk of a tree, and
remained infenfible during the winter months, till the vernal
warmth again waked it to animation, and invited it to its flowery
pafture. T.]

VOL. VI.   B   productions

productions are ever the moſt wonderful; ſhe
has placed it in the order of birds, at the bottom
of the ſcale of magnitude; but all the talents
which are only ſhared among the others, nim-
bleneſs, rapidity, ſprightlineſs, grace, and rich
decoration, ſhe has beſtowed profuſely upon this
little favourite. The emerald, the ruby, the
topaz, ſparkle in its plumage, which is never
ſoiled by the duſt of the ground. It inhabits
the air; it flutters from flower to flower; it
breathes their freſhneſs; it feeds on their nectar, and reſides in climates where they blow in
perpetual ſucceſſion.

It is in the hotteſt part of the new world
that all the ſpecies of Fly-birds are found. They
are numerous, and ſeem confined between the
two tropics *; for thoſe which penetrate in
ſummer within the temperate zones make but
a ſhort ſtay. They follow the courſe of the
ſun; with him they advance or retire; they
fly on the wings of the zephir, to wanton in
eternal ſpring.

The Indians, ſtruck with the dazzle and
glow of the colours of theſe brilliant birds,
have named them the *beams or locks of the ſun*†.
The Spaniards call them *tomineos*, on account
of their diminutive ſize, *tomine* ſignifying a
weight of twelve grains. " I ſaw," ſays Nieremberg, " one of theſe birds weighed with its

* Laet. *Ind. Occid.* Lib. V. 256.　　† Marcgrave.

neſt,

neft, and the whole together did not amount
to two tomines *." The fmaller fpecies do not
exceed the bulk of the great gad-fly, or the
thicknefs of the drone. Their bill is a fine
needle, and their tongue a delicate thread; their
little black eyes refemble two brilliant points;
the feathers of their wings are fo thin as to look
tranfparent †; hardly can the feet be perceived, fo fhort they are and fo flender: and thefe
are little ufed, for they reft only during the
night. Their flight is buzzing, continued, and
rapid; Marcgrave compares the noife of their
wings to the *whirr* of a fpinning-wheel: fo
rapid is the quiver of their pinions, that when
the bird halts in the air, it feems at once de-
prived of motion and of life. Thus it refts a
few feconds befide a flower, and again fhoots
to another like a gleam. It vifits them all,
thrufting its little tongue into their bofom, and
careffing them with its wings; it never fettles,
but it never quite abandons them. Its playful
inconftancy multiplies its innocent pleafures;
for the dalliance of this little lover of flowers
never fpoils their beauty. It only fips their honey, and its tongue feems calculated for that
purpofe: it confifts of two hollow fibres, forming a fmall canal ‡, parted at the end into two

---

* Nieremberg, p. 239. Acofta, Lib. IV. cap. 37.
† Marcgrave. ‡ Macrgrave.

filaments:

filaments* : it refembles the probofcis of infects,
and performs the fame office +.   The bird pro-
trudes it from its bill, probably by a mechanifm
of the *os hyoides*, fimilar to what obtains in the
tongue of wood-peckers.   It thrufts it to the
bottom of the flowers, and fucks their juices.
Such is its mode of fubfifting according to all
the authors who have written on the fubject ‡.
One perfon alone denies the fact; he is Ba-
dier §, who, finding in the œfophagus fome
portions of infects, concludes that the bird lives
on thefe, and not the nectar of flowers.   But
we cannot reject a number of refpectable au-
thorities for a fingle hafty affertion ; though the
Fly-bird fwallow fome infects, does it thence
follow that it fubfifts upon them ?   Nay, muft
it not neceffarily happen, that, fucking the honey from the flowers, or gathering their pollen, it will fometimes fwallow the little infects which are entangled ?   Befides, the rapid
wafte of its fpirits, the confequence of its extreme vivacity and its rapid inceffant motion,
muft continually be recruited by rich nutritious aliments: and Sloane, on whofe obfervations I lay the greateft ftrefs, pofitively avers

* Labat, t. IV. 13.
+ Natural Hiftory of Guiana, p. 165.
‡ Garcilaffo, Gomara, Hernandez, Clufius, Nieremberg, Marc-
grave, Sloane, Catefby, Feuillée, Labat, Dutertre, &c.
§ Journal de Phyfique, *Janvier* 1778, p. 32.

that

that he found the ſtomach of the Fly-bird entirely filled with the pollen, and ſweet juice of
flowers *.

Nothing can equal the vivacity of theſe little creatures, but their courage, or rather audacity; they furiouſly purſue birds twenty times
larger than themſelves, fix in the plumage, and
as they are hurried along ſtrike keenly with the
bill, till they vent their feeble rage†: ſometimes
even they fight obſtinately with each other.
They are all impatience; if upon alighting in a
flower they find it faded, they will pluck the
petals with a precipitation that marks their diſ-
pleaſure. Their voice is only a feeble cry *ſcrep*,
ſcrep, which is frequent and reiterated‡. They
are heard in the woods at the dawn of the morning §, and as ſoon as the ſun begins to gild the
ſummits of the trees, they take wing and diſ-
perſe in the fields.

They are ſolitary‖; and indeed, fluttering
irregular in the breeze, they could hardly aſſo-
ciate. But the power of love ſurmounts the
elements, and, with its golden chains, it binds
all animated beings. The Fly-birds are ſeen to
pair in the breeding ſeaſon; their neſt correſ-

* Nat Hiſt. Jamaica, p. 307.

† Browne, p. 475; Charlevoix, *Nouvelle France*, t. III. p. 158;
Dutertre, *t*. II. *p* 263.

‡ Marcgrave compares this note, for its continuance, to that of
the ſparrow, *p*. 196.

§ Marcgrave, *p*. 196.

‖ Philoſophical Tranſactions, *No*. 200, *art*. 5.

B 3 ponds

ponds to the delicacy of their bodies; it is formed with the foft cotton or filky down gathered
from flowers, and has the confiftency and feel
of a thick fmooth fkin. The female performs
the work, and the male collects the materials*.
She applies herfelf with ardour; felects, one by
one, the fibres proper to form the texture of this
kindly cradle for her progeny; fhe fmooths the
margin with her breaft, the infide with her tail;
fhe covers the outfide with bits of the bark of
the gum tree, which are ftuck to fhelter from
the weather, and give folidity to the fabric † :
the whole is attached to two leaves, or a fingle
fprig of the orange or citron ‡, or fometimes
to a ftraw hanging from the roof of an hut §.
The neft is not larger than the half of an apri-
cot ||, and it is alfo fhaped like a half cup. It
contains two eggs, which are entirely white,
and not exceeding the bulk of fmall peafe. The
cock and hen fit by turns twelve days ; on the
thirteenth the young are excluded, which are
then not larger then flies. "I could never
perceive," fays Father Dutertre, "how the
mother fed them, except that fhe prefented the
tongue covered entirely with honey extracted
from flowers."

We may eafily conceive that it is impoffible
to raife thefe little flutterers. Thofe who have

* Dutertre, t. II, p. 262.   † Id. Ibid.
‡ Browne.   § Dutertre.
|| Feuillée *Journal d'Obfervations*, t, I. p. 413.

tried

tried to feed them with fyrups could not keep
them alive more than a few weeks; thefe aliments, though of eafy digeftion, are very different from the delicate nectar collected from
the frefh bloffoms. Perhaps honey would have
fucceeded better.

The method of obtaining them is to fhoot with
fand, or by means of the *trunk-gun*; they will
allow one to approach within five or fix paces
of them *. They may be caught by placing a
twig fmeared over with a clammy gum in a
flowering fhrub. It is eafy to lay hold of the
little creature while it hums at a bloffom. It
dies foon after it is caught †, and ferves to
decorate the Indian girls, who wear two of
thefe charming birds, as pendants from their
ears. The Peruvians had the art of forming
their feathers into pictures, whofe beauty is
perpetually extolled in the older narratives ‡.
Marcgrave, who faw fome of thefe pieces of
workmanfhip, admires their brilliancy and delicacy.

With the luftre and glofs of flowers, thefe
pretty birds have been fuppofed to have alfo the
perfume ; and many authors have afferted that

* They are fo numerous, fays Marcgrave, that a fowler may
eafily take fixty in a day.

† Dutertre and Marcgrave.

‡ *See* Ximenes, who attributes the fame art to the Mexicans :
Gemelli Carreri, Thevet, Lery, Hern.mdez, &c.

they

they have the fragrance of mufk. The miftake
originated probably from the name applied by
Oviedo, of *paſſer moſquitus*, which would eaſily
be changed into *paſſer moſcatus* *. But this is
not the only marvellous circumſtance with
which their hiſtory has been clouded †; it has
been ſaid that they are half birds, half flies, and
produced from a fly ‡ ; and a Provincial of the
Jeſuits gravely affirms in Cluſius, that he was
witneſs to this transformation §. It has been
alledged that during the winter ſeaſon they remain torpid, ſuſpended by the bill from the
bark of a tree, and awakened into life when
the flowers begin to blow. Theſe fictions
have been rejected by intelligent naturaliſts ||;
and Cateſby aſſures us, that he ſaw them
through the whole year at St. Domingo and
Mexico, where nature never entirely loſes her
bloom ⊥. Sloane ſays the ſame of Jamaica,
only that they are more numerous after the
rainy ſeaſon; and prior to both, Marcgrave

* Geſner very juſtly remarks that this epithet is derived rather
from *muſca* (a fly), than from *moſchus* (the name in modern Latin
for mufk.)

† Dutertre corrects very judiciouſly many puerile exagger-
ations, and detects, as uſual, the miſtakes of Rochefort, *t.* II.
p. 263.

‡ *See* Nieremberg, *p.* 240.

§ This Jeſuit, ſays Cluſius, made ſtrange relations in natural
hiſtory. *Exotic, p.* 96.

|| See *Willoughby.*

⊥ Nat. Hiſt. of Carolina, *Vol.* I. *p.* 65.

mentioned

mentioned their being frequent the whole year
in the woods of Brazil.

We are acquainted with twenty-four fpecies in the genus of the Fly-bird; and it is probable fome have been overlooked. We fhall
diftinguifh them by their different denominations, drawn from the moft obvious characters.

The LEAST FLY-BIRD.

Le plus petit Oiseau-Mouche, Buff.

FIRST SPECIES.

Trochilus Minimus, Linn. Gmel. and Klein.
Mellisuga, Briss.
Guainumbi septima species, Marcq.
Guainumbi minor, corpore toto cinereo, Ray.
Polythmus minimus variegatus, Brown.
The *Least Humming-bird,* Sloane, Edw. and Lath.

IT is congruous to begin with the smallest species, in enumerating the smallest genus. This Least Fly-bird is scarce fifteen lines in length; its bill is three and a half, its tail four: so that there remains only nine lines for the head, the neck, and the body. It is smaller, therefore, than some of our flies. All the upper side of the head and body is of a gold green changing brown, and with reddish reflections; all the under side is of a white grey. The feathers of the wing are brown, inclining to violet, and this is the general colour of the wings in all the Fly-birds, as well as in the colibris. The bill also and the feet are commonly black, the legs are clothed pretty low with little downy plumules; and the toes are furnished with little sharp curved nails. All of them have six feathers in the tail; Marcgrave mentions only four, which

3   is

is probably a miftake of the tranfcriber. The
colour of thefe tail-feathers is, in moft of the
fpecies, bluifh black, with the luftre of burnifhed fteel. In the female the colours are generally not fo bright; it is acknowledged too
by the beft obfervers to be rather fmaller than
the male *. The bill of the Fly-bird is equally
thick throughout, flightly fwelled at the tip,
compreffed horizontally and *ftraight*. This laft
character diftinguifhes the Fly-bird from the colibris, which moft naturalifts, and even Marcgrave, have confounded.

This firft and leaft fpecies is found in Brazil
and the Antilles. The bird was fent to us from
Martinico with its neft; Edwards received it
from Jamaica.

* Grew in the Philofophical Tranfactions, *No.* 200, *art.* 5.—
Labat, Dutertre.

[A] Specific character of the *Trochilus Minimus:* " Its lateral
tail-feathers white at their outer edge, its body of fhining green,
below whitifh." It weighs between twenty and forty-five grains.

The R U B Y.

Le Rubis, Buff.

SECOND SPECIES.

Trochilus-Colubris, Linn. Gmel. and Klein.
Mellisuga Carolinensis gutture rubro, Briss.
Tomineo Virescens, gutture flammeo, Petiv.
Guainumbi, Ray. and Will.
The *American Tomineius* or *Humming-bird,* Catesby and
    Kalm.
The *Red-throated Honey-sucker,* Penn.
The *Red-throated Humming-bird,* Edw. and Lath.

IN observing the scale of magnitude, many
would occupy the second place. We shall
take the Carolina Fly-bird, and denominate it
the *Ruby.* Catesby feebly expresses the lustre
and beauty of the colour of its throat, when he
calls it a *crimson enamel:* it has the brilliancy
and fire of the ruby. In a side view, it has a
gold tinge, and, seen from below, it appears a
dull garnet. We may remark that the feathers
of the throat are fashioned and disposed like
scales, round and detached; which arrangement multiplies the reflections, that play both
on the neck and the head of the Fly-birds,
among all their sparkling feathers. In the present, all the upper side of the body is gold-
green, changing into red copper; the breast
and the fore part of the body are mixed with
white,

white, grey, and blackifh; the two feathers in the middle of the tail are of the colour of the back, and the lateral feathers are purple-brown; Catefby fays *copper colour*. The wing is brown, tinged with violet, which, as we have already obferved, is the common colour of the wings in all thefe birds; fo that we may omit them in the fubfequent defcriptions. The form of the wings is fingular: Catefby compares it to the blade of a Tuɹkifh fcimeter. The four or five firft outer quills are long, the next much lefs fo, and thofe neareft the body are extremely fmall; which, joined to another circumftance, that the largeft are curved outwards, makes the two wings when fpread refemble a drawn bow, of which the little body of the bird reprefents the arrow in the middle.

The Ruby appears in fummer in Carolina, and even in New England; it is the only Fly-bird that penetrates into the northern provinces [*]. Some narratives tranfport it to Gaf-pefia [†], and Charlevoix fays that he faw it in Canada. But he appears little acquainted with it when he fays, that the bottom of its neft *is interwoven with fmall bits of wood, and that it lays five eggs* [‡] ; and in another place that *its*

[*] Catefby and Edwards.

[†] *Nouvelle relation de la Gafpefie,* par le R. P. Chretien le Clerque, *Paris,* 1691, *p.* 486. The Gafpefians, according to this account, call it *virido,* bird of heaven.

[‡] Hift. and Defcrip. de la Nouv. France, *Paris,* 1744, *t.* III. 158.

feet

*feet are like its bill, very long**. Little ſtreſs
can be laid on ſuch evidence. The winter retreat is ſaid to be in Florida†; it breeds in Carolina in ſummer, and departs when the flowers begin to fade. It extracts its nouriſhment
from the flowers only; " and I have always
obſerved," ſays Cateſby, " that it never feeds
on inſects, but entirely on honey-juices ‡."

* Hiſt. de St. Domingue, *Paris*, 1730, p. 31.
† *See* Hiſt. Gen. des Voy. *t*. XIV. *p*. 456.
‡ Carolina, Vol. I. p. 65.

[A] Specific character of the *Trochilus-Colubris:* " It is gold-
green; its tail-feathers black, the three lateral ones ferruginous
tipt with white; its throat flame coloured." It is three inches and
one third long.

The AMETHYST, Buff.

Trochilus Amethyſtinus, Gmel.
The *Amethyſtine Humming-bird*, Lath.

THIRD SPECIES.

ALL the throat and the fore part of the neck
are of a brilliant amethyſt, which it is
impoſſible to figure or paint. This is one of
the ſmalleſt of the Fly-birds; its ſize and form
the ſame with thoſe of the Ruby; its tail is alſo
forked. The fore ſide of the body is marbled
with white, grey, and brown; the upper ſide
is gold-green; the amethyſt colour of the throat
changes into purple brown, when the eye is
placed lower than the objeƈt. The wings ſeem
rather ſhorter than in the other Fly-birds, and
reach not the two middle feathers of the tail,
which are however the ſhorteſt, and give it a
forked ſhape.

[A] Specific charaƈter of the *Trochilus Amethyſtinus*: " It is
gold-green, below variegated with aſh and brown, its throat amethyſtine." It is found in Cayenne.

The GOLD GREEN.

L'Orvert, Buffon.

FOURTH SPECIES.

Trochilus Viridissimus, Gmel.
The *All-green Humming-bird,* Lath.

GREEN and gold yellow sparkle more or less in all the Fly-birds; but these fine colours cover the whole plumage of this, with a brilliancy and glofs which the eye cannot enough admire. In certain positions it is pure dazzling gold; in others, it is a glazed green, which is not inferior to the lustre of polished metal. These colours extend over the wings; the tail has the black hue of burnished steel *.

To this we shall refer the *All-green Humming-bird* of Edwards. We shall also refer the second species of Marcgrave; its singular beauty, its short bill, and the dazzle of gold, and of brilliant and resplendent green, distinguish it sufficiently. Brisson makes this his sixteenth species, under the name of the *Forked-tail Bra-*

---

* Specific character: " It is very green, its belly white, its tail steel-coloured.'

zilian

zilian Honey-fucker * ; but he was not aware that Marcgrave reprefents its tail neither long nor forked. Its *tail is like the former*, fays that author; and in the firft fpecies the tail is *ftraight, only an inch long*, and does not exceed the wing.

The TUFTED-NECK.

Le Hupecol, Buff.

FIFTH SPECIES.

Trochilus Auratus, Gmel.
Trochilus Ornatus, Lath. Ind.
The *Tufted-necked Humming-bird*, Lath. Syn.

THIS name marks a very fingular character, which diftinguifhes this bird from all the reft. Not only its head is ornamented with a pretty long rufous tuft, but on each fide of the neck, below the ears, rife feven or eight unequal feathers; the two longeft, being fix or feven lines, are rufous, and narrow throughout,

---

* *Trochilus Glaucopis*, Gmel.
Trochilus Frontalis, Lath.
Mellifuga Brafilienfis caudâ bifurcâ, Briff.
Guainumbi Major, Ray and Will.
The *Blue fronted Humming-bird*, Lath. Syn.

Specific character: " It is gold-green; its front fky-blue, its vent white, its wing-feathers violet-brown, its feet feathered, its tail fteel coloured, and fomewhat forked."

but the ends a little widened, and tipt with a green dot. The bird erects them reclining them back; while at reft they lie flat on the neck, as alfo does the beautiful tuft: but they are all briftled when it flies, and the bird appears quite round. The throat and the fore fide of the neck are of a rich gold green (if the eye be held much lower than the object, thefe brilliant feathers appear entirely brown); the head and all the upper fide of the body green, with dazzling reflections of gold and bronze, as far as the white bar that croffes the rump; beyond this, to the end of the tail, is fpread a fhining gold on a brown ground on the outer webs of the quills, and rufous on the inner ones; the under fide of the body is gold-green brown; the lower belly, white. The Tufted-neck does not exceed the fize of the Amethyft; the female refembles it, except that it has no tuft or external ears: the bar of the rump is rufty, and fo is the throat; the reft of the under fide of the body rufous, fhaded with greenifh; its back and the upper fide of the head are as in the male, green with gold and bronze reflections.

The RUBY TOPAZ, *Buff.*

SIXTH SPECIES.

Trochilus Mofchitus. Linn. and Gmel.
Mellifuga Brafilienfis gutture topazino, Briff.
The *Trochilus with a ferruginous tail,* Bancroft.
The *Ruby-necked Humming-bird,* Lath.

OF all the birds of the genus, this is the moft beautiful, fays Marcgrave, and the moft elegant: it has colours and the fparkling fire of the Ruby and the Topaz : the upper fide of its head and neck is as brilliant as a ruby; the throat, all the fore fide of the neck, in the front view, dazzle like the Aurora Topaz of Brazil. The fame parts, feen a little lower, refemble unburnifhed gold, and ftill lower, change into a dull green ; the top of the neck and the belly are of a velvet black brown; the wing is violet brown; the lower belly white ; the inferior coverts of the tail and its quills are of a fine gold-rufous, and tinged with purple; it is edged with brown at the end; the rump is brown, heightened with gold-green ; the wings, when clofed, do not extend beyond the tail, whofe quills are equal. Marcgrave remarks that it is broad, and that the bird difplays it gracefully in flying. It is pretty large for its kind. Its total length is three inches and from four to fix lines ; its bill is feven or eight; Marcgrave calls this *half an*

inch.

inch. This beautiful fpecies feems numerous, and has become common in the cabinets of naturalifts: Seba fays that he received many of them from Curaçoa. We may notice a character which all the Fly-birds and Colibris have, viz that the bill is thick feathered at its bafe, and fometimes as far as the fourth or third of its length.

The female has only a ftreak of gold or topaz, on the throat and fore-part of the neck; the reft of the under fide of the body is white-gray [A].

We conceive that the Fly-bird reprefented, No. 640, fig. 1 *, *Planches Enluminées*, is a proximate fpecies, or perhaps the fame with this; for the only difference confifts in its having a creft, but which is not much raifed. In other refpects, the refemblance is ftriking; and, from a comparifon of the figures, the latter appears rather fmaller and its colours not fo deep, though the tints and diftributions are effentially the

[A] Specific character of the *Trochilus Mofchitus:* " It is gold-green; its tail-feathers equal and ferruginous, the outermoft tipt with brown; its wing-feathers black."

* *Trochilus Elatus*, Gmel.
  Mellifuga Americana gutture topazino, Briff.
  Florifuga aut Mell fuga, Seba.
  The *Ruby crefted Humming bird*, Edw. and Lath.

Specific character: " It is gold-greenifh, its tail-feathers equal and ferruginous, tipt with black; a red crefted cap."

fame:

fame: fo that the one feems to be young, the other adult. Or perhaps it is a variety of climate; fince the one comes from Cayenne, the other from Brazil. The *Ruby-crefted Humming-bird* given in Edwards's Gleanings correfponds exactly with the above-mentioned coloured figure. Frifch has alfo given the head of this Fly-bird, *pl.* 24, on which Briffon has formed his fecond fpecies, taking, for the female, another figure inferted by Frifch in the fame place, and which reprefents a Little Goldgreen Fly-bird. But the female of the Topazbreafted Fly-bird, whofe body is brown, cannot furely be green. In this, as in every other genus of birds, the colours of the female are always duller than thofe of the male. We may, therefore, with the higheft probability, refer the *fecond all-green Fly-bird* of Frifch to the gold-green.

The CRESTED FLY-BIRD.

L'Oiseau-Mouche Huppé, Buff.

SEVENTH SPECIES.

Trochilus Cristatus, Linn. Gmel. and Borowsk.
Mellisuga Cristata, Briss. and Klein.
The *Green strait-billed Humming-bird*, Bancroft.
The *Crested Green Humming-bird*, Lath.

DUTERTRE and Feuillée have taken this bird
for a *Colibri* ; but it is one of the smallest
of the Fly-birds, since it does not exceed the
Ruby. Its crest resembles the most brilliant
emerald ; and this distinguishes it, for the rest
of its plumage is dull. The back has green and
gold reflexions, on a brown ground; the wing
is brown; the tail blackish, and shining like
polished steel; all the fore-side of the body is velvet brown, mixed with a little gold-green near
the shoulders; the wing, when closed, does
not exceed the tail. The under side of the bill
is covered with little green brilliant feathers as
far as the middle. Edwards has delineated the
nest. Labat observes that the female has no
crest.

[A] Specific character of the *Trochilus Cristatus* : " It is green,
its wings brown, its belly brown cinereous, its crest bluish, its legs
feathered."

The RACKET FLY-BIRD.

L'Oiseau-Mouche a Raquettes, Buff.

EIGHT SPECIES.

Trochilus Longicaudus, Gmel.
Trochilus Platurus, Lath. Ind.
The *Racket-tailed Humming-bird,* Lath. Syn.

Two naked shafts, extending from the two middle feathers of the tail, are terminated with little fans, which gives them the form of rackets. The ribs of all the quills of the tail are very thick, and of a rusty white; the rest is brown, like the wings. The upper side of the body is of a bronze green, which is the colour common to all the Fly-birds; the throat is of a rich emerald-green. The point of the bill is about thirty lines from the end of the true tail; the two shafts extend ten lines farther. This species is not well known, and seems very rare. We have described it from a specimen in Mauduit's cabinet. It is one of the smallest Fly-birds, and, exclusive of the tail, it exceeds not the Tufted-neck.

[A] Specific character of the *Trochilus Longicaudus:* " It is gold green, its throat emerald; its wings, and its tail-feathers, brown, the two mid-ones very long."

The PURPLE FLY-BIRD.

L'Oiseau-Mouche Pourpré, Buff.

NINTH SPECIES.

Trochilus Ruber, Linn. and Gmel.
Mellisuga Surinamensis, Briss.
Mellisuga Alis Fuscis, Klein.
The *Little Brown Humming-bird*, Edw. Banc. and Lath.

ALL the plumage of this bird is a mixture of orange, purple, and brown; and it is, perhaps, as Edwards observes, the only one of the genus that has not the gold-green on the back. Klein has therefore discriminated it imperfectly by the epithet of *brown-winged*; since brown, with more or less of violet and purple, is the general colour of the Fly-birds. The bill is ten lines, which is nearly one third of its length.

[A] Specific character of the *Trochilus Ruber*: " Its lateral tail feathers are violet; its body of a brown brick-colour, somewhat spotted."

The GOLD CRAVAT,

TENTH SPECIES.

Trochilus Leucogafter, Gmel.
Mellifuga Cayanenfis ventre albo, Briff.
Guainumbi prima fpecies, Ray and Will.
The *Larger Humming-bird*, Sloane.
The *Gold-throated Humming-bird*, Lath.

THIS feems to be the firft fpecies of Marc-grave; for it has a gold ftreak on the throat, which that author thus defcribes, " the fore-fide of the body is white, mixed under the neck with fome feathers of a fhining colour." Briffon omits that circumftance in his eighth fpecies, though it is formed upon the defcrip-tion of Marcgrave's firft. Its length is three inches and five or fix lines; all the under fide of the body, except the gold ftreak on the fore fide of the neck, is white-grey, and the upper fide gold-green [B]. We fhall reckon Briffon's ninth fpecies* the female of this, there being no material difference between them.

[B] Specific charaﬅer of the *Trochilus Leucogafter:* " It is gold-green, below white, its legs feathered."

* *Trochilus-Pegafus*, Gmel.
Mellifuga Cayanenfis, ventre grifeo, Briff.
The *Grey-bellied Humming bird*, Lath.

It is thus defcribed by Briffon: "Above gold-green, varying with a pure copper colour; the feathers of the tail gold-green on their firft-half, varying with a pure copper colour, and dark purple on their other half, the lateral ones tipt with grey; the feet feathered."

The S A P P H I R E.

Le Saphir, Buff.

ELEVENTH SPECIES.

Trochilus Saphirinus, Gmel.
The *Sapphire Humming-bird,* Lath.

IT is rather above the middle fize; the fore-
fide of the neck and breaft is of a rich fap-
phire-blue, with violet reflections; the throat
is rufous; the upper and under fides of the body
dull gold-green; the lower belly white; the inferior coverts of the tail rufous; the fuperior ones
of a fhining gold-brown; the quills of the tail
are gold-rufous, edged with brown; thofe of
the wings brown; the bill is white, except the
point, which is black.

The EMERALD-SAPPHIRE, *Buff.*

TWELFTH SPECIES.

Trochilus Bicolor, Gmel.
The *Sapphire and Emerald Humming-bird,* Lath.

THE two rich colours which decorate this
bird defervedly confer upon it the names
of thofe precious ftones. A fapphire blue covers the head and throat, and melts admirably
8   into

into the glazed emerald green, with gold reflexions that cover the breaft, the ftomach, the
circle of the neck, and the back. The bird is
middle fized; it comes from Guadeloupe, and,
we believe, has not hitherto been defcribed.
We have feen another, brought from Guiana,
of the fame bulk; but it had not the fapphire
throat, and the reft of its body was of a very
brilliant glazed green. Both thefe are depofited with the firft in the excellent cabinet of
Mauduit. The laft appears to be a variety, or
at leaft a fpecies nearly related to the firft. In
both, the lower belly is white; the wing is
brown, and exceeds not the tail, which is cut
equally and rounded: it is black, with blue reflexions; their bill is pretty long, its lower half
whitifh, and upper black.

[A] Specific charaƈter of the *Trochilus Bicolor:* " It is gold-
emerald; its head and throat fky-blue."

The AMETHYST EMERALD.

THIRTEENTH SPECIES.

Trochilus-Ouriffia, Linn. and Gmel.
Mellifuga Surinamenfis pectore cœruleo, Briff.
The *Green and Blue Humming-bird,* Edw. and Lath.

THIS Fly-bird is above the middle fize; it is
near four inches long, and its bill is eight
lines. Its throat and the fore part of its neck
are

are emerald green, brilliant and golden; its
breaſt, its ſtomach, and the top of its back, are
purple blue amethyſt of the utmoſt beauty: the
lower part of the back is gold-green, on a brown
ground; the belly is white; the bill blackiſh;
the tail velvet black, ſhining like poliſhed ſteel.
To the ſame ſpecies we may refer the *Green and
Blue Humming-bird* of Edwards, and the *Blue-
breaſted Surinam Honey-ſucker* of Briſſon. It is
figured rather larger in Edwards [B].

The CARBUNCLE.

L'Eſcarboucle, Buff.

FOURTEENTH SPECIES.

Trochilus Carbunculus, Gmel.
The *Carbuncle Humming-bird,* Lath.

A CARBUNCLE red, or deep ruby, is the colour of the throat and breaſt; the upper
ſide of the head and neck is of a duller red; a
velvet black envelopes the reſt of the body; the
wing is brown, and the tail of a deep gold-rufous. The bird is rather above the middle ſize;
the bill, both above and below, is beſet with
feathers, through almoſt one half of its length.

[B] Specific charaƈter of the *Trochilus Ouriſſia:* " It is golden-
green, the feathers of its tail ſomewhat equal and gold-brown,
the feathers of the wings black, its belly blue."

It

It was fent from Cayenne, and feems to be very rare. Mauduit, in whofe poffeffion it is, would refer it as a variety to the *Topaz-ruby*; but the difference between the topaz-yellow and the deep ruby on the throat of thefe two birds, feems too great to admit this claffification. In all other refpects, they are very fimilar.—The preceding fpecies, except the thirteenth, are new, and not defcribed by any naturalift [A].

The GOLD-GREEN, *Buff.*

FIFTEENTH SPECIES.

Trochilus Mellifugus, Linn. and Gmel.
Guainumbi Nona Species, Ray and Will.
Mellifuga Cayanenfis, Briff.
The *Cayenne Humming-bird,* Lath.

THIS is the ninth fpecies of Marcgrave; the whole body, fays he, is of a brilliant green, with gold reflections; the upper mandible is black, the lower rufous; the wing is brown; the tail pretty broad, and fhines like polifhed fteel. The total length of the bird exceeds fomewhat three inches. The under fide of the body has not fo much green as the back, and is

[A] Specific character of the *Trochilus-Carbunculus:* " It is black; its head, neck, and breaft, red; its wings brown; its tail gold-rufous."

only

only marked with fpots or waves of that colour.
The female is rather fmaller, as ufual in this
tribe of birds [A].

THE
SPOTTED-NECKED FLY-BIRD.

L'Oifeau-Mouche a Gorge Tachetée, Buff.

SIXTEENTH SPECIES.

Trochilus Fimbriatus, Gmel.
Mellifuga Cayanenfis gutture nævio, Briff.
The *Spotted necked Humming-bird*, Lath.

THIS fpecies is much related to the preceding. It is larger, and, but for that difference, we fhould have affigned it the fame
place. Briffon fays that it is four inches long,
and its bill eleven lines. Its plumage is exactly
like that of the preceding *.

[A] Specific character of the *Trochilus Mellifugus*: " It is gold-
green; its tail-feathers equal and blue, its wing-feathers dark
bluifh; its legs feathered."

* Specific character: " It is gold-green, below gray; its
tail fteel coloured, tipt with gray; the feathers of its breaft fringed
with white."

The EMERALD RUBY, *Buff.*

SEVENTEENTH SPECIES.

Trochilus Rubineus, Lath. Ind.
Mellisuga Brasiliensis gutture rubro, Briss.
The *Ruby-throated Humming-bird*, Lath. Syn.

THIS is much larger than the Carolina Ruby, being four inches four lines in length;
its throat is of a sparkling ruby, or, in certain
positions, rose colour; its head, its neck, the
anterior and upper parts of its body, emerald
green, with gold reflections; the tail is rufous.
It is found both in Brazil and in Guiana [A].

The EARED FLY-BIRD.

L'Oiseau-Mouche a Oreilles, Buff.

EIGHTEENTH SPECIES.

Trochilus Auritus, Gmel.
Mellisuga Cayanensis Major, Briss.
The *Violet-eared Humming-bird*, Lath.

WE apply the epithet *eared* to this Fly-
bird, both on account of the remarkable
colour of the two pencils of feathers, which

[A] Specific character of the *Trochilus Rubineus:* " It is gold-
green, its throat gold-red, its wings and tail rufous."

extend

extend behind the ears, and on account of
their great length, which is twice or thrice
that of the fmall adjoining feathers that cover
the neck.   They feem only the production
of what, in all birds, cover the *meatus audi-
torius*; they are foft, and their downy fibres
not glued together.   Thefe are the remarks of
Mauduit, and well agree with his ingenious obfervation, which we formerly had occafion to
mention, viz. that all the feathers which appear fuperabundant, or, fo to fpeak, parafite,
in birds are not peculiarities of ftructure, but
merely the extenfion and developement of parts
common to all the others.   The Eared Fly-
bird is of the firft magnitude, being four inches
and a half long.   Of the two pencils which diftinguifh the ears, and which confift each of
five or fix feathers, the one is emerald-green,
and the other amethyft-violet; a ftreak of velvet
black ftretches under the eye ; all the fore part
of the head and body is of a bright gold-green,
which changes on the coverts of the tail into
a very lively bright green ; the throat and under fide of the body are of a fine white ; of the
tail quills, the fix lateral ones are of the fame
white, the four mid-ones black, inclining to
deep blue ; the wing is blackifh, and the tail
projects beyond it nearly one-third of its length!
In the female, the pencils and the black ftreak
under the eye are lefs diftinct ; in other refpects
it refembles the male.

The COLLARED FLY-BIRD,

Called the *Jacobine*.

NINETEENTH SPECIES.

Trochilus Mellivorus, Linn. Gmel. and Browſk.
Melliſuga Surinamenſis Torquata, Briſſ.
The *White-bellied Humming-bird,* Edw. and Lath.

THIS Fly-bird is of the firſt magnitude ; it is four inches eight lines in length ; its bill ten lines ; its head, throat, and neck, of a fine obſcure blue, gloſſed with green ; on the back of the neck, and near the back, is a white half collar ; the back is gold-green ; the tail white at the end, and edged with black ; its two middle quills, and their coverts, gold-green ; the breaſts and ſides the ſame ; the belly white. It is probable, on account of this diſtribution, it has been called *Jacobine.* The two middle feathers of the tail are ſhorter than the reſt, and the wing, when cloſed, does not project beyond it. The ſpecies is found at Cayenne and Surinam.

[A] Specific character of the *Trochilus Mellivorus :* " Its tail-feathers are black, the lateral ones white ; the head blue ; the back green ; the belly white."

The BROAD-SHAFTED FLY-BIRD.

L'Oiseau-Mouche a Larges Tuyaux, Buff.

TWENTIETH SPECIES.

Trochilus Campylopterus, Gmel.
Trochilus Latipennis, Lath. Ind.
The *Broad-shafted Humming-bird,* Lath. Syn.

THIS bird and the preceding are the two
largest of the genus. The present is four
inches eight lines long; all the upper side of
the body is of a faint gold-green; the under
side grey; the middle feathers of the tail are
like those of the back; the lateral ones white
at the tip, the rest of a brown, resembling polished steel. It is easily distinguished from the
other Fly-birds by the protuberance of three
or four great wing-quills, whose shafts appear
swelled and dilated, bent near the middle,
which gives the wing the shape of a broad
sabre. This species is new and apparently rare,
and has not hitherto been described. We saw
the specimen in the cabinet of Mauduit, who
received it from Cayenne.

[A] Specific character of the *Trochilus Campylopterus:* " It is
gold-green; below grey; its lateral tail-feathers brown, tipt with
white; the shafts of three or four of the middle feathers of the
wings curved in the middle.''

The LONG-TAILED STEEL-CO-LOURED FLY-BIRD. *Buff.*

TWENTY-FIRST SPECIES.

Trochilus Macrourus, Gmel.
Trochilus Forcipatus, Lath. Ind.
Mellifuga Cayanenfis cauda bifurca, Briff.
The *Cayenne fork-tailed Humming-bird,* Lath. Syn.

THE beautiful violet blue, which covers the head, throat, and neck, would feem to indicate an analogy to the fapphire, did not length of the tail exhibit too great a difference. The two exterior quills are two inches longer than the two mid-ones; the lateral ones continually diminifh, which makes the tail very much forked. The bird is dark blue, gliftening like burnifhed fteel; all the body, both above and below, is of a fhining gold-green; there is a white fpot on the lower belly; the wings, when clofed, reach only to the middle of the tail, which is three inches and three lines; the bill is eleven lines, and the total length is fix inches. The entire refemblance between this defcription and that which Marcgrave gives of his third fpecies, convinces us that they are the fame, contrary to the opinion of Briffon, who makes it his twentieth fpecies. But the third fpecies of Marcgrave has *a tail more than three inches long*; whereas the twentieth Honey-

D 2   fucker

sucker of Briffon has it only *an inch and six lines*: and this is too wide a difference to occur in the same species. We shall consider the bird of Briffon in the following article [A].

The FORKED-TAIL VIOLET FLY-BIRD.

L'Oiseau-Mouche Violet a Queue Fourchue, Buff.

TWENTY-SECOND SPECIES.

Trochilus Furcatus.
Mellifuga Jamacienſis Violacea cauda bifurca, Briff.
The *Leſſer fork-tail Humming-bird*, Lath.

BESIDES the difference of size, which, as we have already remarked, obtains between this and the preceding species, there is also a difference of colours. The upper parts of the head and neck are brown, gloffed with gold-green, whereas these glisten with blue in Marcgrave's third species. In the present, the back and breaft are of a shining violet blue; in that of Marcgrave they are gold-green. The throat and the lower part of the back are brilliant gold-green; the small coverts below the wings

[A] Specific character of the *Trochilus Macrourus:* " It is gold-green, its head and throat violet, its belly marked with a white space, its tail forked and steel-coloured."

are

are of a fine violet, the great ones gold-green;
their quills black: thofe of the tail the fame;
the two exterior ones are the longeft, which
makes it forked; it is only an inch and half
long; the bird meafures four inches.

The LONG-TAIL FLY-BIRD,
Of Gold, Green, and Blue. *Buff.*

TWENTY-THIRD SPECIES.

Trochilus Forficatus, Lath. Gmel. and Browſk.
Falcinellus vertice caudaque cyaneis, Klein.
Mellifuga Jamaicenfis cauda bifurca, Briſſ.
The *Long-tailed Green Humming-bird,* Edw.
The *Fork-tailed Humming-bird,* Lath.

THE two exterior feathers of the tail of this
Fly-bird are near twice as long as the body,
and project above four inches. Thefe feathers,
and all thofe of the tail, of which the two
middle ones are very fhort, and not exceeding eight lines, are wonderfully beautiful and
mingled, fays Edwards, with reflections of
green and of gold blue; the body is green; the
wing is purple brown.—This fpecies occurs in
Jamaica.

[A] Specific character of the *Trochilus Forficatus:* " It is green,
the lateral feathers of the tail very long, its cap and its tail feathers blue."

D 3

The BLACK LONG-TAILED FLY-BIRD. *Buff.*

TWENTY-FOURTH SPECIES.

Trochilus-Polytmus, Linn. and Gmel.
Falcinellus cauda septem unciarum, Klein.
Mellisuga Jamaicensis Atricapilla cauda bifurca, Briss.
The *Long tailed Humming-bird*, Albin.
The *Long-tailed Black-cap Humming-bird*, Edw. & Ban.
The *Black-capped Humming bird*, Lath.

THIS Fly-bird has a longer tail than any of the rest; the two great feathers are four times as long as the body, which is fcarcely two inches; thefe are alfo the two outermoft; their webs confift of parted downy fibres, and they are black like the crown of the head; the back is gold brown-green; the forefide of the body green; the wings purple-brown. Albin's figure is a very bad one, and he was much miftaken in fuppofing this to be the fmalleft fpecies in the genus; though he fays, that he found it in Jamaica in its neft, which confifted of cotton [A].

We find in the Effay on the Natural Hiftory of Guiana, mention of a little Humming-bird

---

[A] Specific character of the *Trochilus Polytmus:* " It is greenifh, the lateral feathers of its tail very long, its cap and tail feathers brown."

with

with a blue creſt. We are unacquainted with
it; and the account of it, and indeed of two
or three others, is inſufficient to aſcertain their
ſpecies. We may, however, be convinced that
the genus of theſe handſome birds is ſtill richer
and more multiplied in nature than we have delineated it.

The COLIBRI*.

WHEN nature beftowed beauty fo lavifh-
ly on the fly-birds, fhe neglected not
their kindred tribe, the Colibris. Both inha-
biting the fame climate, fafhioned after the fame
model, and decorated by the fame brilliancy of
plumage : the fame vivacity, the fame perpetual flutter of action, and the fame habits and
economy. As their refemblance is fo entire,
they have often been confounded under the fame
name : that of *Colibri* is adopted from the language of the Caribbees. Marcgrave applies to
both indifferently the Brazilian appellation,
Guainumbi. But they are diftinguifhed by an
obvious and permanent character : in the Colibris the bill is equal and taper, inflated flightly near the end, and not ftraight, as in the fly-
birds, but curved throughout, and longer alfo
in proportion. Further, the neat and flender
form of the Colibris feems to be more length-
ened than that of the fly-birds ; and they are
in general larger : yet there are fome little Colibris fmaller than the great fly-birds. The

*+In the Brazilian language, the Fly-bird and the Colibri have
the common name of *Guainumbi:* in Guiana, the Colibri is called
in the dialect of Garipana *Toukouki:* and, according to Seba, certain tribes of Indians term it *Ronckjes.*

Colibris

COLIBRIS, OF THE NATURAL SIZE.

Colibris fhould be ranged below the creepers, though they differ in the fhape and length of their bill; in the number of the feathers of their tail, there being ten in the former and twelve in the latter; and in the ftructure of their tongue, which is fimple in the latter, but in the former divided into two femi-cylindrical portions, as in the fly-bird.

All naturalifts agree that the Colibris and fly-birds have the fame manner of living. It has, indeed, been denied that either of thefe tribes feed on the honey of flowers *. But the rea-fons already adduced convince us that this af-fertion is unfounded; and the general refem-blance of thefe birds corroborates the evidence that their mode of fubfifting is the fame.

It is no lefs difficult to breed the young of the Colibri than thofe of the fly-bird; they are as delicate, and confinement proves equally fa-tal to them. The parents have been feen, hur-ried on by the audacity of affection, to rufh with food for their progeny into the very hands of the plunderer. Labat relates an inftance of this, which deferves to be quoted. " I fhow-ed," fays he, " to Father Montdidier a neft of Colibris, which was placed on a fhed near the houfe. He carried it off with the young, when they were about fifteen or twenty days old, and put them in a cage at his room window, where

* Journal de Phyfique, *Janvier* 1778.

the

the cock and hen continued to feed them, and grew fo tame, that they fcarcely ever left the room; and though not fhut in the cage, nor fubjeƈted to any reftraint, they ufed to eat and fleep with their brood. I have often feen all the four fitting upon Father Montdidier's finger, finging, as if they had been perched upon a branch. He fed them with a very fine and almoft limpid pafte, made with bifcuit, Spanifh wine, and fugar. They dipt their tongue in it, and when their appetite was fatisfied they fluttered and chanted . . . . I never faw any thing more lovely than thofe four pretty little birds, which flew about the houfe, and attended the call of their fofter-father *."

Marcgrave, who does not difcriminate the Colibris from the fly-birds, mentions them as having only a feeble cry, and no travellers afcribe fong to them. Thevet and Lery alone affert of their *gonambouch* that it chants fo as to rival the nightingale †; for it is from them that

---

* " He preferved them in this way five or fix months, and we hoped foon to fee them breed, when Father Montdidier, having one night forgotten to tie the cage in which they roofted by a cord that hung from the ceiling, to keep them from the rats, had the vexation in the morning to find that they were difappeared; they had been devoured." Labat, *Nouveau Voyage aux Iles de l'Amerique.* Paris, 1722, t. IV. p. 14.

† " But, as a fingular curiofity, and as a mafter-piece of littlenefs, we muft not omit a bird which the favages call *gonambouch*, of a whitifh and fhining plumage, which, though not larger than a hornet, excels in fong; infomuch that this diminutive creature,

fcarce

that *Coreal* and some others have repeated the same. But it is most likely a mistake; the gonambouch, or little bird of Levy, which has a *whitish shining plumage, and a clear distinct voice*, is the *sugar bird*, or some other, and not the Colibri, whose notes form, according to Labat, only a sort of pleasant hum.

It does not appear that the Colibris advance so far into North America as the fly-birds; at least, Catesby says that he saw only one species of these in Carolina. And Charlevoix, who pretends that he found a fly-bird in Canada, confesses that he never saw there a Colibri *. Yet it is not the cold that prevents it from visiting that province in the summer, since it seeks a cool temperature at a considerable height among the Andes. M. de la Condamine never saw Colibris more numerous than in the gardens of Quito †, where the climate is not hot. They prefer, therefore, a warmth of twenty or twenty-one degrees ‡ : there, in a perpetual round of pleasures and joys, they fly from the

---

scarce stirring from the great millet, which the Americans name *avati*, or other great plants, has its bill and throat always open. If one did not repeatedly see and hear, he would hardly be persuaded that from so slender a body could proceed notes so clear, so liquid, and so loud, as not to yield to those of the nightingale." *Voyage au Bresil, par Jean de Lery.* Paris, 1578, p. 175. The same fact is mentioned by Thevet. *Singularités de la France Antartique.* Paris, 15 8 p. 94.

* Hist. de Saint Domingue. *Paris,* 1730, *t.* I. *p.* 32.
† Voy. de la Condamine. *Paris,* 1745, *p.* 171.
‡ *i. e.* 77° or 79° of Farenheit.

expanded

expanded bloffom to the opening bud, and where
the harmonious year for ever invites them, by
its enchanting mildnefs, to love and fruition.

The TOPAZ COLIBRI, *Buff.*

FIRST SPECIES.

Trochilus Pella, Linn. and Gmel.
Polytmus Surinamenfis Longicaudus Ruber, Briff.
Falcinellus gutture viridi, Klein.
The *Long-tailed Red Humming-bird*, Edw.
The *Topaz Humming-bird*, Lath.

As fmallnefs was the moft ftriking character
of the fly-birds, we began with the fmalleft : but that property, not being fo confpicuous in the Colibris, we fhall refume the natural
order of magnitude. The Topaz appears, ex-
clufive of the two long fhafts that extend from
its tail, to be the largeft of the genus; we
fhould alfo call it the moft beautiful, did not all
thefe brilliant birds rival each other, and be-
wilder the imagination amid the blaze of their
charms. Its form is delicate, flender, elegant,
and rather fmaller than the common creeper,
its total length, from the point of the bill to the
end of the true tail, being near fix inches ; the
two long fhafts project two inches and a half
beyond it ; the throat, and the fore fide of the
                                        neck,

neck, decorated by the moſt brilliant topaz
mark; that colour viewed obliquely changes
into gold-green, and from below it appears pure
green; a hood of ſoft black covers the head,
a thread of the ſame black incloſes the topaz
mark; the breaſt, the neck, the top of the
back, are of a finer deep purple; the belly is
of a ſtill richer purple, and dazzling with red
and gold reflections; the ſhoulders and the
lower part of the back, are orange rufous; the
great quills of the wing, violet-brown; the little quills, rufous; the colour of the ſuperior and
inferior coverts of the tail, gold-green; the lateral quills rufous, the two middle ones, purple brown; theſe project into two long ſhafts,
which are webbed with a ſmall edging a line
broad on each ſide; theſe long ſhafts, in their
natural poſition, croſs each other a little beyond
the tail, and then diverge; they drop in moulting, and the male to which they belong would
then reſemble the female, were he not diſcri-
minated by other characters. The female has
not the topaz breaſt, but only a ſlight trace of
red; and in place of the fine purple and flame
rufous of the male's plumage, almoſt all that
of the female is gold-green: in both the feet
are white.

[A] Specific character of the *Trochilus Pella:* " It is red, its
middle tail-feathers very long, its head brown, its throat golden,
and its rump green."

The G A R N E T.

Le Grenat, Buff.

SECOND SPECIES.

Trochilus Auratus var, Gmel.

THE cheeks as far as under the eye, the fides
and lower part of the neck and throat to
the breaft, are of a fine brilliant garnet ; the
upper fide of the head and back, and the under
fide of the body, are of a foft black ; the tail
and wings of the fame colour ; but ornamented
with gold-green.   The bird is five inches long,
and the bill ten or twelve lines.

The W H I T E S H A F T.

Le Brin Blanc, Buff.

THIRD SPECIES.

Trochilus Superciliofus, Linn. and Gmel.
Polytmus Cayanenfis Longicaudus, Briff.
The *Supercilious Humming-bird,* Lath.

OF all the Colibris, this has the longeft bill,
which is twenty lines ; the feathers of the
tail, next the two long fhafts, are alfo the
longeft, and the lateral ones continually de-
creafe, to the two outermoft, which are the
fhorteft,

shortest, and this gives the tail a pyramidical shape; its quills have a gold gloss on a grey and blackish ground, with a whitish edge at the point, and the two shafts are white through the whole projecting portions; all the upper side of the back and head, gold colour; the wing violet-brown; and the under side of the body white-gray [A].

ZITZIL, or DOTTED COLIBRI.

Le Zitzil, ou Colibri Piqueté, Buff.

FOURTH SPECIES.

Trochilus Punctulatus, Gmel.
Polytmus Punctulatus, Briss.
Hoitzitziltototl, Fernandez.
The *Spotted Humming-bird*, Lath.

ZITZIL is contracted for *Hoitzitzil*, which is the Mexican name of this bird. It is pretty large; its wings blackish, marked with white points on the shoulders and back; the tail is brown, and white at the tip. This is all we can gather from an ill-written description of Hernandez' editor *. He subjoins that he got his information from one Father Aloaysa; and

[A] Specific character of the *Trochilus Superciliosus*: " It is glossy brown; its middle tail-feathers very long; its belly somewhat flesh-coloured; its eye-brows white."
* *Jo. Fab. Linceus.*

that

that the Peruvians call the fame bird *pilleo*,
and that living upon the juice of flowers, it
prefers that of the thorny tribes *.

The BLUE SHAFT.

Le Brin Blue, Buff.

FIFTH SPECIES.

Trochilus Cyanurus, Gmel.
Polytmus Mexicanus Longicaudus, Briff.
Yayauquitototl, Seba and Klein.
The *Blue-tailed Humming-bird*, Lath.

ACCORDING to Seba, whom Klein and Briffon have followed in reckoning this a fpecies of Colibri, the two long projections of feathers which decorate its tail are of a fine blue;
the fame colour, only deeper, covers the ftomach and fore part of the head ; the upper fide
of the body and of the wings is light green; the
belly cinereous. It is one of the largeft Colibris, and almoft equal to the epicurean warbler.
Seba's figure reprefents it as a creeper, and that
author feems to have never obferved the three

* In another part of his work, Hernandez gives the names of
feveral fpecies of fly-birds and colibris, without characterizing
any : thefe names are, *Quetzal Hoitzitzillin*, *Zochio Hoitzitzillin*,
Xiulks Hoitzitzillin, *Tozcacoz Hoitzitzillin*, *Yotac Hoitzitzillin*, *Te-
noc Hoitzitzillin*; whence it appears that *Hoitzitzillin* is the generic name.

fhades

fhades in the form of the bill which difcriminate thefe three tribes, the fly-birds, the colibris,
and the creepers. Nor is he more fortunate in
difplaying his erudition ; he applies to this Colibri the Mexican name *yayauhquitototl*, which,
in Fernandez, denotes a bird of the fize of a
ftare. But fuch errors are trifling in comparifon of thofe into which naturalifts are led by
the collectors of curiofities, who value nothing
but the glitter of their cabinets. To find an
inftance we need not ftep afide : Seba mentions
Colibris from the Moluccas, from Macaffar,
and from Bali, not knowing that this tribe of
birds is peculiar to the new world. Briffon
copies the miftake, and defcribes three fpecies
of *Colibris from the Eaft Indies*. Thefe are undoubtedly creepers, the brilliancy of whofe colours, and the names *tfioei* and *kakopit*, which
Seba tranflates *little kings of flowers*, have fug-
gefted the Colibri. No traveller acquainted
with natural hiftory has found Colibris in the
old continent ; and what Francis Cauche fays of
the fubject, is too obfcure to merit attention *.

* In his account of Madagafcar, *Paris*, 1651, *p.* 137, borrow-
ing the name and the habits of the Colibri, he afcribes them to a
little bird of this ifland. It is probably by a fimilar abufe of names,
that *fly-bird* occurs in the voyages of the Company, applied to a
bird of the Coromandel coaft, which is indeed very fmall, and is
elfewhere called *tati*. *Recueil de Voyages qui ont fervi a l'établiffe-
ment de la Compagnie des Indes*. Amfterdam, 1702, t. VI. p. 513.

[A] Specific character of the *Trochilus Cyanurus*: " It is green,
below cinereous ; its front, its throat, and the two middle feathers
of the tail longer than the reft, and blue."

The GREEN and BLACK COLIBRI.

SIXTH SPECIES.

Trochilus Holofericeus, Linn. Gmel. and Borowſk.
Polytmus Mexicanus, Briſſ.
Avis Auricoma Mexicana, Klein.
The *Black-bellied Humming-bird*, Edw. Bancr. and Lath.

I T is rather more than four inches long; its
bill thirteen lines; its head, neck, and back,
are gold colour and bronze; the breaſt, the belly, the ſides of the body, and the legs, are ſhining black, with a light reddiſh reflection; a little white bar croſſes the lower belly, and another of gold-green, gliſtening with lively blue,
interſects tranſverſely the top of the breaſt; the
tail is velvet black, with the blue gloſs of poliſhed
ſteel.   It is ſaid that the female may be diſtinguiſhed in this ſpecies by the want of the white
ſpot on the lower belly.   The bird is found both
in Mexico and in Guiana.   Briſſon refers to
this ſpecies the *Avis auricoma Mexicana* of Seba,
which is indeed a Colibri; but his deſcription is
ſo vague and indefinite, as to apply equally to
them all.

[A] Specific character of the *Trochilus Holofericus:* "It is
green; the quills of its tail equal, and black above; a blue bar on
the breaſt; its belly black."

The TUFTED COLIBRI, *Buff.*

SEVENTH SPECIES.

Trochilus Paradifeus, Linn. Gmel. and Borowfk.
Polytmus Mexicanus Longicaudus ruber criftatus, Briff.
The *Paradife Humming-bird,* Lath.

BRISSON finds this alfo in Seba's catalogue. I am generally averfe to form fpecies on the indications, fo often defective, of that compiler; but the characters of the prefent feem fufficiently diftinct to be adopted. " This little bird," fays Seba, " has a fine red plumage, blue wings; two long feathers project from the tail; and on its head there is a tuft which is very long in proportion to its thicknefs, and falls back on the neck ; the bill is long and curved, including a fmall *bifid* tongue, which ferves to fuck the flowers."

Briffon meafuring Seba's figure, which is not of much account, found near five inches fix lines to the end of the tail.

[A] Specific character of the *Trochilus Paradifeus:* " It is red, its wings blue, its head crefted; its middle tail-feathers very long."

THE

VIOLET-TAILED COLIBRI, *Buff*.

EIGHTH SPECIES.

Trochilus Albus, Gmel.
Trochilus Nitidus, Lath. Ind.
The *Violet-tailed Humming bird*, Lath. Syn.

THE bright pure violet which paints the tail
of this Colibri, difcriminates it from the
reft ; the four middle feathers of the tail are of
a violet colour, melted under brilliant reflections of gold-green ; the fix outer ones, viewed
from below, prefent a white point, with a violet fpot that furrounds a fpace of dark blue like
burnifhed fteel ; all the under-fide of the body
is richly gilded in the front view, and when
held obliquely it appears green ; the wing, as
in all thefe birds, is brown, verging on violet ;
the fides of the throat are white, and, in the
middle, there is a longitudinal ftreak of brown,
mixed with green ; the fides are coloured with
the fame ; the breaft and belly are white. This
fpecies is pretty large, it being five inches ; and
has one of the longeft bills, which is fixteen
lines.

[A] Specific charaƈter of the *Trochilus Albus*: " It is gold-
green ; its under furface, the fides of the neck, and the tips of the
fix outer tail feathers, white ; its tail violet."

7

GREEN-THROATED COLIBRI, *Buff.*

NINTH SPECIES.

Trochilus Maculatus, Gmel.
Trochilus Gularis, Lath. Ind.
The *Green-throated Humming-bird,* Lath. Syn.

A STREAK of very bright emerald-green is
traced on the throat of this Colibri,
which falls, spreading on the fore-side of the
neck ; there is a black spot on the breast ; the
sides of the throat and neck are rufous, mixed
with white ; the belly is pure white ; the upper side of the body, and of the tail, dull gold-
green ; below the tail, are the same violet,
white and burnished steel spots, as in the *Violet-
tailed Colibri*.   These two species appear analogous, and they are of the same size, but the
bill of the Green-throated Colibri is not so long.
We saw in Mauduit's cabinet a Colibri of the
same dimensions, with the upper side of the
body faintly tinged with green and gold on a
blackish grey ground, and all the fore-part of
the body rufous, which seems to us the female.

THE
CARMINE-THROATED COLIBRI,
Buff.

TENTH SPECIES.

Trochilus Jugularis, Linn. and Gmel.
The *Red-breasted Humming-bird*, Edw. and Lath.

IT is four inches and a half in length ; its bill
thirteen lines, much curved, and therefore
analogous to that of the creepers, as Edwards
remarks; the throat, the cheeks, and all the
fore-part of the neck, carmine red, with a ru-
by-luftre; the upper fide of the head, body,
and tail, of a foft blackifh brown, with a flight
fringe of blue on the edge of the feathers ; a
deep gold-green fhines on the wings ; the inferior and fuperior coverts of the tail are of a
fine blue.   This bird was brought from Surinam into England.

[A] Specific character of the *Trochilus Jugularis*: " It is bluifh,
its tail feathers equal, its neck below blood-coloured."

The VIOLET COLIBRI, *Buff.*

ELEVENTH SPECIES.

Trochilus Violaceus, Gmel.
Polytmus Cayanenfis Violaceus, Briff.
The *Violet Humming-bird*, Lath.

IT is four inches and two lines in length; its
bill eleven lines; the whole head, neck,
and belly, covered with purple violet, which is
brilliant on the throat and on the fore-fide of
the neck, and diluted on all the reft of the
body with a mixture of velvet black; the wing
is gold green; the tail the fame, with a changing reflection of black. It is found in Cayenne; its colours refemble thofe of the *garnet*
Colibri; but the difference of fize is too great
to admit of their being claffed together.

[A] Specific character of the *Trochilus Violaceus:* " It is violet, its wings and tail gold-green."

E 4

The GREEN GORGET.

Le Hauſſe-Col Vert, Buff.

TWELFTH SPECIES.

Trochilus Gramineus, Gmel.
The *Black-breaſted Humming-bird*, Lath.

IT is rather larger than the Violet-tailed Colibri, but its bill is not ſo long; all the fore-
part and ſides of the neck, with the lower part
of the throat, emerald green; the top of the
throat, or the ſmall portion beneath the bill, of
a bronze colour; the breaſt velvet black, tinged with dull blue; green and gold appear on the
flanks, and cover all the upper ſide of the body;
the belly white; the tail purple blue, with the
reflection of burniſhed ſteel, and exceeds not
the wing. We conceive the female to be another Colibri of the ſame ſize and diſtribution
of colours, except that the green, on the fore-
part of the neck, is interſected by two white
ſtreaks, and that the black, on the throat, is
neither ſo broad nor ſo deep. Theſe two birds
are in the admirable ſeries of Colibris and Fly-
birds in Dr. Mauduit's cabinet.

[A] Specific character of the *Trochilus Gramineus:* " It is gold-
green, below white; its throat emerald; its breaſt black; its tail
purple."

The **RED COLLAR**, *Buff*.

THIRTEENTH SPECIES.

Trochilus Leucurus, Linn. and Gmel.
Polytmus Surinamenfis, Briff.
The *White-tailed Humming bird*, Lath.

THIS is of the middle fize, being four inches and five or fix lines in length; on the lower and fore-part of the neck, there is a handfome red half collar, of confiderable breadth; the back, the neck, the head, the throat, and the breaft, are of a bronze and gold green; the two middle feathers of the tail are of the fame colour; the eight others are white, and this is the character by which Edwards difcriminates the bird.

[A] Specific character of the *Trochilus Leucurus*: " It is gold-green, its tail feathers equal, its collar red."

The BLACK PLASTRON.

FOURTEENTH SPECIES.

Trochilus Mango, Linn. and Gmel.
Guainumbi minor, rostro nigro, Ray and Will.
Polytmus Jamaicensis, Briss.
The *Mango Humming-bird*, Lath.

THE throat, the fore-side of the neck, the breast, and the belly of this Colibri, are of the most beautiful velvet black; a streak of brilliant blue rises from the corners of the bill, and, descending over the sides of the neck, separates the black plastron, or breast-piece, from the rich gold-green, with which all the under surface of the body is covered; the tail is of a purple brown, glossed with shining violet, and each quill is edged with the blue of burnished steel. These colours resemble those of Marcgrave's fifth species, only the bird is rather smaller; it is four inches long; the bill one inch; the tail eighteen lines. It is found equally in Brazil, in St. Domingo, and in Jamaica.

[A] Specific character of the *Trochilus Mango*: " It is glossy green; its tail-feathers somewhat equal and ferruginous; its belly black."

The WHITE PLASTRON.

FIFTEENTH SPECIES.

Trochilus Margaritaceus, Gmel.
The *Grey-necked Humming-bird,* Lath.

ALL the under fide of the body, from the throat to the lower belly, is white pearl gray; the upper fide of the body is gold-green; the tail is white at the tip, then croffed by a bar of black burnifhed fteel, and after that by one of purple brown; and it is black with a blue fteel caft at its origin. It is four inches long, and its bill an inch.

The BLUE COLIBRI, *Buff.*

SIXTEENTH SPECIES.

Trochilus Venuftiffimus, Gmel.
Trochilus Cyaneus, Lath. Ind.
Polytmus Mexicanus Cyaneus, Briff.
The *Crimfon-headed Blue Humming-bird,* Lath. Syn.

IT is ftrange that Briffon, who never faw this bird, fhould follow the vague, inaccurate account of Seba, inftead of the defcription of Dutertre. The wings and tail are not blue, as Briffon reprefents, but black, as Father Duter-tre mentions, and indeed according to the analogy

logy of all the birds of this tribe. The whole of the back is azure; the head, the throat, and the fore-part of the body, as far as the middle of the belly, are velvet crimfon, which, if held in different pofitions, is enriched with a thoufand beautiful reflections. Dutertre only adds, that it is about *half the fize of the little crowned wren.* The figure of Seba, which Briffon feems to take, reprefents a creeper [A].

The PEARL GREEN.

SEVENTEENTH SPECIES.

Trochilus Dominicus, Linn. and Gmel.
Polytmus Dominicenfis, Briff.
The *St. Domingo Humming-bird,* Lath.

THIS is one of the fmalleft of the tribe, and hardly exceeds the crefted fly-bird; all the upper fide of the head, body, and tail, are of a faint gold-green, which is intermixed, on the fides of the neck, and more and more on the throat, with pearl white-gray; the wing is brown, as in the reft, and tinged with violet; the tail is white at the end, and of the co-lour of polifhed fteel below [B].

[A] Specific character of the *Trochilus Venuftiffimus:* " It is red; its back blue; its wings black."
[B] Specific character of the *Trochilus Dominicus:* " It is fhin-ing green, below fomewhat cinereous; its tail-feathers ferrugi-nous in the middle, and white at the tips."

The RUSTY BELLIED COLIBRI.

EIGHTEENTH SPECIES.

Trochilus Hirſutus, Gmel.
Polytmus Braſilienſis, Briſſ.
Guainumbi minor, roſtro incurvo, Ray and Will.
The *Rufous-bellied Humming-bird,* Lath.

THIS is the fourth ſpecies of Marcgrave, and muſt be very ſmall, ſince he ſays that it is inferior to the third, which he had formerly ſtated as the leaſt. All the upper ſide of the body is gold-green; all the under ſide ruſty blue; the tail is black, with green reflections, and the point is white; the lower mandible is yellow at its origin, and black to the extremity; the feet are yellowiſh white.

The LITTLE COLIBRI, *Buff.*

NINETEENTH SPECIES.

Trochilus-Thaumantias, Linn. and Gmel.
Guainumbi minor toto corpore aureo, Ray and Will.
Polytmus, Briſſon.
Melliſuga Ronckje dicta, Klein.
Avicula Americana Colubritis, Seba.
The *Admirable Humming-bird,* Lath.

THIS is the laſt and ſmalleſt of all the Co-libris; it is only two inches and ſix
lines

lines in length; its bill eleven lines, and its
tail twelve or thirteen; it is entirely gold-
green, except the wing, which is violet or
brown: there is a ſmall white ſpot on the
lower belly, and a ſmall border of the ſame
colour on the feathers of the tail, broader on
the two outer ones, which it half covers.
Marcgrave again ſtops to admire the brilliant
plumage with which nature has decked theſe
charming birds.   The little Colibri in particular, he obſerves, dazzles like the ſun*.

* *In ſummâ ſplendet ut ſol.*

[A] Specific charaƈter of the *Trochilus Thaumantias:* " It is
ſhining green; its tail-feathers equal and fringed with white, the
outermoſt white exteriorly."

The P A R R O T*.

Le Parroquet, Buff.

THE animals which man has the moſt ad-
mired, are thoſe that ſeem to participate of his nature. He is ſtruck with wonder
as often as he traces his external form in the
ape, or hears his voice imitated by the Parrot ;
and, in firſt moments of his ſurpriſe, he is diſpoſed to rank them above the reſt of the brutes.
Theſe animals have fixed even the ſtupid attention of ſavages, who behold the magnificent
ſcene of nature and her exquiſite productions
with the moſt perfect inſenſibility : they ſtop
the progreſs of their canoes, and linger gazing
whole hours at the capers of the marmoſet.
Parrots are the only birds which they are fond
of raiſing and educating, and which they are even
at pains to improve ; for they have diſcovered
the art, which is ſtill unknown to us, of vary-

* In Greek Ψιϑλακη; in modern Greek Παπαγας; in Latin
Pſittacus. In German the Parrot is called *Pappengey,* the Parakeet *Sittick,* or *Sickuſt :* in Spaniſh the Parrot is named *Popagio* ;
in Italian *Papagallo,* and the Parakeet *Peroquetto :* in Poliſh *Papu-*
ga : in Turkiſh *Dudi :* In Mexican *Tuznene :* in Brazilian *Ajuru,*
and the Parakeet *Tui.* In old French *Papegaut.* According to
Aldrovandus, moſt of theſe names are derived from *Papa,* and denote *the pope of the birds.*

ing

ing and heightening the colours that deck the plumage *.

The power of ufing the hand, and of walk-ing on two feet, the refemblance, how faint foever, to the face, the want of a tail, the naked hams ; the fimilarity of the fexual parts, the pofition of the breafts, and the menftrual flux in the females ; the ardent paffion of the males for women : all thefe circumftances have procured to the ape the name of *wild man* from thofe who themfelves are indeed only half-men, and who can compare only the exterior charac-ters.   Had what was equally poffible taken place, had the voice of the Parrot been beftow-ed on the ape ; the human race would have been ftruck dumb with aftonifhment, and the philofopher could hardly have been able to de-monftrate that the ape was ftill a brute.   It is fortunate, therefore, that nature has feparated the faculties of imitating our fpeech and our geftures, and fhared them between two very different fpecies ; and while fhe has conferred on all animals the fame fenfes, and on fome the fame members and organs, with man, fhe has referved for him alone the power of improving

---

* Thofe Parrots to which the favages give artificial colours are termed *tapirés*.   This is effected, it is faid, by means of the blood of a frog, which they drop into the fmall wounds made in young Parrots by plucking their feathers : thofe which fprout again change their green or yellow tints into orange, rofe colour, or variegated hues, according to the medicaments employed.

them ;

them; that noble mark of our pre-eminence, which conftitutes our empire over the animated world.

There are two kinds of improvement; the one barren, and confined to the individual; the other prolific, and extending through the fpecies, and cultivated in proportion as it is encouraged by the inftitutions of the fociety. Among brutes, the experience of one race is never tranfmitted to the fucceeding; their acquifitions are merely individual; they are the fame now that they ever were—ever will be. But man is progreffive; he receives the inftructions of paft ages, he reaps the benefit of the difcoveries of others, and, by a proper ufe of his time, he may continually advance in knowledge. And who can, without regret and indignation, view that long gloomy night of ignorance and barbarifm, which overfpread Europe, and which not only arrefted our improvement, but thruft us back from that elevation which we had attained? But for thefe unfortunate viciffitudes, the human fpecies would invariably approach towards the point of perfection.

The mere favage, who fhuns all fociety, and receives only an individual education, cannot improve his fpecies, and will not differ, even in underftanding, from thofe animals on which he has beftowed his name. Nor will he acquire even fpeech, if the family be difperfed, and the

children

children abandoned foon after birth. The firft rudiments of the focial difpofition are therefore unfolded by the tender attachment and the watchful folicitude of the mother; the helplefs ftate of the infant requires conftant and affiduous attention; its claimant cries are anfwered by foothing expreffions, which begins the formation of language, and, during the fpace of two or three years, this grows in fome degree fixed and regular. But, in other animals, the growth is much more rapid; the parental endearments laft only fix weeks or two months; and the impreffions are flight and tranfitory; and, after feparation, they entirely ceafe. It is not, therefore, to the peculiar ftructure of our organs that we are indebted for the attainment of fpeech; the Parrots can articulate the fame founds, but their language is mere prattle, and void of fignification.

The power of imitating our difcourfe or our actions, confers no real fuperiority on an animal. It never incites to the cultivation of talents; it never tends to the improvement of the fpecies. The articulation of the Parrot implies only the clofe analogy of its organs of hearing and of voice to thofe in man; and that fimilarity of ftructure obtains, though in a lefs degree, in many other birds, whofe tongue is thick, round, and nearly of the fame form. The ftares, the blackbirds, the jays, the jack-daws, &c. can imitate words. Thofe whofe tongue is forked

(and

(and almoſt all the ſmall birds may be ranged
in that claſs), whiſtle more eaſily than they
prattle; and if, with this ſtructure, they have
alſo ſenſibility of ear, and can accurately retain
the impreſſions made on that organ, they will
learn to repeat airs: the canary, the linnet, the
ſiſkin, and the bulfinch, ſeem natural muſicians.
The Parrot imitates every ſort of noiſe, the
mewing of cats, the barking of dogs, and the
notes of other birds, as well as the human
voice; yet it can only ſcream or pronounce
very ſhort phraſes; and, though capable of
even articulating ſounds, it is unable to mo-
dulate theſe, or ſupport them by intermingling
gentle cadences. It has therefore leſs acute-
neſs of perception, leſs memory, and leſs flexi-
bility of organs.

There are alſo two different kinds of imitation; the one is acquired from reflection; the
other is innate and mechanical: the latter pro-
ceeds from the common inſtinct diffuſed through
a whole ſpecies, which prompts or conſtrains
each individual to perform ſimilar actions; and
the more ſtupid the animal, the more entire
will be this influence, and the cloſer will be
the reſemblance. A ſheep has invariably the
ſame habits with every other ſheep; the firſt
cell of a bee is preciſely like the laſt. The
knowledge of the individual is equal to that of
the ſpecies;—ſuch is the diſtinction between
reaſon and inſtinct. The other kind of imita-

F 2   tion

tion, which should be regarded as artificial, is
the acquifition of the individual, and cannot be
communicated. The moft accomplifhed Parrot
will never tranfmit his talent of prattling to his
offspring. When an animal is inftructed by
man, the improvement refts with it alone.
This imitation depends as well as the former on
the peculiar ftructure ; but it alfo implies fenfibility, attention, and memory ; and thofe fpecies which are fufceptible of education, rank
high in the order of organized beings. If the
animal be eafily trained, and each individual receive a certain degree of inftruction, as in the
cafe of the dogs, the whole fpecies will acquire
fuperiority under the direction of man ; but
when abandoned to nature, the dog will relapfe
into the wolf or the fox, and would never of
itfelf emerge from that ftate.

All animals may therefore be improved by
affociating with man ; but they cannot be inftructed to improve each other ; for they never can communicate the ideas and knowledge which they have acquired. Even birds
whofe fhape and proportions are fo different
from thofe of quadrupeds, are fufceptible of the
fame degrees of education. The *agamis* can
be trained to perform nearly all the actions of
the dogs ; a canary, properly bred, fhews its
attachment by careffes that are equally animated, and more innocent and more fincere
than thofe of the cat. There are many inftances

ftances of the wonderful effects of education on
the rapacious birds *, which feem the moft favage and the moft averfe to bend to inftruction.
In

* " In 1763," fays M. Fontaine, " a buzzard was brought to
me that had been taken in a fnare : it was at firft extremely favage
and even cruel. I undertook to tame it, and I fucceeded by leaving it to faft, and conftraining it to come and eat out of my hand.
By purfuing this plan, I brought it to be very familiar ; and after
having fhut it up about fix weeks, I began to allow it a little liberty, taking the precaution, however, to tie both pinions of its
wings. In this condition it walked out into my garden, and returned when I called it to feed. After fome time, when I judged
that I could truft to its fidelity, I removed the ligatures, and faftened a fmall bell, an inch and a half in diameter, above its talon,
and alfo attached on the breaft a bit of copper having my name
engraved. I then gave it entire liberty, which it foon abufed ;
for it took wing, and flew as far as the foreft of Belefme. I gave
it up for loft ; but four hours after I faw it rufh into my hall,
which was open, purfued by five other buzzards, which had conftrained it to feek its afylum . . . . After this adventure it ever
preferved its fidelity to me, coming every night to fleep on my
window ; it grew fo familiar with me, as to feem to take fingular
pleafure in my company. It attended conftantly at dinner, fat on
a corner of the table, and very often careffed me with its head and
bill, emitting a weak fharp cry, which however it fometimes foftened. It is true that I alone had this privilege. It one day followed me, when I was on horfeback, more than two leagues, fail-
ing above my head . . . . It had an averfion both to dogs and cats,
nor was it in the leaft afraid of them ; it had often tough battles
with them, and always came off victorious. I had four very
ftrong cats, which I collected into my garden befide my buzzard ;
I threw to them a bit of raw flefh, the nimbleft cat feized it, the
reft pu fued ; but the bird darted upon her body, bit her ears with
his bill, and fqueezed her fides with his talons, with fuch force
that the cat was obliged to relinquifh her prize. Often another
cat fnatched it the inftant it dropt, but fhe fuffered the fame treatment, till the buzzard got entire poffeffion of the plunder. He
was fo dexterous in his defence, that when he perceived himfelf af-

failed

In Afia, the pigeon is taught to carry letters
between places an hundred leagues diftant : and
the art of falconry proves that, by directing the
inftinct of birds, they may be as much improved as the other animals. On the whole,
it appears that if man beftowed equal time and
attention upon any animal as upon a child, it
would acquire a mechanical imitation of the
fame actions ; the effects only would differ. In
the

failed at once by the four cats, he took wing, and uttered a cry of
exultation.   At laft, the cats, chagrined at their repeated difap-
pointment, would no longer contend.

" This buzzard had a fingular antipathy ; he would not fuffer a
red cap on the head of any peafant, and fo alert he was in whip-
ping it off, that they found their head bare without knowing what
was become of their cap.   He alfo fnatched wigs without doing
any injury, and he carried thefe caps and wigs to the talleft tree
in a neighbouring park, which was the ordinary depofit'of his
booty . . . . He would fuffer no other bird of prey to enter his domain ; he attacked them very boldly, and put them to flight.   He
did no mifchief in my court-yard, and the poultry, which at firft
dreaded him, grew infenfibly reconciled to him.   The chickens
and ducklings received not the leaft harfh ufage, and yet he bathed
among the latter.   But what is fingular, he was not gentle to my
neighbours' poultry ; and I was often obliged to publifh that I
would ay for the damages which he might occafion.   However,
he was often fired at, and he received fifteen mufket-fhots, without fuffering any fracture.   But once early in the morning, hover-
ing over the fkirts of a foreft, he dared to attack a fox ; and the
keeper feeing him on the fhoulders of the fox, fired two fhots at
him ; the fox was killed and the buzzard had his wing broken ; yet
notwithftanding this fracture he efcaped from the keeper, and was
loft feven days.   This man having difcovered, from the noife of
the bell, that he was my bird, came next morning to inform me ;
I fent to make a fearch near the fpot ; but the bird could not be
found, nor did it return till feven days after.   I had been ufed to
call

the one cafe, reafon extends and diffufes the attainments; in the other, they continue ftationary, and perifh with the poffeffor.

But that education which feems to unfold the faculties, and meliorate the difpofitions of quadrupeds or birds, renders them odious to the reft of their fpecies. When a buzzard, for inftance, a magpie, or a jay, efcapes to the woods, its favage kindred flock around it to gaze at the novelty. Their wonder is foon converted into rage; and they furioufly attack and drive off the intruder: nor is it admitted into their fociety till it relinquifhes its artificial habits, and adopts the manners of the tribe.

Birds are deftined by nature to enjoy the completeft independence, and exult in the moft unbounded freedom. Other animals are condemned to crawl on the furface; thefe foar aloft

call him every evening with a whiftle, which he anfwered not for fix days; but, on the feventh, I heard a feeble cry at a diftance, which I judged to be that of my buzzard: I repeated the whiftle a fecond time, and I heard the fame cry. I went to the part whence the found came, and, at laft, found my poor buzzard with his wing broken, which had travelled more than half a league on foot to regain his afylum, from which he was then diftant about 120 paces. Though he was extremely reduced, he gave me many careffes. It took near fix weeks till he was recruited, and his wounds healed; after which he began to fly as before, and follow his old habits for about a year: he then difappeared for ever. I am convinced that he was killed by accident; and that he would not have forfaken me from choice."

Letter of M. Fontaine, Curé de Saint-Pierre de Belefme, to M. le Comte de Buffon, bearing date 28 January, 1778.

in

in the air. No obstacle can oppose their progress; no spot can fix their residence: the sky
is their country, and their course is on the
wings of the breeze. They foresee the viciffi-
tude of the seasons, and watch their return.
They generally appear when the mild influence
of spring has clothed the forests with verdure;
there they nestle, concealed under the foliage.
Heaven and earth seem to conspire to their felicity. But solicitude soon arises: they dread
the cruel visits of the same animals on which
they before looked down with contempt. The
wild cat, the marten, the weazel, seek to devour the objects of their tenderest affection:
the adder clambers to gain their eggs, or devour
their progeny: and children, that amiable portion of human kind, but who, from want of em-
ployment, are ever in mischief, wantonly plunder the sacred deposits of love. Often the mother rushes into danger in defence of her young;
and sacrifices to the ardor of her attachment,
her love, her liberty, and her life.

Why is the season of the highest pleasures
also the season of the greatest solicitude? Why
are the most delicious enjoyments always damp-
ed, even in the freest and most innocent of beings, by the cruellest anxieties? May we not
complain of harshness in nature, the common
mother of all? Her benevolence is never pure,
or of long continuance. No sooner the happy pair united, by choice and by their mu-
                                              tual

tual labours, have fabricated the manfion of love,
than they dread the plunderer's attack. The
feathered race has alfo its tyrants; and the rapacious birds are the more formidable, as they
are more independent. The eagle fnatches
with impunity the prey from the lion; all
dread his afpect; the feebler birds fcream at
his approach, and feek immediate fhelter. Perhaps the eagle would have occupied a large portion of the earth, if man had not driven him
to the fummits of mountains and inacceffible
tracks, where in folitude he ftretches out his
gloomy dominion.

From this curfory view, it would appear that
birds rank next to man in the great fcale of exiftence. Nature has accumulated and concen-
trated more ftrength in their little bodies than
fhe has communicated to the huge limbs of the
moft powerful quadrupeds; agility is combined
with folidity; their empire extends over the inhabitants of the air, the earth, and the waters.
The whole of the infect tribes are exclufively
fubject to their dominion, and feem only deftined to feed thefe deftroyers: they alfo feize
the noxious reptiles on the ground, and fnatch
the fifh from their element. They even attack the quadrupeds; the buzzard fometimes
darts on the fox, and the falcon ftops the ante-
lope; the eagle preys on the fheep, murders the
dog equally with the hare, and tranfports their
carcafes to his eyry. The birds walk on two
feet,

feet, imitate fpeech, and repeat mufical airs ;
in thefe refpects too they refemble man, while
their power of flying marks a decided fuperiority above all the other terreftrial animals.

But from this general view of the nature of
birds, let us defcend to furvey the genus of the
Parrots. That tribe, the moft numerous of all,
affords ftriking illuftrations of a new propofition: that, in birds, as in quadrupeds, thofe
which inhabit the tropical regions are confined
exclufively to their refpective continents. This
principle ferves to fix their nomenclature ; the
fpecies are much diverfified and multiplied;
above 100 are known, and yet of thefe not one
is common to both continents. What can be
a more decifive proof of this general propofition
which we explained in the Hiftory of Quadrupeds ? The two continents were never joined,
except towards the north, and therefore no animal incapable of fupporting the intenfe cold of
the frozen regions could migrate from the one
into the other. Birds alfo, fuch as the Parrots,
which live and propagate only in warm climates, have remained indigenous ; fome inhabit the tropical regions of the new continent,
others thofe of the old, and occupy in each a
zone extending twenty-five degrees on both
fides of the equator.

But it will be faid that if the elephants and
other large quadrupeds, which at prefent are
peculiar to Africa and India, inhabited origi-

5   nally

nally the northern tracts in both continents,
might not this have alfo been the cafe in regard
to the Parrots? And as the earth gradually
cooled, thefe might continually advance towards the tropics; and neither the lofty mountains, nor the narrow pafs of the ifthmus of
Panama, could prevent their migration.

This objection, though plaufible, is only a
new queftion, which, in whatever way it be
refolved, cannot affect our hypothefis, that the
north was the primæval refidence of animals,
and that they afterwards removed to the regions
of the fouth. But thofe birds whofe conftitution is adapted to a hot climate could never rife
to the frozen fummits of mountains; and the
cold that prevails in the elevated regions of the
air would as effectually ftop their flight, as the
various obftacles to be furmounted would limit
the progrefs of the elephant. Thus what
appears at firft an objection, is really a con-
firmation of the theory; fince not only the
quadrupeds, but alfo the birds, which are natives of the torrid tracts in the old world, have
never penetrated or fettled in the infulated continent of South America. In the cafe of the
birds, however, this principle has fome exceptions; for a few fpecies are found equally in the
equatorial parts of both continents. But this
is owing to particular circumftances; their vigorous wings and their power of refting on the
furface of the water by means of the broad mem-
branes

branes of their feet. The Parrots can neither
foar to a vaſt height, nor fly to a great diſtance,
and their feet are not webbed. Accordingly,
none of theſe have ever migrated from the one
continent to the other, unleſs tranſported by
men acroſs the intervening ocean *. This will
be better perceived after viewing the arrangement, and comparing the deſcriptions of the
ſeveral ſpecies. It was perhaps as difficult to
claſs them as the monkeys; ſince all the preceding naturaliſts have confounded them together.

The Greeks were acquainted at firſt with
only one ſpecies of Parrot, or rather of Parrakeet; it is what we now call the *Great Ring
Parrakeet*, and comes from India. They were
brought from the iſland of Taprobane into
Greece by Oneſicrites, who commanded Alexander's fleet. They were ſo new and uncommon that Ariſtotle himſelf appears not to have
ſeen them, and mentions them only from report †. But the beauty of theſe birds, and
their power of imitating ſpeech, ſoon made

* The Parrots have a laborious ſhort flight, ſo that they cannot
croſs an arm of the ſea ſeven or eight leagues broad. Each iſland
of South America has its particular Parrots; thoſe of St. Lucia,
of St. Vincent, of Dominica, of Martinico, are different from
each other: thoſe of the Caribbee iſlands do not reſemble them, nor
are theſe Caribbee Parrots found near the Oronooco, which is the
part of the continent neareſt theſe iſlands. *Note communicated by
M. de la Borde, King's Phyſician at Cayenne.*

† *Hiſt. Anim.* Lib. VIII. 12. " There is an Indian bird called
pſittace, which is ſaid to ſpeak."

3   them

them the objects of luxury among the Romans, and the prevalence of that practice provoked the indignation of the rigid Cato*. They were lodged in cages of filver, of fhells, and of ivory; and the price of a Parrot often exceeded that of a flave †.

No Parrots were known at Rome, but thofe brought from India ‡, until the time of Nero; the emiffaries of that prince found them in the ifland of the Nile, between Syene and Meroe, which is exactly in the limit that we affigned of twenty-four or twenty-five degrees latitude §. Pliny tells us that the Latin name *pfittacus* was derived from the Indian appellation *pfittace*, or *fittace* ‖.

The Portuguefe, who firft doubled the Cape of Good Hope, and explored the fhores of Africa, found the country of Guinea, the iflands fcattered in the Indian ocean, and alfo the continent, inhabited by various kinds of Parrots, all unknown in Europe. So numerous they were

* This auftere cenfor exclaimed in the midft of the affembled fenate, " O! fenators! O! unhappy Rome! what forebodings! in what times do we live, to fee the women feed dogs on their knees, and the men carry Parrots in their hands!" *Columella, Dict. Antiq.* Lib. III.

† Statius.

‡ Pliny, *Lib. X.* 42.—Paufanias.

§ *Id.* Lib. VI. 29.

‖ Lib. X. 42. They were brought alfo in the fifteenth century from the countries through which Alexander marched. *Relation de Cadamofto.* See *Hift. Gen. des Voyages,* t. II. 305.

at Calicut *, in Bengal, and on the African coasts, that the Indians and negroes were obliged during harvest to watch their fields of rice and maize, and to repel the destructive havock of these birds †.

This vast multitude of Parrots in all countries which they inhabit ‡, seems to prove that they breed several times annually, since the product of one hatch is inconsiderable. Nothing could equal the variety of the species which navigators found on every part of the coast of South America. Many islands were called the *Parrot Islands.* They were the only animals that Columbus met with in the one where he first landed §. They were the early articles of traffic between the Europeans and Americans ||. The American and African Parrots were imported in such numbers, that the Parrot of the ancients was forgotten; it was known only by description in the time of Belon ╪.

* Recueil des Voyages qui ont servi à l'etablissement de la Compagnie des Indes, &c. *Amsterdam,* 1702, *t. III. p.* 195.

† *See* Mandeslo, at the end of Olearius, *t. II. p.* 144.

‡ " Among the many remarkable animals, the Parrots of Malabar excite the admiration of navigators, by their prodigious numbers, and by the variety of their species. Dellon avers that often he had the pleasure of seeing two hundred taken in one draw of a net." *Hist. Gen. des Voy. t. XI. p.* 454.

§ Guanahani, one of the Lucayos.

|| First Voyage of Columbus in the beginning of the *Hist. Gen. des Voy. t. XII.*

╪ *Nat des Oiseaux,* p. 296.

We

We shall range the Parrots in two great claffes; the firft comprehending thofe of the old continent, the fecond thofe of the new. The firft will be fubdivided into five families; the Cockatoos, the Parrots properly fo called, the Lories, the long-tailed Parrakeets, and the fhort-tailed Parrakeets. Thofe of the new world will include fix other families; the Maccaws, the Amazonians, the Creeks, the Popinjays, the long-tailed Paroquets, and the fhort-tailed Paroquets.

PARROTS

OF THE OLD CONTINENT.

The COCKATOOS.

Les Kakatoes, Buff.

THE largeſt Parrots of the old continent are the Cockatoes. They are all natives of the ſouth of Aſia, where they ſeem indigenous. We are uncertain whether they are alſo found in Africa, but they are undoubtedly not found in America. They are ſpread through the ſouthern parts of India *, and in all the iſlands of the Indian ocean, at Ternate †, at Banda ‡, at Ceram §, in the Philippine iſlands ‖, and in

---

* "The trees of this city (Amadabat, capital of Guzarat), and thoſe on the road from Agra to Brampour, which is 150 German leagues, breed an inconceivable number of Parrots . . . Some are white, or pearl grey, and capped with a carnation tuft; theſe are called *kakatous,* becauſe they diſtinctly articulate that word. Theſe birds are very common through all India, where they neſtle in the towns on the roofs of houſes, like the ſwallows in Europe." *Voyage de Mandeſlo,* t. II. p. 144.

† *Voyage autour du Monde,* par Gemelli Carreri, *Paris,* 1719, *t. V. p. 5.*

‡ Recueil des Voyages qui ont ſervi à l'etabliſſement de la Compagnie des Indes, &c. *Amſterdam,* 1702, *t. V. p. 26.*

§ Dampier.   ‖ Gemelli Carreri.

thoſe

thofe of Sunda *. Their name *kakatoes, cata-
cua*, and *cacatou*, is formed from their cry +.
They are eafily diftinguifhed from the other
Parrots, by their white plumage, by the round-
er and more hooked fhape of their bill, and
particularly by a creft of long feathers, which
they can raife or deprefs at pleafure ‡.

It is difficult to teach the Cockatoos to prattle, and fome fpecies can never acquire the
imitation. But they are more eafily bred § ;
they all grow tame, and in fome parts of India they feem domefticated, for they build
their nefts on the roofs of the houfes. And
this facility of education feems to refult from
their fuperior underftanding; they are more
attentive and obedient than other Parrots, and
they ftrive, though without fuccefs, to repeat what they hear. Their defects are com-
penfated by other expreffions of feeling, and
by affectionate careffes. All their motions have
a gentlenefs and grace which adds new charms
to their beauty. Two of thefe birds, a male
and a female, were fhewn in March 1775 at
the fair of St. Germain at Paris. They difco-

---

* Voyage de Siam, par le P. Tackard, *Paris*, 1686, *p.* 130.

+ " We made feveral tacks to double the ifle of Cacatoua, fo
called becaufe of the white Parrots that refide in it, and which inceffantly repeat that name. This ifle is very near Sumatra." *Ibid.*

‡ The crown of the head, which is covered by the long reclined
feathers, is entirely bald.

§ " At Ternate, thefe birds are domeftic and docile; they
fpeak little, but fcream much." *Gemelli Carreri.*

　　vered

vered great docility, raifed their creft, made a falute with their head, touched with their bill or their tongue, anfwered their keeper's queftions with a fign of affent, as they were defired; they marked by repeated motions the number of perfons in the room, the colour of their clothes, the hour of the day, &c.; they billed each other without being directed, an evident token of their inclination to couple, and their keeper told us that they had often commerce together even in our climate.—Though the Cockatoos, like the other Parrots, ufe their bill in climbing, they have not the fame heavy unpleafant gait; they are, on the contrary, very agile, and walk gracefully, tripping with fhort quick fteps.

THE
WHITE-CRESTED COCKATOO.

Le Kakatöos a Huppe Blanche, Buff.

FIRST SPECIES.

Pfittacus Criftatus, Linn. Gmel. and Borowfk.
Cacatua, Briff.
Kakatocha tota alba, Klein.
Pfittacus albus Criftatus, Ray and Will.

I⊤ is nearly as large as a hen. Its plumage is entirely white, except a yellow tinge on the
under

THE GREAT WHITE COCKATOO.

under fide of the wings, and of the lateral quills
of the tail; the bill and feet are black. Its noble creft is very remarkable, confifting of ten
or twelve feathers, not of the foft downy kind,
but of the nature of quills, tall and broad
webbed; they are inferted in two parallel lines
running back from the face, and form a double
fan [A].

THE

YELLOW-CRESTED COCKATOO.

Le Kakatoes a Huppe Jaune, Buff.

SECOND SPECIES.

Pfittacus Sulphureus, Gmel.
Cacatua Luteo-criftata, Briff.
The *Crefted Parrot or Cockatoo,* Albin.
The *Leffer White Cockatoo,* Edw. and Lath.

OF this fpecies, there are two branches, differing in fize. In both the plumage is
white, with a yellow caft under the wings and
the tail, and fpots of the fame colour round the
eyes; the creft is yellow citron, confifting of
long foft ragged feathers, which the bird ele-
vates and projects; the bill and feet are black.
It was a Cockatoo of this fpecies, and probably
the firft ever feen in Italy, that Aldrovandus

[A] Specific character of the *Pfittacus Criftatus:* " It is white,
its creft pliant and yellow."

G 2 defcribes;

defcribes; and he admires its elegance and beauty. It is as intelligent, gentle, and docile, as the preceding.

We faw this beautiful Cockatoo alive. It expreffes joy by fhaking its head brifkly feveral times upwards and downwards, making a flight cracking with its bill, and difplaying its elegant creft. It returns the careffes; touches the face with its tongue, and feems to lick it; the kiffes are foft and gentle. When the one hand is laid flat under its body, and the other refts on its back, or only touches its bill, it preffes firm-ly, claps its wings, and with its bill half open it blows and pants, and feems to feel the moft intoxicating delight. It repeats this as often as one choofes. It is alfo very fond of being fcratched; holds its head, and raifes its wing to be ftroked: it often whets its bill, by gnaw-ing and breaking bits of wood. It cannot bear the confinement of the cage, but it never roves out of its mafter's fight. It anfwers its call, and retires when he commands; in which cafe it difcovers anxiety, often looking back for the fign of invitation. It is exceedingly neat; all its motions are graceful, delicate, and pretty. It feeds on fruits, pulfe, all the farinaceous grains, on paftry, eggs, milk, and whatever is fweet, but not too fugary.

[A] Specific character of the *Pfittacus Sulphureus*: " It is white, its creft pliant and drawn to a point; and this, with a fpot below the eyes, is brimftone colour."

The RED-CRESTED COCKATOO.

Le Kakatoës a Huppe Rouge, Buff.

THIRD SPECIES.

Pfittacus Moluccenfis, Gmel.
Pfittacus Rofaceus, Lath. Ind.
Cacatua Rubro-criftata, Briff. and Gerini.
The *Greater Cockatoo*, Edw.
The *Great Red-crefted Cockatoo*, Lath.

IT is one of the largeft of the genus, being near a foot and half long ; the upper part of its creft, which reclines backwards, confifts of white feathers, and covers a bundle of red ones [A].

The LITTLE FLESH-BILLED COCKATOO.

Le Petit Kakatoës a Bec Couleur de Chair, Buff.

FOURTH SPECIES.

Pfittacus Erythroleucus, Linn. Gmel. Ray, and Will.
The *Red and White Parrot*, Lath.

THE plumage is entirely white, except fome tints of pale red on the temples, and on

[A] Specific chara&er of the *Pfittacus Moluccenfis :* " It is white, inclining to a dilute rofe colour ; its creft is red above ; the lateral feathers of its tail below, from the bafe to the middle, brimftone coloured."

the

the feathers of the upper part of the creſt, which red caſt is deeper on the coverts of the lower ſurface of the tail.   There is a little light yellow at the origin of the ſcapular feathers and of thoſe of the creſt, and on the inſide of the quills of the wing and of moſt of thoſe of the tail;   the feet are blackiſh;   the bill reddiſh brown,   which is peculiar to this ſpecies,   the bills of the other Cockatoos being all black.   It is alſo the leaſt of the genus; Briſſon makes it of the ſize of the Guinea Parrot, but it is much ſmaller.   It has a creſt, which lies flat, and is erected at pleaſure.

We may obſerve that the bird termed by Brif-ſon the *Cockatoo with red wings and tail* does not appear to belong to the ſame genus, fince he makes no mention of the creſt, which is the diſtinguiſhing character.   Beſides,   he borrows his account from Aldrovandus, who deſcribes it in the following terms.   " This Parrot ought to be reckoned among the largeſt;   it is equal in ſize to the capon ;   all its plumage is cinereous white ;   its bill is black and much in-curvated ;   the lower part of the back,   the rump, all the tail, and the quills of the wings, are vermilion."   Theſe characters would cor-reſpond to thoſe of the Cockatoos,   if the creſt were added ;   and this great red and white Par-rot of Aldrovandus might perhaps form a fifth ſpecies, or a variety of one of the preceding.

[A] Specific character of the *Pſittacus Erythroleucus:* " It is cinereous ; the quills of its wings are white crimſon."

The BLACK COCKATOO,
Buff. and *Lath.*

FIFTH SPECIES.

Pſittacus Aterrimus, Gmel.

EDWARDS, who deſcribes this Cockatoo, aſſerts that it is as large as a maccaw. Its
plumage is entirely bluiſh black, which is deeper on the back and the wings than under the
body; the creſt is brown or blackiſh, and the
bird has, like the other Cockatoos, the power
of erecting it high, and of reclining it almoſt
cloſe on the head; the cheeks below the eye
are covered by a red, naked, wrinkled ſkin,
which covers the inferior mandible of the bill,
whoſe colour, as well as that of the feet, is
blackiſh brown; the eye is fine black. The
bird may be reckoned the negro of the Cockatoos, which are generally white; the tail is
long, and conſiſts of tapered feathers. The figure delineated from nature was ſent from Ceylon to Edwards, and that naturaliſt recogniſed
the ſame bird in a collection publiſhed by *Vander
Meulen* at Amſterdam, in 1707, and termed by
Peter Schenk the *Indian Crow.*

[A] Specific character of the *Pſittacus Aterrimus:* " It is black,
its creſt large and lighter coloured, its cheeks red and naked."

The PARROTS

PROPERLY SO CALLED.

WE fhall apply the name of Parrot to thofe
of the old continent whofe tail is fhort,
and confifts of quills nearly equal in length.
We may reckon eight fpecies, all natives of
Africa or India, and none of them found in
America.

THE

JACO or CINEREOUS PARROT, *Buff.*

FIRST SPECIES.

Pfittacus Erithacus, Linn. Gmel. and Kram.
Pfittacus Guineenfis Cinereus, Briff.
Ufchgraver Papagey, Wirs.
The *Afh-coloured Parrot,* Albin. Will. and Lath.

THIS fpecies is now the moft commonly
brought into Europe, and generally preferred, as well on account of the mildnefs of
its difpofition, as of its fagacity and docility, in
which it at leaft equals the green Parrot, without the difagreeable cries. It feems to pronounce
the word *Jaço,* and hence its ufual appellation.
All the body is of a fine pearl and flaty gray,
which is deeper on the upper furface, lighter
on the lower, and inclined to white on the
belly.

belly. The tail, which is vermilion, terminates and heightens this plumage, which is
gloffed and powdered with a fnowy colour,
that gives it conftantly a frefh appearance.
The eye is placed in a white, naked, mealy
íkin, that covers the check; the bill is black;
the feet gray; and the iris gold colour. The
total length of the bird is a foot.

Moft of thefe Parrots are imported from the
coaft of Guinea *, and come from the interior
parts of Africa † : they are alfo found at Congo ‡,

* Willughby.

† " They are found on the whole of this coaft (of Guinea), but
in fmall numbers, and moft of them even come from the interior
parts of the country. Thofe of Benin, of Calbari, of Cabolopez,
are moft efteemed, for which reafon they are brought from thofe
places; but they are much older than fuch as can be obtained
here, and confequently are not fo docile, nor fo eafily trained. All
the Parrots here on the coaft, and alfo near the angle of Guinea,
and in the above-mentioned places, are of a blue colour . . . Thefe
birds are fo common in Holland, that they are lefs efteemed there than
here, and not fold fo dear." *Voyage en Guinée*, par Bofman, *Utrecht*,
1705.—Albin is miftaken when he fays that this fpecies comes
from the Eaft Indies; it appears confined to Africa, and *à fortiori*
it occurs not in America, though Briffon places it at Jamaica, probably from the indication of Browne and Sloane; but without having confulted them, fince Sloane (Jamaica, *Vol. II. p.* 297) fays
exprefsly that the Parrots, which are numerous in Jamaica, were
all brought thither from Guinea. This fpecies is not a native of any
part of the new world. " Among the multitude of Parrots found
at Para, we cannot perceive the gray fpecies, which is fo common
in Guinea." *Voyage de la Condamine, p.* 173.—In Antarctic France
there is no gray kind found, as in Guinea, and in upper Africa.
Thevet, Singularités de la France Antarctique, *Paris*, 1558, *p.* 92.

‡ Recueil des Voyages qui ont fervir à l'etabliffement de la Compagnie des Indes. *Amflerdam*, 1702, *t. IV. p.* 321.

and

and on the coaſt of Angola *. They are very
eaſily taught to ſpeak †, and ſeem fondeſt of
imitating the voice of children, who are alſo the
moſt ſucceſsful in training them. It has indeed
been remarked by the older writers ‡ that the
birds moſt ſuſceptible of imitating the human voice
are eager to liſten to children, whoſe articulation is imperfect and unequal, and therefore
more correſpondent to their own. But the cinereous Parrot copies alſo the deep tones of the
adult; though the effort is laborious, and the
words are leſs diſtinct. One of theſe Guinea
Parrots was ſo completely drilled by an old
ſailor, that it acquired exactly his hoarſe voice
and cough; and though it was afterwards given
to a young perſon, and was in no other company, it never forgot the leſſons of its firſt
maſter, and it was diverting to obſerve its tran-
ſitions from a ſoft gracious tone to its former
hoarſeneſs and coarſe ſea tones.

* Hiſtorie Generale des Voyages, t. V. p. 76.
† They inhabit likewiſe the iſles of France and Bourbon, whither they have been tranſported. _Lettres Edifiantes, Recueil_ 18,
p. 11. " This iſle (of Mauritius or France) breeds tortoiſes, turtles, grey parrots, and other game, which are caught by the hand
in the woods. Beſides the profit derived from this exerciſe, it affords much diverſion. Sometimes when a great Parrot is taken,
it is made to ſcream, and inſtantly hundreds flock round it, which
are felled with ſticks." _Recueil des Voyages qui ont ſervir a l'éta-_
bliſſement de la Compagnie des Indes. Amſterdam, 1702, t. III. p. 195.
‡ Albertus, _Lib. XXIII._

But

But not only has this bird a facility, it has
alſo an eagerneſs, in imitating the human voice.
It liſtens with attention, and ſtrives to repeat;
it dwells conſtantly on ſome ſyllables which it
has heard, and ſeeks to ſurpaſs every voice by
the loudneſs of its own. We are often ſur-
priſed at its repeating words or ſounds, which
we never taught it, and which we ſhould not
ſuppoſe it to have noticed *. It ſeems to ſet
itſelf taſks, and tries every day to retain its leſ-
ſon †. This engages its attention even in ſleep,
and, according to Marcgrave, it prattles in its
dreams ‡. They are moſt capable of improvement when young; then they ſhew more ſa-
gacity, more docility: and their memory, if
early cultivated, becomes ſometimes aſtoniſhing. Rhodiginus § mentions a Parrot which a
Cardinal purchaſed for 100 crowns, becauſe it
recited correctly the Apoſtles' Creed ||. But when

* Witneſs that Parrot of Henry VIII. which, as Aldrovandus
relates, having fallen into the Thames, called to the boatmen for
aſſiſtance, as it had heard the paſſengers call from the beach.

† Cardan goes ſo far as to aſcribe to it meditation and inward
reflection on what it has been taught, and this, ſays he, through
emulation and the love of glory .. *Meditatur ob ſtudium gloriæ* ...
The love of the marvellous muſt have had mighty influence upon
this philoſopher, to make him advance ſuch abſurdities.

‡ Ariſtotle had propoſed a quære, whether animals hatched
from eggs ever dream *(Lib. V. 10. Hiſt. Anim.)* Marcgrave an-
ſwers, that " his Parrot Laura often roſe in the night, and prattled
half aſleep."

§ *Lib. III.* 32.

|| M. de la Borde tells us that he ſaw one, which ſerved as al-
moner on board a veſſel; it recited the ſailors prayer, then the
roſary.   it

it grows older, it becomes ſtubborn, and will hardly be taught. Olina recommends; the evening, after their meal, as the proper time to inſtruct them; for their wants being ſatiſ-fied, they are moſt docile and attentive.

The education of the Parrot has been com-pared to that of the child *. At Rome, the perſon who trained a Parrot held in his hand a ſmall rod, with which he ſtruck it on the head. Pliny ſays that its ſkull is very hard, and that it requires ſmart blows to make it feel †. How-ever, the bird which we mentioned feared the rod more than a child that has been often whip-ped. If after remaining perched the whole day, it anticipated the hour of walking out into the garden, and deſcended too ſoon (which ſeldom happened), threats and the ſight of the rod drove it with precipitation to its rooſt; there it continued, but ſhowed its impatience by flap-ping its wings and ſcreaming.

" We ſhould ſuppoſe that the Parrot does not perceive when he ſpeaks himſelf, but fancies that ſome perſon addreſſes him. He often aſked his paw, and anſwered by holding up the paw. Though he liked to hear the voice of children, he ſeemed to have an antipathy to them; he purſued and bit them till he drew blood. He had alſo his objects of attachment, and though his choice was not very nice, it was conſtant.

* Ælian.   † Pliny, *Lib. X.* 42.

He

He was exceffively fond of the cook-maid; followed her every where, fought for her, and
feldom miffed finding her. If fhe had been
fome time out of his fight, the bird climbed
with his bill and claws to her fhoulders, lavifh-
ed his careffes, and would, on no account, leave
her. His fondnefs had all the marks of clofe
and warm friendfhip. The girl happened to
have a very fore finger, which was tedious in
healing, and fo painful as to make her fcream.
While fhe uttered her moans, the Parrot
never left her chamber. The firft thing he
did every day was to pay her a vifit; and this
tender condolence lafted the whole time of
the cure, and he again returned to his former
calm fettled attachment. Yet this ftrong predilection for the girl feems to have been more
directed to her office in the kitchen, than her
perfon; for when another cook-maid fucceeded
to her, the Parrot fhewed the fame degree of
fondnefs the very firft day *."

But Parrots of this kind not only imitate
difcourfe; they alfo mimic geftures and actions. Scaliger faw one that performed the
dance of the Savoyards, at the fame time repeating their fong. The one already mentioned
liked to hear a perfon fing, and, when he faw
him dance, he alfo tried to caper, but with the
worft grace imaginable, holding in his toes, and

* Note communicated by Madame Nadault, my fifter, to whom
this Parrot belonged.

tumbling

tumbling back clumfily. He was then the
moft cheerful; but he had alfo an extravagant
joy, and an inceffant prattling when in the
ftate of intoxication: for all Parrots love wine,
particularly the Spanifh and the mufcadine.
Even in the time of Pliny it was remarked that
the fumes of that liquor gave the Parrots a flow
of fpirits *. He crept near the fire in winter, and
his greateft pleafure, in that feafon, was to get on
the chimney; and when warmed he gave many
figns of his comfortable feelings. He had equal
pleafure in the fummer fhowers; he continued
whole hours expofed, and fpread his wings the
better to receive the rain, and did not feek for
cover till he was wet to the fkin. After he had
returned to his rooft, he ftripped all the feathers
one after another through his bill. If the weather
was dry, he liked to bathe in a ciftern of water,
and entered into it repeatedly, though always
very careful not to wet his head. But he was
as averfe to plunge in winter; and if then
fhewn a veffel full of water, he would run off,
and even fcream.

Sometimes he was obferved to yawn, and
this was almoft always the fymptom of weari-
nefs. He whiftled with more force and clear-
nefs than a man; but, though he expreffed
many tones, he could never be taught to copy
an air. He imitated perfectly the cries of wild

* In vino præcipue lafciva. Lib. X. 42.

and

and domeftic animals, particularly the crow,
which he mimicked fo well, that he might
have been taken for one. He feldom prattled
in a room with company; but if alone in the
adjacent room, he was noify in proportion to
the loudnefs of the converfation which he over-
heard; he feemed prompted to repeat precipi-
tately all that he had learnt, and was never fo
animated or fo clamorous. In the evening he
retired of his own accord to his cage, which he
fhunned during the day: there with one foot
concealed in the plumage, or hooked to the bars
of the cage, and his head beneath his wings, he
flept until he perceived the dawn of the morning; but he often wakened to the blaze of can-
dles. Then he ftepped down to the bottom of
the cage, and fharpened his claws, ufing the
fame motion with the fcratching of a hen.
Sometimes he whiftled or prattled in the night
when expofed to light; but in the dark he was
filent and tranquil *.

That fort of fociety which the Parrot forms
with man, is, by means of language, more inti-
mate and pleafing than what the monkey can
claim from its antic imitation of our geftures and
actions. If the ufeful and amiable qualities of the
dog, the horfe, or the elephant, command our
attention and efteem, the fingular talents of the

* Reft of the note communicated by Madame Nadault.

4

prattling

prattling bird sometimes engage more power-
fully our curiosity. It diverts and amuses; in
solitude it is company; it takes part in conver-
sation, it laughs, it breathes tender expressions,
or mimics grave discourse; and its words uttered indiscriminately please by their incongru-
ity, and sometimes excite surprise by their apt-
ness *. This play of language without meaning is uncommonly whimsical, and though not
more empty than much other talk, it is always
more amusing. The Parrot seems also to receive a tincture of our inclinations and manners;
it loves, or it hates; it has particular attach-
ments, predilections, and caprices; it is the
object of its own admiration and applause; it
becomes joyous or sad; it is melted by caresses,
and bills tenderly in return: in a house of
mourning, it learns to moan +, and often accustomed to repeat the dear name of a mistress

---

* Willughby speaks from Clusius of a Parrot, which, when a
person said to it, *Laugh, Poll, laugh,* laughed accordingly, and
the instant after screamed out; *What a fool to make me laugh!* We
have seen another which grew old with its master, and shared with
him the infirmities of age. Being accustomed to hear scarce any
thing but the words *I am sick* (Je suis malade); when a person
asked it, *How d'ye, Poll, how d'ye* (Qu'as-tu, perroquet, qu'as-
tu)? *I am sick,* it replied with a doleful tone, stretching itself over
the fire, *I am sick* (Je suis malade).

+ *See,* in the Annals of Constantine Manasses, the story of the
young Prince Leo, son of the Emperor Basil, condemned to death
by his implacable father, whom the cries of the persons around
him could not move, till the accents of the bird, which had learnt
to deplore the fate of the Prince, at last stung his barbarous heart.

whose

whofe lofs is bewailed, it awakens, in feeling hearts, the memory of paft joys *.

The power of imitating exactly articulate difcourfe implies in the Parrot a peculiar and more perfect ftructure of organ ; and the accuracy of its memory, though independent of the underftanding, manifefts a clofenefs of attention and a ftrength of mechanical recollection that no bird poffes in fo high a degree. Accordingly, all the naturalifts have remarked the fingular form of its bill, its tongue, and its head : its bill, found on the outfide and hollow within, has, in fome meafure, the capacity of a mouth, and allows the tongue to play freely ; and the found, ftriking againft the circular border of the lower mandible, is there modified as on a row of teeth, while the concavity of the upper mandible reflects it like the palate ; and hence it does not utter a whiftling, but a full, articulation. The tongue, which modulates all the founds, is proportionally larger than in man, and would be more voluble, were it not harder than flefh, and invefted with a ftrong horny membrane.

But this organization, though adjufted with fkill, is ftill inferior to the ftructure contrived to give an eafy and powerful motion to the upper mandible, and, at the fame time, not to hinder its opening. The mufcles are not fixed

* *See*, in Aldrovandus (p. 662), a pleafing and affecting piece, which a poet, who grieves for his miftrefs, addreffes to his Parrot, that inceffantly repeats her name.

to the root, where they would have exerted no force; nor to the fides, where they would have clofed the aperture. Nature has adopted a different plan; at the bottom of the bill are fixed two bones, which, extending on both fides, and under the cheeks, form a continuation of it, fimilar in form to the *pterygoid* bones in man, except that their hinder extremity is not concreted into another bone, but loofe. Thick layers of mufcles, fent off from the back of the head, and inferted in thefe bones, move them and the bill. For a fuller defcription of this fingular contrivance, I fhall refer to Aldrovandus *.

This naturalift properly obferves, that, between the eye and the lower jaw, there is a fpace, which deferves better the name of cheek than in any other bird; it is alfo more protuberant, occafioned by the number of mufcles that extend over it to the bill.

The bill is very ftrong; the Parrot eafily cracks the nuts of the red fruits; it gnaws the wood, and even bends or wrenches the bars of its cage, if they be flender, or if it be tired of confinement. It ufes its bill, oftener than its claws, in climbing and fufpending itfelf; it alfo holds by the bill in defcending, as if it were a third foot, which fteadies its motion; it alfo ferves to break its fall †. It is a fecond organ

---

* *Tom. I. pp.* 640 *and* 641.   † Pliny, *Lib. X.* 42.

          of

of touch, and is equally ufeful with its toes, in fcrambling and clenching.

The mobility of its upper mandible gives it a power which no other birds have, of chewing its food. In thofe, whether of the granivorous or carnivorous tribes, the bill is like a hand which throws the food into the gizzard, or an arm which fplits or tears it. The Parrot feizes the piece fideways, and gnaws deliberately*. The lower mandible has little motion, but that from right to left is moft perceptible; and this is often performed when the bird is not eating, which has made it be fuppofed to ruminate. In fuch cafes it probably only whets the edge of this mandible, with which it cuts and bites its aliments.

The Parrot difcovers hardly any choice in its food: it lives in its native country on almoft every fort of fruit or grain. The feeds of the baftard faffron † have been found to fatten it, though they act on man as a violent purge ‡. In the domeftic ftate, it eats whatever is pre-

---

* We muft remark that the external hind toe is moveable, and that the bird draws it fidewife and forward, to feize and handle what is given to it; but only in this fingle cafe does it ufe that power, and at other times, whether it walks or perches, it conftantly carries two toes before and two behind. Apuleius and Solinus fpeak of Parrots with five toes; but this was owing to their miftaking a paffage of Pliny, where that naturalift afcribes that uncommon property to a family of magpies (Lib. X. 42.)

† Carthamus Carduncellus, Linn.

‡ The Spaniards call this feed Seme de Papagey, Parrot-feed.

fented;

fented; but flefh, which it would rather prefer, is extremely hurtful to it, and occafions an un-natural longing, which prompts it to fuck and gnaw its feathers, and pluck them one by one from every part that its bill can reach. This cinereous Guinea Parrot is particularly fubject to that difeafe; it tears the feathers from its body, and even from its beautiful tail, which never afterwards recovers the fame bright red as at firft.

Sometimes after moulting this Parrot is ob-ferved to become marbled with white and rofe colour; occafioned either by fome diftemper, or by advanced age.

What Briffon reckons as varieties, under the names of the *Red-winged Guinea Parrot*, and the *Red variegated Guinea Parrot*, are owing to fuch accidental changes of plumage. In the one figured by Edwards, the red feathers are mingled at random with the gray, as if the bird had been dreffed out *(tapired)*. The cinereous Parrot is like others of the genus, fubject to the epilepfy and the cramp *; yet is it very hardy and lives to a great age †. Salerne fays that he faw one at Orleans which was above fixty years old, and ftill cheerful and lively ‡.

It

* Olina. *Occelleria, p.* 23.
† " I knew one at the Cape of St. Domingo, which was averred to be forty-fix years old." *Note communicated by M. de la Borde.*
‡ Vofmaer fays that he knew a Parrot which had lived in a fa-mily

It is uncommon for Parrots to propagate in our temperate climates; but they frequently lay addle eggs. There are fome inftances, however, of Parrots being reared in France. M. de le Pigeoniere had a cock and hen in the town of Marmande in Agenois, which hatched regularly each fpring for five or fix years, and the young Parrots lived, and were educated by the parents. Each hatch confifted of four eggs, three of which fucceeded. The birds were fhut in a room with nothing but a barrel open at top and filled with faw duft; fticks were faftened both on the outfide and infide, that the male might fcramble upwards and downwards, and fit befide the hen. In entering the room it was neceffary to have boots; for the male, fired by jealoufy, bit furioufly whatever he perceived to approach·his female *. Father Labat alfo mentions two Parrots that had feveral hatches at Paris †.

mily for an hundred years, having defcended from father to fon: but Olina, more credible and better informed, afcribes only twenty years for the average term of the Parrot.

* Letter dated from *Marmande en Agenois*, 25th Auguft, 1774.

† Nouveaux Voyages aux îles de l'Amerique. *Paris*, 1722, *t. II. p.* 160.

[A] Specific character of the Afh-coloured Parrot, *Pfittacus Erithacus:* " It is hoary, its temples white and naked, its tail crimfon."

The GREEN PARROT,

SECOND SPECIES.

Pſittacus Sinenſis, Gmel. and Briſſ.
The *Green and Red Chineſe Parrot,* Edw. and Lath.

EDWARDS deſcribes this bird as brought from China. But it is not found in moſt of the provinces of that vaſt empire ; it is confined to the moſt ſouthern, ſuch as Quanton and Quangſi *, which are near the tropic, the uſual limit of the climate of Parrots. This is probably one of thoſe which travellers have fancied were the ſame both in China and in America. But that notion, which is contrary to the general order of nature, is overturned by comparing each ſpe-cies in detail. The preſent is unlike any of the Parrots of the new world : it is as large as a middle-ſized hen ; the whole of its body is bright ſhining green ; the great quills of the wing and the ſhoulders are blue ; the flanks, and the under ſide of the top of the wing, bril-liant red ; the quills of the wings and tail are lined with brown.—Edwards ſays that it is very

* " The ſouthern provinces, ſuch as Quanton, and eſpecially Quangſi, have Parrots of all kinds, which differ in nothing from thoſe of America : their plumage is the ſame, and they have no leſs facility in learning to ſpeak." *Hiſtoire Generale des Voyages,* t. VI. p. 488.

rare.

rare. It is found in the Moluccas, and in New
Guinea, whence it was fent to us [A].

The VARIEGATED PARROT.

THIRD SPECIES.

Pfittacus Accipitrinus, Linn. Gmel. Gerini, & Borowſk.
Pfittacus Varius Indicus, Briſſ.
Pfittacus Elegans, Cluſius.
The *Hawk-headed Parrot,* Edw. and Lath.

IT is of the ſize of a pigeon. The feathers
round the neck, which it briſtles when angry, but which Cluſius overdoes in his figure,
are purple, edged with blue. The head is covered with feathers mixed with ſtreaks of brown
and white, as in the plumage of the hawk, and
hence Edwards applies the epithet of *Hawk-
headed.* There is ſome blue on the great quills
of the wing, and at the point of the lateral ones
of the tail, of which the two middle ones are
green, and ſo are the feathers on the upper ſide
of the body.

The mailed Parrot, *No.* 526, *Pl. Enl.* appears to be the ſame with the one juſt deſcribed;
and we preſume that the ſmall number of theſe

[A] Specific character of the *Pfittacus Sinenfis:* " It is green;
the lower coverts of its wings red; ſome of the greater ones, and
the margin, blue; the tail brown below."

birds

birds which have been brought from America
to France were introduced from India into the
new world, and that if they are found in the
interior parts of Guiana, they have been naturalized there like the canaries, finches, the Guinea
pig, and some other animals, that were carried
thither by navigators from the old continent.
That this species is not a native of America seems
evinced, because no traveller mentions it. Besides, its voice, which is shrill and acute, is different from that of all the other Parrots indigenous in that continent; and we may therefore conclude that it originated from a few individuals carried accidentally from India [A].

The VAZA, or BLACK PARROT*.

FOURTH SPECIES.

Psittacus Niger, Linn. Gmel Klein. and Gerin.
Psittacus Madagascarensis, Briss.
The *Black Parrot of Madagascar*, Edw. and Lath.

VAZA is the name which this species bears in
Madagascar, according to Flaccourt, who

[A] Specific character of the *Psittacus Accipitrinus*: " It is
green; its head gray; its neck and breast somewhat violet and variegated; the quills of its wings and tail tipt with blue."
* *Vaza* is the black Parrot of this country; some of the young
are brown red, but they are difficult to be had." *Voyage au Madagascar*, par Flaccourt. *Paris*, 1661.

THE BLACK PARROT.

THE MASCARINE PARROT.

adds that it imitates the human voice. Rennefort alſo mentions it*; and it is the ſame with
what Francis Cauche calls *Woures-meinte* †;
which, in the Madagaſcar dialect, ſignifies the
black bird. Aldrovandus likewiſe takes notice of
black Parrots that inhabit Ethiopia ‡. The Vaza
is as large as the cinerous Guinea Parrot, and is
uniformly black over its whole plumage ; the
colour is not indeed intenſe, but inclined to
brown, and tinged faintly with violet. It has
a remarkably ſmall bill ; its tail is, on the contrary, of conſiderable length. Edwards, who
ſaw it alive, ſays that it is a very familiar and
lovely bird [A].

The MASCARINE.

FIFTH SPECIES.

Pſittacus Maſcarinus, Linn. Gmel. and Briſſ.

THIS Parrot is ſo called, becauſe, round its
bill, there is a kind of black maſk which

* " At Madagaſcar the large Parrots are black." *Relation de
Rennefort, Hiſt. Gen. des Voy.* t. VIII. p. 606.

† *Voyage au Madagaſcar* par Fr. Cauche, *Paris,* 1651.

‡ *Tom. I. p* 636.

[A] Specific character of the *Pſittacus Niger* : " Its tail elongated and equal ; its body bluiſh black ; its bill and orbits whitiſh."

envelopes

envelopes the forehead, the throat, and the
border of the face. Its bill is red; a gray hood
covers the back of the head and neck; all the
body is brown; the quills of the tail, which are
brown two thirds of their length, are white at
their origin. The total length of this Parrot is
thirteen inches. The Vifcount Querhoent affures us, that it is found in the ifland of Bourbon, whither it has probably been carried from
Madagafcar. We have one in the King's Cabinet of the fame fize and colour, except that
it has not the black mafk, nor the white colour on the tail, and that all its body is equally
brown; its bill is alfo fmaller, and, in that refpect, it refembles the Vaza, of which it would
appear to be a variety, if it does not form an
intermediate fpecies between that bird and the
mafcarine. To the fame fpecies we would refer the *brown Parrot* of Briffon.

[A] Specific character of the *Pfittacus Mafcarinus :* " It is cinereous, with the bridle black below; its orbits naked and reddifh; its lateral tail-feathers whitifh at their bafe."

The BLOODY-BILLED PARROT.

Le Parroquet a Bec Couleur du Sang, Buff.

SIXTH SPECIES.

Pſittacus Macrorhyncos, Gmel.
The *Great-bellied Parrot*, Lath,

THIS Parrot is found in New Guinea. It is remarkably large. Its bill is blood-coloured, thicker and broader in proportion than that of any of the other Parrots, and even than that of the American maccaws. The head and neck are of a brilliant green with gold reflections; the fore part of the body is yellow ſhaded with green; the tail is yellow below and green above; the back is ſky blue; the wing appears tinged with a mixture of the ſame ſky blue and green, according to its different poſitions; the coverts are black, edged and ſprinkled with ſtreaks of gold yellow; this Parrot is fourteen inches long.

[A] Specific character of the *Pſittacus Macrorynchos*: " It is green, inclining below to yellow; its wings mixed with ſky blue and green; its coverts black."

The GREAT BLUE-HEADED GREEN PARROT.

SEVENTH SPECIES.

Pfittacus Gramineus, Gmel.
The *Amboyna Parrot,* Lath.

THIS is one of the largeſt of the Parrots; it is near ſixteen inches in length, though its tail is rathei ſhort. The face and the upper ſide of the head, are blue; all the upper ſurface is meadow-green, mixed with blue on the great quills; all the under ſurface is olive-green: the tail is green above, and dirty yellow below [A].

The GRAY-HEADED PARROT.

EIGHTH SPECIES.

Pfittacus Senegalus, Linn. Gmel. and Briſſ.
The Senegal Parrot, Lath.

THIS bird has a ſhort tail, which excludes it from the family of the Parrakeets; and though only ſeven inches and a half long, it is

[A] Specific character of the *Pfittacus Gramineus:* " It is green, below olive; its front and top blue; its tail yellow below."

thick

thick and round fhaped. Its head and face are of a glofly bluifh gray; its ftomach and all the under fide of its body are of a full marigold-yellow, fometimes mixed with aurora red; its breaft and all its upper furface green; except the quills of the wings, which are only edged with that colour on a brown gray ground.

Thefe Parrots are frequent in Senegal; they fly in fmall flocks of five or fix, and perch on the ftraggling trees in the burning, fandy plains of that country, and utter a fhrill, difagreeable cry. They keep clofe together, fo that a perfon may kill feveral at once; and it often happens that a fingle fhot levels with the ground the whole of the little flock. Le Maire affirms that they never fpeak *; but perhaps they have been neglected in their education.

* " The Parrots are thefe of two kinds (at Senegal); fome fmall and entirely green, others larger, having the head gray, the belly yellow, the wings green, and the back mixed with gray and yellow; the latter never fpeak, but the fmaller have a fweet, clear voice, and prattle whatever they are taught." *Voyage de le Maire. Paris, 1695, p. 107.*

[A] Specific character of the *Pfittacus Senegalus:* " It is green, below yellow; its head cinereous; its orbits black and naked."

The L O R I S.

THIS name has been applied in the Eaſt Indies to a family of Parrots whoſe cry reſembles the ſound of the word *lori*. They are
hardly diſtinguiſhed from the reſt of the genus,
except by their plumage, which is chiefly red,
and of various intenſity. Their bill is alſo ſmaller, not ſo much hooked, but ſharper than that
of the other Parrots. Their aſpect is lively,
their voice ſhrill, and their motions quick.
They are, according to Edwards, the moſt
nimble of all the Parrots, and the only ones
that can leap to the height of a foot. Theſe
well aſcertained facts confute the aſſertion of
a traveller, that they brood in ſilent melancholy *.

They are taught with great eaſe to whiſtle
and articulate words; they ſoon grow tame,
and, what is uncommon in all animals, they
retain their cheerfulneſs in captivity. But they
are in general very delicate, and difficult to tranſ-
port; and, in our temperate climates, they are
ſhort lived. Even in their native regions, they
are ſubject to epileptic fits, like the maccaws
and other Parrots; yet it is probable that this
diſorder attacks only the domeſticated birds.

* *Hiſt. Gen. des Voy*. t. X. p. 459.

" Ornithologiſts

" Ornithologifts have improperly," fays Sonnerat *, " difcriminated the Loris by the epithets of the *Philippine*, the *Eaft Indian*, the
Chinefe, &c. Thefe birds inhabit only the
Moluccas and New Guinea, and thofe found in
other parts have been carried thither." But
thefe nomenclators are guilty of a greater im-
propriety in reckoning fome fpecies of Loris as
natives of America, fince none exift there;
and, if travellers have feen a few individuals,
they muft have been introduced from the Afiatic iflands.

Sonnerat adds too, that he conftantly found
the Loris in one ifland to be of a different fpecies from thofe in another, though at a fhort
diftance only. A fimilar obfervation has been
made in regard to the iflands of the Weft Indies.

The N O I R A - L O R I.

FIRST SPECIES.

Pfittacus Garrulus, three Varieties, Gmel.
Lorius Moluccenfis, Briff. and Gerini.

T HIS bird is found at Ternate †, at Ceram,
and at Java, where it is called *Noira*, a
name

* *Voyage a la Nouvelle Guinee*, p. 173.
† " There are many beautiful Parrots in the ifle of Ternate,
which are red on the back, with little feathers on the fore fide of
the

name which the Dutch have adopted. It is
held in fuch high eftimation in India, that ten
reals are readily offered for one Noira. In the
account of the firft voyage from Holland to Java, it is faid that feveral of thefe beautiful birds,
which were tried to be brought home, all died
on the paffage *. In the fecond voyage, however, one was carried to Amfterdam; and, fince
that time, they have been more frequent.

The Noira fhews ftrong attachment, and
even affection, to its mafter; it careffes him
with its bill, and ftrokes his hair with furprifing gentlenefs and tamenefs. At the fame
time it cannot bear ftrangers, and bites them
with a fort of rancour. The natives of Java
breed many of thefe birds †. In general the
cuftom of keeping tame Parrots feems to have
been very ancient in India, fince Ælian mentions it [A].

the wings. They are fomewhat fmaller than thofe of the Weft Indies, but they learn much better to fpeak." *Argenfola,* Conquêtes
des Moluques. *Paris,* 1706, *t. III. p.* 21.

   * Linfcot *apud Clufium. Auct. p.* 364.

   † " The Dutch paffed into the apartment of the Parrots, which
appeared to them much more beautiful than what they had feen in
other places, but of a moderate fize. The Portuguefe give them
the name of *noyras:* they have a bright gloffy red on the throat and
under the ftomach, and a beautiful gold plate on the back." *Hift.
Gen. des Voy. t. VIII. p.* 136.

   [A] Briffon thus defcribes his *Lorius Moluccenfis :* " It is fcarlet ;
the fpot on the upper part of its back and the upper coverts of the
wings, yellow ; the quills of the wings green externally and above,
below pale rofe colour, within faffron tirt with black ; the lateral
quills of the tail above, crimfon on their firft half, and green on
the other; the two outermoft mixed externally on their laft half
with deep violet."

VARIETIES of the NOIRA.

I*. To the Noira we ought perhaps to refer the Java Parrot mentioned by Aldrovandus, and which the inhabitants of that ifland
term *nor*, which means brilliant. The whole
of the body is of a deep red; the wings and the
tail are of a deep green; there is a yellow fpot
on the back, and a fmall border of the fame colour on the fhoulder. Of the feathers of the
wings, which when clofed appear entirely
green, the coverts only, and the fmall quills,
are yellow, and the large ones are brown.

II†. The Lori defcribed by Briffon under
the name of the *Ceram Lori*, and to which he
applies what we have afcribed to the Noira, is
only a variety, and in no refpect different, except that its legs are green, while thofe of the
former are red, like the reft of the body.

* This is the fecond variety of Linnæus' *Pfittacus Garrulus.*
 † *Pfittacus Garrulus*, Linn. Gmel. and Borowfk.
  Lorius Ceramenfis, Briff.
  Pfittacus Rufus, femoribus alifque viridibus, Fris. & Klein.
  The *Purple Parrot*, Charlton.
  The *Scarlet Parrakeeto with green and black wings*, Will.
  The *Ceram Lory*, Lath.
[A] Specific character: " It is red; its orbits cinereous; its
cheeks and wings green; its tail-quills blue on their pofterior half."

The COLLARED LORY.

Pſittacus-Domicella, Linn. Gmel. and Borowſk.
Lorius Orientalis Indicus, Briſſ.
The *Second Black-capped Lory*, Edw.
The *Purple-capped Lory*, Lath.

ALL the body, including the tail, is of a deep blood colour; the wing is green, the top of the head is black, terminated with violet on the nape; the legs and the fold of the wing are of a fine blue; the lower part of the neck is furniſhed with a yellow collar, which we have adopted as the ſpecific character.

The bird figured in the *Planches Enluminées* under the name of the *Eaſt Indian Lory*, and which Briſſon deſcribes by the ſame appellation, appears to be the female of this; for the only difference is that it wants the yellow collar, and that the blue ſpot on the top of its wing is not ſo broad; it is alſo ſomewhat ſmaller. This Lory is like all the reſt of the kind, very gentle and familiar; but it is alſo very delicate and difficult to breed. None more eaſily learns to ſpeak, and even with diſtinctneſs. " I have ſeen one," ſays Aublet, " which repeated every thing it heard the firſt time *." Though this

---

* " It had come from India to the Iſle of France, and had been given to me by the Count d'Eſtaing; it was aſtoniſhing." *Note communicated by M. Aublet.*

capacity

capacity is very aftonifhing, there is no reafon
to doubt of it*. This bird is valued very high;
Albin fays that he faw one fold for twenty gui-
neas.—We may regard the *Eaft Indian collared
Lory* as a variety of this fpecies [A].

The TRICOLOR LORY.

THIRD SPECIES.

Pfittacus-Lory, Linn. Gmel. and Borow&k.
Lorius Philippenfis, Briff.
The *Firft Black-capped Lory,* Edw. and Lath.

THE fine red, the azure, and the green,
which are difpofed in large fpots on the
plumage of this Lory, have induced us to give
it the epithet of *Tricolor.* The forepart and
the fides of the neck, the flanks, the lower
part of the back, the rump and half the tail,
are red. The under fide of the body, the legs,
and the top of the back, are blue; the wing is

---

* " The Dutch had one that in a moment imitated the cries of
the other animals which it heard." *Second Voyage des Hollandois.*
Hift. Gen. des Voy. t. VIII. p. 377.—" All voyagers fpeak with
admiration of the facility with which the Parrots of the Moluccas
can repeat what they hear. Their colours are variegated, and
form an agreeable mixture; they fcream much, and very loud."—
Ibid.

[A] Specific charaFter of the *Pfittacus-Domicella:* " It is red;
its cap violet: its wings green; its fhoulders and knees blue; its
orbits brown."

I 2   green,

green, and the point of the tail, blue; the crown of the head is covered by a black cap. The bird is near ten inches long. Few are fo beautiful, both on account of the brilliancy of the colours, and their elegant contraft. Edwards faw it alive, and terms it the *Little Lory*; it whiftled pleafantly, he fays, and pronounced feveral words diftinctly; and, leaping brifkly on its rooft or on the finger, it called with a foft clear voice, *Lory, lory*. It played with the hand, and ran after perfons, hopping like a fparrow. This charming bird lived but a few months in England. The fpecimen which we have defcribed was brought by Sonnerat from the ifland Yolo, which the Spaniards claim as one of the Philippines, and the Dutch as one of the Moluccas [A].

The CRIMSON LORY.

FOURTH SPECIES.

Pfittacus Puniceus, Gmel.
Lorius Amboinenfis, Briff.

THIS Lory is near eleven inches long. We term it *crimfon*, becaufe the red of its plumage, the face except, is not fo brilliant as

[A] Specific character of the *Pfittacus-Lory:* " It is purple; its cap violet; its wings green; its breaft, its cheeks, and its tail, blue; its orbits fomewhat carnation."

in

THE LORY.

in the others, and has a dull brown caft on the wing. The blue of the top of the neck and of the ftomach is weak, and inclined to violet; but, on the fold of the wing, it is bright and azure, and, at the edge of the great quills, it is loft in their blackifh ground; the tail is of a fmoky red below, and of the fame tile-red above, as the back. This is not the only fpecies feen at Amboyna, and from Gemelli Carreri the following alfo appears to be found there [A].

The RED LORY.

FIFTH SPECIES.

Pfittacus Ruber, Gmel.
The *Molucca Lory*, Lath.

IT is entirely red, except the tip of the wing, which is blackifh, and two blue fpots on the back, and one of the fame colour on the under coverts of the tail. It is ten inches long, and appears to be a new fpecies. It is improperly termed the *Chinefe Lory* in the *Pl. Enlum.* for voyagers never mention the Lories as found in China *, and one of our beft obfervers,

[A] Specific charaƈter of the *Pfittacus Puniceus*: " It is deep fcarlet, below white; the leffer and inferior coverts of its wings, and the inner and under fide of its wing-quills are blackifh brown."

* " At Amboyna there are many fpecies of Parrots, and among
others

fervers, M. Sonnerat, affures us, on the contrary, that they are all inhabitants of the Moluccas and of New Guinea. In fact, the *Gilolo
Lory* * of this author feems to be exactly the
fame with the prefent [A].

The RED and VIOLET LORY.

SIXTH SPECIES.

Pfittacus Guebienfis, Gmel.
The *Gueby Lory,* Lath.

THIS bird has hitherto been found only at
Gueby. All the body is of a fhining red,
regularly fcaled with violet brown from the
back of the head, paffing by the fides of the
neck, as far as the belly ; the wing is broken
by red and black, in fuch a manner that this
laft colour terminates all the points of the quills,
and marks a part of their webs; the fmall quills
and their coverts neareft the body are dun-vio-
let ; the tail is copper red. The total length
of the bird, eight inches [B].

others is one which has all its feathers carnation." *Voyage autour
du monde,* par Gemelli Carreri, t. V. p. 236.
  * Voyage à la nouvelle Guinee, p. 177.
  [A] Specific character of the *Pfittacus Ruber:* " It is red ; the
fpace about the eyes and the wing-quills black ; the fpot on the wings
and the lower coverts of the tail fky-blue ; the tail tipt with bay."
  [B] Specific character of the *Pfittacus Guebienfis:* " It is brilliant
red ; its wing-quills ftriped tranfverfely with black ; its tail brown-
red."

The GREAT LORY, *Buff.*

SEVENTH SPECIES.

Pſittacus Grandis, Gmel.
The *Grand Lory,* Lath.

THIS is the largeſt of all the Loris; it is thirteen inches long. The head and neck are of a fine red; the lower part of the neck, where it joins the back, is violet blue; the breaſt is richly clouded with red, blue, violet, and green, and the mixture of green and fine red is continued under the belly; the great quills, and the edge of the wing, from the ſhoulder, are ſky blue; the reſt of the upper ſurface is dull red. Half of the tail is red, the tip yellow.

It appears that Voſmaer deſcribes the ſame ſpecies by the name of *Ceylon Lory.* It was probably carried from a greater diſtance to that iſland, and thence brought into Holland; but it lived there only a few months.

The PARRAKEET LORIS.

THE following fpecies are almoft entirely
red, like the Loris, but their tail is longer,
though not fo long as that of the Parrakeets.
We fhall therefore confider them as forming
the intermediate gradation.

———————

The RED PARRAKEET LORY, *Buff*.

FIRST SPECIES.

Pfittacus Borneus, Linn. and Gmel.
Pfittaca Coccinea Bonarum Fortunarum Infulæ, Briff.
The *Long-tailed Scarlet Lory*, Edw. and Lath.

THE plumage of this bird is almoft wholly
red, except fome of the coverts and the
tips of the quills of the wing and of the tail,
which are partly green and partly blue. The
total length is eight inches and a half. Edwards fays that it is very rare, and that a traveller brought it from Borneo, and gave it to
Sir Hans Sloane.

[A] Specific charaĉter of the *Pfittacus Borneus*, as given by Mr.
Latham: " It is red; the quills of its wings and tail green; a
blue fpot on its wings, its orbits brown."

The VIOLET and RED PARRA-KEET LORY, *Buff*.

SECOND SPECIES.

Pſittacus Indicus, Gmel.
Pſittacus Coccineus, Lath. Ind.
Pſittaca Indica Coccinea, Briſſ.
The *Indian Lory*, Lath. Syn.

THE prevailing colour is red, mixed with violet blue. The bird is ten inches long, and its tail occupies near one third of that ſpace. The tail is entirely of a full blue, which alſo covers the flanks, the ſtomach, the top of the back, and of the head ; the great quills of the wing are yellow ; all the reſt of the plumage is of a fine red, edged with black, which is diſ-poſed in feſtoons on the wings.

[A] Specific character of the *Pſittacus Indicus :* " It is ſcarlet, variegated with brown and violet ; the upper parts of its head and neck, its breaſt, and a band behind the eyes, violet ; the tips of the greater tail quills dilute brown, thoſe of the leſſer and the co-verts, brown violet."

TRICOLOR PARRAKEET LORY.

THIRD SPECIES.

Pſittacus Amboinenſis, Linn. and Gmel.
Pſittaca Amboinenſis Coccinea, Briſſ.
The *Amboyna Red-Parrot,* Lath.

RED, green, and turkey blue, are diſpoſed in large marks over all its plumage; red covers the head, the neck, and all the upper ſide of the body; the wing is deep green; the back and tail are of a full velvet blue. The tail is ſeven inches long; and the whole bird is fifteen inches and a half long, and as large as a turtle. —The tail in theſe three laſt ſpecies, though longer than common in the Loris and Parrakeets properly ſo called, is not tapered as in the long-tailed Parrakeets, but conſiſts of equal quills, with a ſquare termination.

[A] Specific character of the *Pſittacus Amboinenſis:* " It is ſcarlet; its back blue; its wings with a green ſpot."

PARRAKEETS
OF THE OLD CONTINENT.

PARRAKEETS
IN WHICH THE TAIL IS LONG AND EQUALLY TAPERED.

WE fhall diftinguifh the long-tail Parrakeets into two families; into thofe which
have the tail equally tapered, and into thofe which
have the two middle quills much longer than
the reft, and appearing detached from each other.
All thefe Parrakeets are larger than the fhort-
tailed ones, which we fhall afterwards defcribe.

THE
GREAT COLLARED PARRAKEET.
La Grande Perruche a Collier d'un Rouge Vif, Buff.
FIRST SPECIES,
With a long, equally tapered Tail.

Pfittacus-Alexandri, Linn. Gmel. Scop. and Bor.
Pfittacus Torquatus Macrourus, Ray.
Perrocello, Olin.
The *Ring Parrakeet,* Edw. and Will.
The *Alexandrine Parrakeet,* Lath.

PLINY and Solinus have both defcribed the
green collared Parakeet, which was the only
one

one known in their time, and which came from
India. Apuleius delineates it with that elegance
which he ufually affects *, and fays that its
plumage is of a pure brilliant green. The only
interruption of this colour is, according to Pliny, a half collar of bright red on the top of the
neck †. Aldrovandus, who has collected all
the particulars, leaves no room to doubt but
that the *long-tailed and collared* Parrot of the ancients is the fame with the red collared Great
Parrakeet of this article. There are two circumftances fufficient to evince this; the firft
is the breadth of the collar, which, about the
middle, is equal to the *thicknefs of the little finger*, and the fecond, that there is a red fpot
which marks *the top of the wing*. Both thefe
are peculiarly the properties of this Parrakeet.
It is equally beautiful with the reft of the
tribe : its plumage is of a lively light green
on the head, and deeper on the wings and the
back ; the rofy half-collar embracing the back
of the neck, joins, on the fides, to the black
bar that covers the throat ; the breaft is of a
vermilion red, and there is a purple fpot on the
crown of the head : the tail is beautiful, and
longer than the body ; its upper furface mixed
with green and beryl, its under furface of a
delicate yellow. The bird is found, not only

* Florid. *Lib. II.*

† *'Viridem toto corpore, torque tantum miniato in cervice diftinctam.*
Plin. *Lib. X.* 42.

5　　　　　　　　　　　　　on

on the fouth of the continent of Afia, but alfo
in the adjacent iflands, and at Ceylon ; for this
is Taprobana, from which Alexander's fleet
brought the firft Parrot into Greece [A].

The DOUBLE COLLARED PARRAKEET.

SECOND SPECIES,
With a long and equal Tail.

Pfittacus-Alexandri, fecond Variety, Gmel.
The *Double-ringed Parrakeet*, Lath.

Two fmall wings, the one rofe-coloured, and
the other blue, entirely encircle the neck ;
all its plumage is green, which is deeper on the
back, has a yellow caft under the body, and,
in many parts, there is a dufky ftreak on the
middle of each feather ; below the tail, a yellowifh fringe edges the brown-gray, which is
impreffed on each quill. The bird is as large as
a turtle; and, as it is found in the ifland of
Bourbon, it probably inhabits alfo the corre-
fponding continent, either of Africa or India.

[A] Specific chara&er of the *Pfittacus-Alexandri :* "It is green;
its collar and breaft red; its throat black."

The RED-HEADED PARRAKEET.

THIRD SPECIES,

With a long and equal Tail.

Pſittacus Erythrocephalus, Gmel.
Pſittaca Ginginiana, Briſſ.
The *Bloſſom-headed Parrakeet*, Lath.

THIS Parrakeet is eleven inches total length, and the tail is longer than the body. The whole of the upper ſide is dull green, with a purple ſpot on the top of the wing; the face is purpliſh red, which, on the head, runs into blue, and is intercepted, on the nape of the neck, by a ſtreak produced from the black that covers the throat; the under ſide of the body is a dull dirty yellow; the bill is red [A].

THE

BLUE-HEADED PARRAKEET.

FOURTH SPECIES,

With a long and equal Tail.

Pſittacus Cyanocephalus, Linn. Gmel. Briſſ. and Gerini.
The *Blue-headed Parrot*, Lath.

IT is ten inches long; its bill is white, its head blue, its body green; the fore-ſide of

[A] Specific character of the *Pſittacus Erythrocephalus:* " It is green; its head red, mixed with blue; its throat black, with a black and pale green collar."

its

its neck yellow, and the under furface of its belly and tail yellow mixed with green ; the middle quills of the tail have a blue caft above; the legs are bluifh [A].

The LORY PARRAKEET.

FIFTH SPECIES,
With a long and equal Tail.

Pfittacus Ornatus, Linn. Gmel. and Bor.
Pfittaca Indica Varia, Briff.

WE adopt the name which Edwards has be-
ftowed on this fpecies, becaufe it is of a beautiful red, like the loris ; that colour, inter-fected by fmall brown waves, covers the throat, the fore-part of the neck, and the fides of the face, as far as the back of the head, which it furrounds. The crown of the head is purplifh, Edwards terms it blue ; the back, the upper furface of the neck, the wings, and the fto-mach, are emerald green ; the fides of the neck, and the flanks, are irregularly fpotted with orange yellow ; the great quills of the wing are blackifh, fringed at the end with yellow; the tail, which is green above, appears tinged be-low with red, and is yellow at the tip; the

[A] Specific character of the *Pfittacus Cyanocephalus :* " It is green ; its head and throat blue."

bill

bill and feet are white-gray. This Parrakeet is about the middle fize, and is only feven inches and a half long. It is one of the handfomeft, on account of the brilliancy and choice of its colours. It is not the *paradife bird* of Seba, as Briffon fuppofes; for in that bird the tail is unequally tapered [A].

THE

YELLOW PARRAKEET, *Buff.*

SIXTH SPECIES,
With a long and unequal Tail.

Pfittacus Solftitialis, Linn. and Gmel.
Pfittaca Angolenfis Lutea, Briff.
Pfittacus Croceus, Klein.
Pfittacus Luteus, cauda longa, Fris.
The *Angola Yellow Parrot,* Alb. and Lath.

ALL its plumage is yellow, except the belly and the ring of the eye, which are red, and the quills of the wing, with a part of thofe of the tail, which are blue: the former are interfeded near their middle by a yellowifh bar. Albin tells us that it can learn to fpeak, and, though he calls it the Angola Parrot, he fays it comes from the Eaft Indies [B].

[A] Specific charader of the *Pfittacus Ornatus:* " It is yellow-green; the back of its head, its throat, and its breaft, red; its top, and its ears, blue, with cinereous orbits."

[B] Specific charader of the *Pfittacus Solftitialis:* " It is yellow; the coverts of its wings green; its orbits red; its lateral tail-quills blue exteriorly."

THE
AZURE-HEADED PARRAKEET.

SEVENTH SPECIES,
With a long and equal Tail.

Pſittacus-Alexandri, Var. 4, Gmel.
Pſittaca Cyanocephalos Indica, Briſſ.
The *Blue-headed Parrakeet*, Edw. and Lath.

THIS Parrakeet is of the ſize of a pigeon; all
its head, face, and throat, are of a fine
ſky-blue; there is a little yellow on the wings;
the tail is blue, equally tapered, and as long as
the body; the reſt of the plumage is green. It
is brought from the Eaſt Indies, according to
Edwards, who deſcribes it.

The MOUSE-PARRAKEET, *Buff.*

EIGHTH SPECIES,
With a long and equal Tail.

Pſittacus Murinus, Gmel.
The *Grey-breaſted Parakeet*, Lath.

THIS ſpecies appears to be new, and we know
not its native country. Perhaps the following extract from a voyage to the Iſle of
France alludes to it*:—" The green Parra-

* Voyage à Iſle de France, 1772, *p.* 122.

keet

keet with a gray cowl is about the bulk of a
sparrow, and cannot be tamed;" though, however, this Parrakeet is considerably larger than
the sparrow. We have called it the Mouse-
Parrakeet, on account of a large mark of mouse
gray that covers the breast, the throat, the forehead, and the whole of the face; the rest of the
body is olive green, except the great quills of
the wing, which are of a deeper green: the
tail is five inches long, the body as much; the
feet are gray; the bill is white gray. All the
plumage is pale and discoloured, and gives it a
sombre air; and in point of beauty it is the most
inferior of the family [A].

The MUSTACHO·PARRAKEET.

NINTH SPECIES,
With a long and equal Tail.

Psittacus-Pondicerianus, Gmel.

A BLACK streak stretches between both eyes,
and two large mustachoes of the same colour rise from the lower mandible, and spread
over the sides of the throat. The rest of the
face is white and bluish; the tail is green above,

[A] Specific character of the *Psittacus Murinus:* " It is olive;
its face, its throat, and its breast, are mouse-coloured; its wing-
quills green."

and

and ftraw coloured below; the back is deep
green; there is fome yellow on the coverts of
the wings, of which the great quills are of a
deep water-green; the ftomach and breaft are
lilac. The bird is near eleven inches in length,
and the tail occupies the half of it. It has not
been hitherto noticed by any naturalift.

BLUE-FACED PARRAKEET, *Buff.*

TENTH SPECIES,
With a long and unequal Tail.

Pfittacus Hæmatodus, Linn. and Gmel.
Pfittacus Amboinenfis Vária, Briff.
The *Red-breafted Parrot*, Edw. and Lath.

Tʜɪs beautiful Parrakeet is green on the up-·
per fide, and the head is painted with
three different colours; the face and throat with
indigo, the back of the head with brown-green,
and the crown with yellow: the lower part of
the neck and the breaft are red fnuff-colour on a
ground of brown green; the belly is green, the
abdomen confifts of yellow and green, and the
under furface of the tail is yellow. Edwards
has given this bird the name of the *Red-breafted
Parrakeet*; but it appears to have been reprefented from a fpecimen preferved in fpirit of

wine, and its colours were tarnifhed. Our fpecimen was in better condition. The bird is
found at Amboyna.

We fhall regard the *Molucca Parrakeet* as
either a variety of this, or a clofely related fpecies ; its fize and ,colours are nearly the fame,
only the head is entirely indigo, and there is a
fpot of the fame colour on the belly ; and the
aurora-red of the breaft is not waved, but mixed with yellow. The tail of thefe Parrakeets
is as long as the body, which is ten inches;
their bill is reddifh white *.

- - -

THE
LACE-WINGED PARRAKEET,
Buff. and *Lath.*

ELEVENTH SPECIES,
With a long and equal Tail.

Pfittacus Olivaceus, Gmel.

THE wings are laced with blue, yellow, and
orange ; the firft of thefe colours occupying the middle of the feathers, the two others
extending to the border; the great quills are
olive-brown, and the fame colour is extended
over all the body, except a bluifh fpot behind

* It is the firft variety of the *Pfittacus Hæmatodus,* according to
Gmelin, and denominated the *Orange-breafted Parrot* in Latham's
Synopfis.

the

the head. The bird is near eleven lines long, of which the tail is more than a third ; but the wing is alſo very long, and covers near half the tail, whereas in the other Parrakeets the wings are much ſhorter.

We now proceed to enumerate the Parrakeets of the old continent, whoſe tail is alſo long, but unequally tapered.

[A] Specific character of the *Pſittacus Olivaceus* : " It is olive ; a blue ſpot on the back of its head ; its wings variegated with blue, green, and orange."

PARRAKEETS

OF THE OLD CONTINENT,

WHICH HAVE A LONG AND UNEQUAL TAIL.

The ROSE-RINGED PARRAKEET.

FIRST SPECIES,
With a long and unequal Tail.

Pſittacus-Alexandri, Second Variety, Gmel.
Pſittaca Torquata, Briſſ. and Gerini.

SO far is this Parrakeet from being peculiar to
the new world, as Briſſon repreſents it, that
it is there entirely unknown. It inhabits many
parts of Africa, and is brought in great numbers to Cairo by the caravans of Ethiopia. The
veſſels that ſail from Senegal or Guinea, where
it is common, carry it with the negroes into the
Weſt India iſlands. None of theſe Parrakeets
are found on the continent of America; they
are only ſeen near the ſettlements of St. Domingo, Martinico, Guadeloupe, &c. which the
African veſſels perpetually frequent, but at Cayenne, where negroes are very ſeldom imported,
they never occur. All theſe facts, which were
communicated by an excellent obſerver, prove
that the Roſe-ringed Parrakeet is not a native

8

of

of the new world. But it is ftill more fingular that Briffon fhould confider the-fame bird as the Parrot of the ancients; as if the Greeks and Romans went to America to find it. Befides, it is a different fpecies, which we have already defcribed.

The Rofe-ringed Parrakeet is fourteen inches long, but of this extent the tail and its two pro-jecting feathers occupy near two thirds; thefe feathers are of a beryl blue; all the reft of the plumage is a light foft green, which is rather more vivid on the quills of the wings and mix-ed with yellow on thofe of the tail; a fmall rofe collar clafps the back of the neck and joints with the black of the throat; a bluifh tinge ap-pears on the feathers of the nape of the neck, which recline upon the collar; the bill is brown red.

The LITTLE PARRAKEET,

With a Rofe-coloured Head and long Shafts.

SECOND SPECIES,
With a long and unequal Tail.

Pfittacus Erythrocephalus, Firft Variety, Gmel.
Pfittacus bengalenfis, Briff.
The *Parrakeet from Bengal,* Albin.
The *Rofe-headed Ring Parrakeet,* Lath.

THIS little Parrakeet, which is not more than four inches long, meafures twelve, if the

K 4　　　　　two

two long fhafts be included; thefe are blue, the reft of the tail, which is not more than two inches and a half long, is olive green, which is alfo the colour of all the under fide of the body, and even of the upper fide, where it is only deeper; a few fmall red feathers appear through the plumage, on the top of the wing; the head is rofe-colour, mixed with lilac, interfected and bordered by a black ring, which, rifing from the throat, encircles entirely the neck. Edwards fpeaks with rapture of this bird: it is termed, he fays, in Bengal *fridytutab*.

The LONG-SHAFTED GREAT PARRAKEET.

THIRD SPECIES,
With a long and unequal Tail.

Pfittacus Erythrocephalus, Var. 3, Gmel.
The *Malacca Parrakeet*, Lath.

THE colours of this Parrakeet are fo like thofe of the preceding, that they might be viewed as the fame fpecies, were they not confiderably different in refpect to fize. This is fixteen inches long, including the two fhafts of the tail, and the other dimenfions are proportional; the fhafts are blue, as in the preceding; the tail is of the fame olive green, but

deeper

deeper, and of the fame tinge as the wings ; the middle of the wings is rather bluer ; all the green on the body is much diluted with an admixture of yellowifh ; the head is not entirely rofe-coloured, but only the fpace near the eyes, and the back of the head ; the reft is green, and there is no ring about the hood.

The REDDISH-WINGED GREAT PARRAKEET, *Buff*.

FOURTH SPECIES,
With a long and equal Tail.

Pfittacus Eupatria, Linn. and Gmel.
Pfittaca Gin iniana, Briff.
The *Gingi Parrot*, Lath.

THIS Parrakeet is twenty inches, from the point of the bill to the extremity of the two long fhafts of the tail : all the upper furface of the body is deep olive green, the under furface is pale green, intermixed with yellowifh ; on the pinion of each wing is a fmall red mark, and another one of dilute blue on the middle of the long feathers of the tail ; the bill is red, and alfo the feet and nails [A].

[A] Specific chara&ter of the *Pfittacus-Eupatria:* " It is green; its cheeks naked ; its fhoulders fcarlet ; its bill purplifh."

THE
RED-THROATED PARRAKEET.

FIFTH SPECIES,
With a long and unequal Tail.

Pſittacus Incarnatus, Linn. and Gmel.
Pſittaca Indica, Briſſ.
The *Little Red-winged Parrakeet*, Edw. and Lath.

EDWARDS, who deſcribes this bird, ſays that
it is the ſmalleſt Parrakeet which he had
ever ſeen.   It is not larger than a titmouſe, but
its tail is longer than its body ; the tail and
back are of a full green ; the coverts of the
wings and of the throat are red ; the under ſide
of the body is yellowiſh green ; the iris of the
eye is of ſo deep a caſt as to appear black,
though in moſt of the Parrakeets it is gold-coloured.   Edwards ſays that it comes from India.

[B] Specific character of the *Pſittacus Incarnatus:* " It is green ;
its bill, feet, and nails, carnation ; its cere and orbits whitiſh ; its
throat and the coverts of its wings red."

The BLACK-BANDED GREAT PARRAKEET.

SIXTH SPECIES,
With a long and unequal Tail.

Pſittacus Atricapillus, Linn. and Gmel.
Ara Moluccenſis Varia, Briſſ.

THE bird, which Briſſon has termed the *Molucca Maccaw*, is undoubtedly only a Parrakeet ; for no Maccaw inhabits either India,
or any part of the ancient continent. Seba
calls the ſame bird a *Lory*; but the long feathers of its tail ſhew evidently that it is a Parrakeet. The total length of the bird is fourteen inches, of which the tail is near ſeven.
The head has a black band, and the neck a red
and green collar ; the breaſt is of a fine light
red ; the wings and back are of a rich Turkey-
blue; the belly is deep green, ſprinkled with
red feathers ; the tail, of which the middle
quills are the largeſt, is coloured with green
and red, with black edges. This Parrakeet
was, according to Seba, brought from the Pa-
puan iſlands ; a Dutch ſettler at Amboyna purchaſed it of an Indian for five hundred florins.
This price was not extravagant, conſidering the
beauty and gentleneſs of the bird ; it pronounced diſtinctly ſeveral words in different languages,
it made its ſalute in the morning, and ſung its
ſong.

fong. Its attachment equalled its accomplifh-
ments; for having loft its mafter, it died of melancholy * [A²].

THE
RED and GREEN PARRAKEET.

SEVENTH SPECIES,
With a long and unequal Tail.

Pfittaca Japanenfis, Briff.

THIS fpecies has been denominated by Briffon the *Japan Parrakeet* ; but no Parrots
are found in that ifland, or in the northern provinces of China, except fuch as have been carried thither; and probably this bird, of which
Aldrovandus faw only the figure, came from
fome more fouthern part of Afia. Willughby
obferves even that both the figure and its defcription appeared fufpicious. The plumage is
compofed of green, red, and a little blue; the
firft of thefe colours is fpread on the upper fide
of the body, the fecond on the under fide and
the tail, except the two long fhafts, which are
green, and the blue that tinges the fhoulders and
the quills of the wings: there are two fpots of
the fame colour on each fide of the eye.

* Kæmpfer, *t. I. p.* 113.
[A] Specific character of the *Pfittacus Atricapillus :* " Above it
is blue; its throat and breaft red; its belly and vent green; its
top black, with a green and black collar."

The CRESTED PARRAKEET.

EIGHTH SPECIES,

With a long and unequal Tail.

Pfittacus Javanicus, Linn. and Gmel.
Pfittaca Javenfis Criflata Coccinea, Briff.
The *Crefled Red Parrakeet*, Lath.

THIS is the *fmall Parrot of Bontius*, whofe luftre and variety of plumage are extolled by Willughby; the pencil can hardly imitate, he fays, its brilliancy and beauty. It is compofed of bright red, and of rofe colour, mixed with yellow and green on the wings, and with green and blue on the tail, which is very long, and projects beyond the wings ten inches; a remarkable excefs in a bird which is not larger than a lark. The feathers on its head form a creft, which muft be very elegant, fince it is compared to a peacock's tuft in the following extract, which feems to allude to this beautiful fpecies. " This Parrakeet is about the bulk of a fifkin; on its head is a tuft of three or four feathers, refembling that of the peacock. This bird is exceedingly gentle *." Thefe little Parrakeets are found in Java, in the interior parts of the country. They fly in flocks, making a great noife. They chatter much, and, when tamed, they eafily repeat whatever they are taught †.

* Lettres Edifiantes, *fecond recueil*, *p.* 69.   † Willughby.

T H E

SHORT-TAILED PARRAKEETS

OF THE OLD CONTINENT.

THESE are frequent in the fouth of Afia, and in Africa. They are entirely different from the Parrakeets of America, except a few which were probably carried thither. The fpecies of Parrakeets that inhabit the old continent are much more numerous than thofe of the new : their habits are alfo different ; fome, for inftance, cling to a fmall branch, and fleep with their head hanging downwards, which is not remarked of thofe fettled in America.

In general, the Parrots of the new world make their nefts in the hollows of trees, particularly thofe deferted by the woodpeckers, which are called *carpenters* in the Weft India iflands *. In the old continent, on the contrary, many travellers inform us that feveral fpecies of Parrots fufpend their nefts, which are formed with rufhes and roots, to the ends of

---

* Lery pofitively avers that the American Parrots never fufpend their nefts, but make them in cavities of trees. *Apud Clufium Auct. p.* 364.

flexible

flexible boughs *. This difference in the manner of neſtling, if it obtains in a great number of ſpecies, may have been prompted by the diverſity of climate. In America, where the heat never is exceſſive, it is neceſſary to concentrate it; but, in the burning plains of Africa, the neſt is rocked by the tempering breeze.

THE

BLUE - HEADED PARRAKEET.

FIRST SPECIES,
With a ſhort Tail.

Pſittacus-Galgulus, Linn. Gmel. and Bor.
Pſittacula Malaccenſis, Briſſ.
The *Sapphire-crowned Parrakeet*, Edw. and Lath.

THE crown of the head is of a fine blue, and round the neck is an orange half-collar; the breaſt and rump are red, and the reſt of the plumage green.

Edwards ſays that he received this bird from Sumatra: Sonnerat found it in the iſland of Luçon †.

This is one of the ſpecies that ſleep with their head hanging downwards. It lives on

---

* *See* the account given by Cadamoſto. *Hiſt. Gen. des Voy.* t. II. p. 305.—*Voy. à Madagaſcar* par Fr. Cauche, *Paris*, 1651.

† *Voyage à la Nouvelle Guinee, p.* 76.

çallou,

callou, a kind of white liquor, which is obtained in the East Indies from the cacoa tree, by
making an incision near the foot stalks that sup-
port the fruit.   A hollow bamboo is fixed to
the extremity of the branch to collect the sap,
which when first drawn is pleasant, and resembles the taste of new cyder.

It appeared to us that we ought to refer to
the same species the bird mentioned by Aldrovandus, of which the crown of the head was a
fine blue, the rump red, and the rest of the
plumage green.   But as that naturalist does not
take notice of the half-collar, or of the red on
the breast,   and also says that it came from
Malacca, it was perhaps of a different, though
closely related, species.

[A] Specific character of the *Psittacus Galgulus*: " It is green;
its rump and breast scarlet; the top (of the male) blue."

FIG 1.THE RED HEADED GUINEA PARAKEET.FIG. 2 THE PHILIPPINE PARAKEET

The RED-HEADED PARRAKEET,
Or the GUINEA SPARROW.

SECOND SPECIES,
With a short Tail.

Psittacus Pullarius, Linn. Gmel. and Bor.
Psittacula Guineensis, Briss.
Psittacus Pusillus viridis, Ray.
The *Red-headed Guinea Parrakeet,* Lath.

THIS bird is known to bird-fanciers under the name of *Guinea Sparrow* *: it is very common in that country, and is brought into Europe on account of the beauty of its plumage, its tamenefs, and gentlenefs; for it cannot be taught to fpeak, and has only a difagreeable fcream. Many are loft in the paffage, and fcarcely one out of ten reaches Europe †; yet they live to a confiderable age in our climates, when fed on panic and canary feeds, and kept in pairs. They alfo lay fometimes ‡, but the

eggs

---

* " The Parroquets are called *Guinea Sparrows,*" fays Bofman, " though it would not be eafy to affign the reafon, fince the common fparrows are here (on the gold coaft) extremely abundant. Their red bill is a little curved, like that of the Parrots. Thefe fmall creatures are carried to Holland in great numbers: there they fell very dear, though in Guinea a dozen may be purchafed for a crown, of which nine or ten will die in the paffage." *Hiſt. Gen. des Voy. t. IV. p.* 247.

† *Hiſt. Gen. des Voy. t. IV. p.* 64.

‡ There is no doubt but with due care thefe birds might be propagated in the domeftic ftate. Sometimes the force of nature alone,

eggs feldom hatch. If one of the pair die, the other grows melancholy, and hardly ever fur- vives. They are extremely attentive to each other; the male fits befide the female, and dif- gorges into her bill, and he is uneafy if fhe be a moment out of his fight. Thus they fweeten their captivity, by love and gentle manners. Travellers relate that, in Guinea, thefe birds are fo numerous as much to injure the crops *. It would feem that the fpecies is difperfed over al- moft all the fouthern climates in the ancient con- tinent; for it is found in Ethiopia †, in the Eaft Indies ‡, in the ifland of Java §, as well as in Guinea ‖.

Many perfons call this bird very improperly the *Brafilian fparrow*; for it is not a native of America.—The body is entirely green, marked with a fpot of fine blue on the rump, and by a

alone, in fpite of the rigor of the climate and of the feafon, pre- vails in them. Her Highnefs of Bourbon and Vermandois, Ab- befs of Beaumont-les-Tours, had two Parrakeets from Goree, that hatched two young ones in the month of January in a room with- out fire, but which the cold foon killed.

* Barbot. *Hift. de Guinee*, p. 220.

† Clufius, *Exot. Auctuar.* p. 365.

‡ Albin, Vol. III. p. 7.

§ Salerne, *Ornithol.* p. 72.

‖ "All along this coaft they are numerous, but efpecially near the lower part, as at Mourée, Cormantin, and Acra." *Voy. en Gui- nee*, par Bofman, *Utrecht*, 1705, p. 277. "Infinite numbers of Parrots are found at Anamaboe: they are of the bulk of fparrows; their body is of a beautiful green; their head and tail of an admir- able red, and all their figure fo fine, that the author brought fome to Paris, as a prefent fit for the king." *Hift. Gen. des Voy. t. IV.* p. 64.

flame

flame coloured mafk, mixed with a blufh tint, which covers the forehead, enclofes the eye, and defcends under the throat, and, in the middle of it, the bill appears of a reddifh white; the tail is very fhort, and appears all green when clofed, but if difplayed it is perceived to be interfected tranfverfely with three bars, the one red, the other black, and the third green, which borders and terminates the extremity; the pinions of the wings are blue in the male, and yellow in the female, which differs not from the former, except that its head is of a fainter red.

Clufius has very diftinctly defcribed this fpecies under the name of *Pfittacus Minimus.* Edwards, Briffon, and Linnæus, have confounded it with the *little American Parrot painted with various colours* of Seba. But it is undoubtedly a different bird, for Seba fays that his Parrot has not only a collar of fine fky-blue, and a tail magnificently tinged with a mixture of five colours, viz. blue, yellow, red, brown, and deep green; but that its voice and fong are pleafant, and that it eafily learns to fpeak. It is evident that all thefe attributes belong not to the Red-headed Parrakeet. Perhaps the bird, which Seba faw alive, forms a fixth fpecies in the fhort-tailed Parrakeets of the new continent.

A variety, or perhaps a contiguous fpecies, may be found in the bird defcribed by Edwards under the denomination of the *fmalleft green and*

red

red Parrakeet, which differs in no refpect from the preceding, except that its rump is red [A].

The COULACISSI.

THIRD SPECIES
Of fhort-tailed Parrakeet.

Pfittacus Galgulus, Var. Gmel.
Pfittacula Philippenfis, Briff.
The *Philippine Parrakeet*, Lath.

THIS is the name which the bird receives in its native region, the Philippine iflands, and particularly in that of Lucon. The fore-head, the throat, and the rump, are red; there is a half-collar of orange on the upper fide of the neck; the reft of the body, and the fuperior coverts of the wings, are green; the great quills of the wings are deep green on the outfide, and blackifh on the infide; the middle quills of the wings and thofe of the tail are green above and blue below; the bill, the feet, and the nails, are red.

The female is difcriminated from the male by a bluifh fpot on each fide of the head between the bill and the eye, and has no half-collar on the neck, or red on the throat, and the red

---

[A] Specific character of the *Pfittacus Pullarius*: " It is green; its front red   its tail fulvous; its bar black; its orbits cinereous."

tinge

tinge of its forehead is alfo more dilute and narrower.

Briffon and Linnæus* confound it with Edwards's Sapphire-crowned Parrakeet, which is our Blue-headed Parrakeet, and the firft fpecies with a fhort tail.

THE

GOLDEN-WINGED PARRAKEET.

FOURTH SPECIES,

With a fhort Tail.

Pfittacus Chryfopterus, Linn. and Gmel.
Pfittacus Alis Deauratis, Briff.

EDWARDS gives an account of this bird. It was probably brought, he fays, from the Eaft Indies, but he is not quite certain. The head, the fmall fuperior coverts of the wings, and the whole body, are green, only it is deeper on the body than beneath ; the great fuperior coverts of the wings are orange; the four primary quills of the wings are deep blue on the outfide, brown on the infide and at the extremity ; the four next are of the fame colour with the firft ; and laftly, thofe near the body are entirely green, as well as the quills of the tail ; the bill is whitifh; the feet and nails are pale flefh colour [A].

* Syft. Nat. *Edit. XIII.*

[A] Specific charaƈter of the *Pfittacus Chryfopteras:* " It is green; its wings marked with a blue and fulvous fpot; its orbits naked and white."

L 3

THE
GRAY-HEADED PARRAKEET.

FIFTH SPECIES,
With a fhort Tail.

Pfittacus Canus, Linn. and Gmel.
Pfittacula Madagafcarienfis, Briff.

BRISSON is the firft who defcribed this bird, which he fays is found in Madagafcar. The head, the throat, and the lower part of the neck, are gray, inclined fomewhat to green; the body is of a lighter green below than above; the fuperior coverts of the wings, and the middle quills, are green; the great quills are brown on the infide, and green on the outfide, and at the extremity; the quills of the tail are light green, with a broad tranfverfe black bar at their extremity; the bill, the feet, and the nails, are whitifh.

[A] Specific charaƈter of the *Pfittacus Canus:* " It is green; its head and its throat gray-green; its tail rounded, with a broad black bar."

The VARIEGATED WINGED PARRAKEET.

SIXTH SPECIES,
With a fhort Tail.

Pfittacus Melanopterus, Gmel.
The *Black-winged Parrakeet,* Lath.

THIS Parrakeet is fomewhat larger than the preceding ones ; it is found in Batavia, and in the ifland of Luçon. We fhall borrow the defcription of it from Sonnerat *. " The head, neck, and belly, are light green and yellowifh ; there is a yellow bar on the wings, but each feather that forms it is edged exteriorly with blue ; the fmall feathers of the wings are greenifh ; the great ones are of a fine velvet black (fo that the wings are variegated with yellow, blue, green, and black) ; the tail is of a light lilac colour ; and near its extremity there is a very narrow black bar ; the feet are gray ; the bill and iris are reddifh yellow "

* Voyage à la Nouvelle Guinee, *p.* 78.

BLUE-WINGED PARRAKEET.

SEVENTH SPECIES,
With a ſhort Tail.

Pſittacus Capenſis, Gmel.

THIS is a new ſpecies, which we received from the Cape of Good Hope, but without any accounts of its habits or of its climate. It is entirely green, except ſome quills of the wings, which are of a fine blue; the bill and feet are reddiſh.—This ſhort deſcription is ſufficient to diſcriminate it from the other ſhort-tailed Parrakeets.

The COLLARED PARRAKEET.

EIGHTH SPECIES,
With a ſhort Tail.

Pſittacus Torquatus, Gmel.

WE ſhall alſo borrow the account of this bird from Sonnerat. " It is found in the Philippines, and eſpecially in the iſland of Luçon; it is of the bulk of the Braſilian (Guinea) Sparrow; all the body is of a lively pleaſant green, which is deeper on the back, and more dilute

3   under

under the belly, and fhaded with yellow; behind the neck and below the head, there is
a broad collar, which confifts, in the male, of
fky-blue feathers; but in both fexes the feathers
of the collar are variegated tranfverfely with
black; the tail is fhort, equal to the wings, and
terminated in a point; the bill, feet, and iris,
are blackifh gray. This fpecies has no merit,
but in its fhape and colours; for it is devoid of
graces, and cannot be taught to fpeak [A].

BLACK-WINGED PARRAKEET,

NINTH SPECIES,
With a fhort Tail.

Pfittacus Minor, Gmel.
The *Luzonian Parrakeet*, Lath.

THIS fpecies comes likewife from the ifle of
Lucon, and Sonnerat thus defcribes it,
" It is rather fmaller than the preceding; the
upper fide of the neck, the back, the fmall feathers of the wings and of the tail, are of a deep
green; the belly is light green and yellowifh;
the crown of the head is a very bright red in
the male, and the feathers, which border the

[A] Specific charaƈter of the *Pfittacus Torquatus*: " It is green;
a yellow bar on its head, ftriped acrofs with black."

upper

upper margin of the bill in the female are of the fame bright red, and there is alfo a yellow fpot above on the middle of the neck; in the male the throat is blue, in the female it is red; in both, the great feathers of the wings are black, thofe that cover the upper furface of the tail are red; the bill, feet, and iris, are yellow. I conceive, fays Sonnerat, thefe two Parrakeets to be the male and female, becaufe they differ little, and are analogous in their fize, fhape, and colour, and inhabit the fame climate; but I will not affert that they may not be diftinct fpecies. Both fleep hanging from branches with their heads downwards, and are fond of the fap that flows from incifions in the cocoa tree [A].

The A R I M A N O N.

TENTH SPECIES,
With a fhort Tail.

Pfittacus Taitianus, Gmel.
The *Otaheitan Blue Parrakeet,* Lath.

THIS bird is found in the ifland of Otaheite, and lodges in the cocoa trees, whence its name *Arimanon.* We borrow the defcription

[A] Specific character of the *Pfittacus Minor:* "It is green; its top fcarlet; its breaft blue; its greater wing-quills black."

from

from Commerſon. We range it after the ſhort-
tailed Parrakeets, though it has a peculiar character that belongs not to any of the genus; viz.
its tongue is pointed, and terminates in a pencil of ſhort white briſtles.

The plumage is entirely of a fine blue, except the throat and the lower part of the neck,
which are white; the bill and feet are red. It
is very common in Otaheite, where it flutters
about continually ſqualling. It forms flocks,
and feeds on bananas, but it is difficult to do-
meſticate; it dies of melancholy, eſpecially if
kept alone in the cage. It cannot be brought
to eat any thing but the juice of fruits, and
conſtantly rejects every kind of ſolid diet.

[A] Specific character of the *Pſittacus Taitianus*: " It is blue;
the feathers on its head longiſh; its throat white; its bill and feet
red."

P A R R O T S

OF THE NEW CONTINENT.

The A R A S.

OF all the Parrots, the Ara is the largeſt, and
the moſt magnificently decked; purple,
gold, and azure, blaze on its plumage; its aſpect
is ſteady and compoſed; its deportment grave,
and even ſupercilious, and as if conſcious of it's
beauty. But its calm temper ſoon renders it
familiar, and even ſuſceptible of a degree of attachment. It may be domeſticated without being enſlaved; it never abuſes the liberty which
has been granted to it; it contracts a fondneſs
for the family where it is adopted, and conſtantly returns from its rambling.

All theſe Aras are natives of the tropical parts
of the new world, and inhabit both the continent and the iſlands; but none are found in
Africa, or in India. Chriſtopher Columbus, in
his ſecond voyage, touched at Guadeloupe, and
there ſaw Aras, which he named *Guaçamayas* *.
They occur even in the deſert iſlands, and every-

---

* Herrera, *Lib. II. cap.* 10.

where

where conftitute the fineft ornament of thofe gloomy forefts with which the wild luxuriancy of nature clothes the face of the earth*.

When thefe Parrots were introduced into Europe, they were beheld with admiration. Aldrovandus, who, for the firft time, faw an Ara at Mantua in 1572, remarks that they were then quite novelties, and highly efteemed; and that princes received thefe birds from one another as rare and valuable prefents. Even Belon, that curious obferver, had never feen a maccaw, for he fays that the gray Parrots are the largeft of all.

We know four fpecies of Aras, viz. the red, the blue, the green, and the black. Our nomenclators enumerate fix †, but which ought to be reduced to one half, as we fhall afterwards find.

The characters which diftinguifh the Aras from the other Parrots of the new world are, 1. their fize, which is at leaft double that of the others; 2. the length of the tail, which is

* " While Commodore Anfon and his officers were contemplating the natural beauties of this folitude, a flock of Aras flew over their heads; and as if thefe birds wifhed to improve the entertainment, and heighten the magnificence of the fpectacle, they ftopped to make numberlefs circles in the air, which allowed time to obferve the luftre and brilliancy of their plumage. Thofe who witneffed this fcene cannot defcribe it without raptures." *Anfon's Voyage round the World* —" It is the fineft fight in the world, to behold ten or twelve Aras on a very green tree; never are more charming colours difplayed." Dutertre. *Hift. des Antilles, t. II. p. 247.* † Briffon.

alfo

alfo much longer even in proportion to the body; 3. the naked and dirty white fkin, which
covers both fides of the head, furrounds it below, and covers alfo the bafe of the lower mandible. This naked fkin, in which the eyes are
placed, gives them a difagreeable afpect. Their
voice is harfh; they feem to articulate only the
found *ara*, and with a raucous, thick tone,
which is grating to the ear.

The R E D   A R A.

FIRST SPECIES.

1. *Pfittacus-Macao*, Linn. Gmel. Scop. and Bor.
   Ara Brafilienfis, Briff. and Gerini.
   Pfittacus maximus alter, Ray, Will. and Klein.
   The *Red and Blue Maccaw*, Edw. Alb. Banc. and Lath.
2. *Pfittacus-Aracanga*, Gmel.
   Araracangua, Ray and Will.
   Ara Jamaicenfis, Briff.
   Pfittacus capite cæruleo, Klein.
   The *Red and Yellow Maccaw from Jamaica*, Alb. Bancr.
   and Lath.

ALL the nomenclators have followed Gefner
and Aldrovandus in ranging the Red Aras
in two fpecies. But, on the other hand, Marcgrave, and all the travellers, who have had an
opportunity of viewing and comparing the birds,
confider them as belonging to the fame family.
They inhabit only the warm climates of America, the Antilles, Mexico, the ifthmus of Pa-

nama,

nama, Peru, Guiana, Brazil, &c. and are found
in no part of the old continent. It is ftrange,
therefore, that fome authors * have copied Albin in calling them *Macao Parrots*, and have
imagined that they came from Japan. Some may
have been carried thither from America, but they
are undoubtedly not natives of Japan, and it is
probable that thefe authors have confounded the
great Red Lory of the Eaft Indies with the Red
Ara of the Weft Indies.

The great Red Ara is near thirty inches in
length, but of this the tail occupies almoft the
one half; all the body, except the wings, is
vermilion, the four longeft feathers of the tail
are the fame; the great quills of the wing are
Turkey-blue above, and copper-coloured on a
black ground beneath; in the middle quills the
blue and green are admirably melted into each
other; the great coverts are gold-yellow, and
terminated with green; the fhoulders are of the
fame red with the back; the fuperior and inferior coverts of the tail are blue; four of the lateral quills, on each fide, are blue above, and
the whole under furface copper-colour, which
is lighter and has more of the metallic glofs under the four great middle quills: a tuft of fnuff-
coloured velvet feathers projects, like a cufhion,
on the forehead; the throat is brown-red; a white
and naked membranous fkin encircles the eye,

---

* Albin and Willughby.

covers the cheek, and fheathes the lower mandible, which is blackifh, and fo are the legs.
This defcription was made from a living bird,
and one of the largeft and moft beautiful.—
Travellers remark differences both in regard to
colour and fize, according to the different countries, or even iflands, from which they are
brought *. We have feen fome in which the
tail was entirely blue, others in which it was
red, and terminated with blue. Their bulk is
as various; but the little Red Aras are more
rare than the large ones.

In general, the Aras were formerly very
common in St. Domingo. I fee from a letter
of the Chevalier Defhayes that, fince the French
have extended their fettlements to the fummits
of the mountains, thefe birds are become lefs
frequent †. Both the Red and Blue Aras inhabit the fame climates, and their habits and
economy are exactly the fame; and what we

---

* " Thefe birds are fo diffimilar according to the grounds where
they procure their food, that every ifland has its Parrots, its Aras,
and its Parroquets, different in fize of body, in tone of voice, and
in the tints of the plumage." Dutertre, *Hift. des Antilles*. Paris,
1667, t. II. p. 247.—" The Aras are birds of extreme beauty:
they have a long tail, which is compofed of charming feathers of
different colours, according to the iflands which have given them
birth." *Hift. Nat. & Morale des Antilles*. *Rotterdam*, 1658,
p. 134.

† " In all thefe iflands (the Antilles) the Aras have become
very rare, becaufe the inhabitants kill them for eating. They retire to the leaft frequented places, and are no more obferved to
approach the plantations." *Obfervations of M. de la Borde, King's
phyfician at Cayenne.*

fhall

fhall now relate, in regard to the one, will equally apply to the other.

Aras live in the woods that cover the fwampy grounds, where the palm trees abound, and they feed chiefly on the dates of the palmettoes, of which there are immenfe forefts in the over-flowed favannahs. They generally appear in pairs, feldom in flocks; fometimes, however, they affemble together, and their united fcreams are heard at a great diftance. When any thing fcares or furprifes them *, they vent the fame fcreams, and while on the wing they cry in-ceffantly. Of all the Parrots, they fly the beft; they traverfe the cleared lands, but never alight; they perch on the fummit of trees, or on the higheft branch. During the day, they roam to a diftance not exceeding a league in fearch of ripe fruits, but always return in the evening to the fame fpot. Dutertre † fays that they are fometimes compelled by hunger to eat the ap-ples of the manchineel, which is poifonous to man, and probably to moft animals; he adds that the flefh of fuch birds is unwholefome, and even envenomed. Yet the Aras are commonly eaten in Guiana, Brazil, &c. without any bad

* " The Indians were in profound fecurity (at Yubarco, in Da-rien), when the cries of a fort f red Parrots of extraordinary fize, which they call *guacamayas,* gave them notice of the approach of their enemies." *Expedition of Ojeda,* &c. *Hift. Gen. des Voy. t. XII. p.* 156.

† Hift. des Antilles, *t. II. p.* 248.

confequences;

confequences; whether it be that the manchineel does not grow in thofe countries, or that
the Aras, finding abundance of their proper food,
fhun the food of that poifonous tree.

It appears that the Parrots of the new world
had nearly the fame difpofitions with all thofe
animals which inhabit defert tracts; they were
familiar, unfufpecting, and fearlefs of the approach of men, who in thofe regions were feebly
armed and few in number, and never could difplay their dominion *.　Peter of Angleria † affures us, that, on the difcovery of America, the
Parrots were caught with the noofe, and almoft
by the hand of the fportfman; that they were
feldom fcared by the noife of fire arms, and did
not take to flight when they faw their companions drop dead; that they preferred the trees
planted near houfes to the folitude of forefts;
and though the Indians caught them three or
four times a year, to ftrip them of their rich
plumage, this violence did not drive them from
their favourite haunts ‡.　Hence Aldrovandus,
who drew his information from the early accounts of America, fays that thefe birds are naturally attached to man, or, at leaft, fhew no

---

* " The fmall birds which inhabited the woods of New Zealand were fo little acquainted with men, that they calmly roofted
on the branches of trees next us, even on the end of our mufkets.
We were new objects to them, which they regarded with a curiofity equal to our own." *Forfter's Relation of Captain Cook's fecond
Voyage.*

† *Lib. X. dec.* 3.　　　　　‡ *Lery, p.* 174.

fymptoms of fear in his prefence; that they
follow the Indians into their huts, and feem to
contraét an affeétion to the fpots inhabited by
thefe peaceful men. A part of this confidence
and fecurity adheres to the Parrots which have
retired to the forefts. M. de la Borde informs
us that this is the cafe with thofe fettled in Guiana ; they allow one to approach very near them
without betraying fufpicion or fear. And Pifo
fays of the Brazilian birds, what may be applied
to all thofe of the new world, that they are in-
cautious, and eafily decoyed into every kind of
fnare.

The Aras make their nefts in the holes of old
rotten trees, which are very common in their
native regions, and more numerous even than
the rifing vigorous plants: they enlarge the
aperture, when it is too narrow, and line the
infide with feathers. They have two hatches
annually, like all the other American Parrots,
and each confifts of two eggs, which, according to Dutertre, are as large as pigeons' eggs,
and fpotted like thofe of partridges *. He adds
that the young ones have two fmall worms in
their noftrils, and a third on a fmall ball which

* " It often happens that an Ara will lay an egg or two in our
temperate climates; Aldrovandus cites fome inftances. M. le Marquis d'Abzac has informed us that a large Red Ara of his laid three
eggs; they had no germ: however, as the bird was ardent and
clamorous to cover, a hen's egg was given to her, which fhe
hatched." *Letter of M. le Marquis d'Abzac, dated from the Caftle of
Noyac, near Perigueux,* 21 September, 1776.

gathers

gathers above the head; and that thefe little worms die when the birds begin to be fledged*. Such worms in the noftrils are not peculiar to the Aras; the other Parrots, the caffiques, and many other birds, are fubject to them when in the neft: many quadrupeds alfo, the monkeys in particular, have worms in the nofe, and in other parts of the body. Thefe infects are well known in America, and in the French fettlements are called *macaques worms*. They creep into the flefh, and occafion dangerous abfceffes both in men and the other animals; horfes fometimes die of fuch diforders, owing to their negligent treatment in thofe countries, where they are never ftabled nor dreffed.

The male and female Aras fit alternately on their eggs, or cherifh their young, and both equally carry food; they never defert the charge as long as their affiftance is needed, and always perch together near the neft.

The young Aras are eafily tamed, and, in many parts of America, thefe birds are never taken but in the nefts, becaufe the adults are too difficult to educate. Yet Dutertre relates, that the Carribs had a fingular way of catching them alive; they obferved when the Aras were on the ground, eating fallen fruit; they endeavoured to furround them, and on a fudden hallooed, clapt their hands, and made fo great a

* Hift. des Antilles, *t. II. p.* 249.
noife,

noife, that the birds, in the moment of furprife,
loft the ufe of their wings, and turned on their
back to defend themfelves with their bill and
nails; the favages then held out a ftick, on
which they clenched, and were immediately
tied by a fmall ftring. He alfo pretends that
they could be tamed, though old, and caught
in this violent way. But this account appears
rather fufpicious; particularly, as all the Aras
in fact fly from man, a loud noife muft have a
greater effect in driving them away. Wafer
fays that the Indians who inhabit the ifthmus
tame the Aras as we do magpies, and allow
them to make excurfions into the woods during
the day, but that they conftantly return home
in the evening; that they imitate their mafter's
voice, and the fong of a bird called *chicali* *.
Fernandez relates that they can be taught to
fpeak, but that their articulation is coarfe and difagreeable; that when they are kept in the houfe,
they rear their young like other domeftic birds.
It is certain, that they never prattle fo well as
the other Parrots; and after they are tamed,
they never try to efcape.

The Indians work the plumage into feftival
caps, and other ornaments; they alfo ftick fome
of the rich feathers into the cheek, and through
the noftrils, or the ears. The flefh of the Aras,
though commonly hard and black, is not un-

* Wafer, in Dampier's Voyage,

M 3 pleafant

pleafant food, and makes excellent foup; and, in general, the Parrots are ufually eaten as the moft plentiful game in the fettlement at Cayenne.

The Ara is, more than any bird perhaps, fubject to the cramp, which is more violent, and more quickly mortal, in the hot countries, than in the temperate climates. I kept one of the largeft and handfomeft of the kind, which was given to me by the Marchionefs of Pompadour in 1751. It was feized with an epilepfy two or three times every month, and yet it lived feveral years at my feat in Burgundy, and might have lived much longer, if it had not been killed. But in South America, thefe birds commonly die of the falling ficknefs, and this is alfo the fate of all the other Parrots, which are equally fubject to that diforder, in the domeftic ftate. Perhaps the caufe which we affigned in treating of the canary finches, obtains here, viz. the feparation from the female, and the fuperabundance of nutrition. The Indians, who rear the Aras in their huts, with the view to fell their feathers, have a remedy for the epilepfy; they cut the point of the toe, and the difcharge of a fingle drop of blood works an immediate cure. The fame operation fucceeds equally in procuring relief to the other diforders incident to the domeftic ftate. I have formerly remarked that the canaries die when the blood does not form a drop at the bill. Nature feems to point out the

the fame remedy which the Indians have difco-vered.

This epilepfy, or cramp, as it is called in the colonies, invariably happens to domefticated Parrots when they perch on a bit of iron, fuch as a nail, a curtain rod, &c. ; fo that great care is taken that they alight only on wood. This fact feems to fhew that the fit, which is a violent convulfion of the nerves, is analogous to electricity, whofe action, it is well known, is more violent when tranfmitted through iron than through wood *.

* This analogy feems to be rather fanciful. The effect muft be referred to the irritability of the mufcular fibres, excited by the fharpnefs or afperity of the touching fubftance, and by the fudden cold fhot through the bird's toes by the contact of metals, which is a rapid conductor of heat. T.

[A] Specific character of the *Pfittacus-Macao* : " It is red; its wing-quills blue above, rufous below; its fcapular feathers variegated with blue and green; its cheeks naked and wrinkled."— That of *Pfittacus Aracangua* : " It is dilute fcarlet; its fcapular feathers yellow, terminated with green ; its wing-quills blue above, rufous below; its cheeks naked and wrinkled."

The B L U E A R A.

SECOND SPECIES.

Pſittacus-Ararauna, Linn. Gmel. Scop. and Bor.
Pſittacus maximus cyano-croceus, Ray, Sloane, and Will.
1. *Ara Jamaicenſis cyano-crocea,* Briſſ.
2. *Ara Braſilienſis cyano-crocea,* Id.
    Pſittacus vertice viridi cauda cyanea, Klein.
    The *Blue and Yellow Maccaw,* Edw. Alb. and Lath.

NOMENCLATORS have divided this into two
ſpecies; the firſt is *the blue and yellow Ara
of Jamaica,* and the ſecond is *the blue and yellow
Ara of Braſil.* But theſe birds are not only of
the ſame ſpecies, but inhabit the ſame parts of
South America.

It is eaſy to deſcribe the Blue Ara: the upper ſide of the body, the wings, and the tail,
are entirely azure, and the under ſide of the
body is fine yellow*; this yellow is rich and
vivid,

---

* " The other called *Canidé,* having all the plumage under its
belly and round its neck as yellow as fine gold; the upper ſide of
the back, the wings, and the tail, are of the pureſt blue: you would
ſay that it is attired below with a cloth of gold, and mantled above
with violet figured damaſk." Lery, *Voyage au Breſil,* 1578, *p.* 171.
Thevet characterizes equally well the two ſpecies of Aras: " Nature has delighted to pourtray this beautiful bird, called by the ſavages *carinde,* clothing it with a plumage ſo beautiful and charming, that it is impoſſible not to admire the workmanſhip. This
bird exceeds not in bulk the raven, and its plumage, from the belly
to the throat, is yellow like fine gold; the wings and the tail,
which is very long, are of a fine azure-colour. There is another
bird

THE BRASILIAN GREEN MACCAW.

vivid, and the blue is gloffy and fparkling. The
favages admire thefe Aras, and celebrate their
beauty; the ufual burthen of their fongs is, *Yellow bird, yellow bird, how charming* *!

The Blue Aras never mingle with the Red
Aras, though they frequent the fame fpots, and
live in harmony. Their voice is fomewhat different, for the favages can diftinguifh them by
the cry alone. It is faid that the blue ones do
not pronounce the word ara fo diftinctly [A].

The GREEN ARA.

THIRD SPECIES.

Pfittacus Severus, Linn. Gmel. Scop. and Bor.
1. *Ara Brafilienfis Viridis,* Briff.
2. *Ara Brafilienfis Erythrochlora,* Id.
   Maracana, Ray and Will.
   The *Brafilian Green Maccaw,* Edw. Sloane, and Lath.

THE Green Ara is much rarer than the two
preceding; it is alfo much fmaller. It
confifts of only one fpecies, though nomencla-

bird refembling this in fize, but different in its colours; for its
plumage, inftead of being yellow, is of a red like fine fcarlet, and
the reft azure.' *Singularites de la France Antarctique,* par Thevet,
Paris, 1558, *p.* 32.
   * *Canidé jouve, canidé jouve, heura oncèbe* Lery, *p.* 173.
   [A] Specific character of the *Pfittacus Ararauna:* " Above it
is blue, below yellow; its cheeks naked, with feathery lines."

tors have divided it into two ; becauſe they confound it with the Green Parrakeet, which they
call the *Ara Parrakeet,* on account of the length
of its tail, and its diſtinctly articulating the word
ara : but notwithſtanding theſe properties it is
ſtill a Parrakeet, and very common in Cayenne;
whereas the Green Ara is there altogether unknown. Sloane ſays that the Little Maccaw,
or Little Green Ara, is very frequent in the
woods of Jamaica. But Edwards properly obſerves that this is certainly a miſtake, becauſe,
though he made ſeveral applications, he could
procure none from his correſpondents in that
iſland. Perhaps Sloane confounded the long-
tailed Green Parrakeet with the Green Ara.

We had a Green Ara alive : it was preſented
by M. Sonini of Manoncour, who procured it at
Cayenne from the ſavages of Oyapoc, where it
was caught in the neſt. Its length, from the tip
of the bill to the end of the tail, is about ſixteen
inches ; its body, both above and below, is
green, which according to the poſition is golden
and ſparkling or deep olive ; the great and ſmall
quills of the wing are beryl blue, on a brown
ground, and the under ſide copper coloured ;
the under ſide of the tail is the ſame, and the
upper ſide painted with beryl blue, melting into
olive green ; the green on the head is brighter
and leſs mixed with olive than that on the reſt
of the body ; at the baſe of the upper mandible,
on the face, there is a black border of ſmall
                                              linear

linear feathers that refemble briftles; the white
naked fkin that furrounds the eyes is fprinkled
with fmall pencils of the fame black briftles
ranged in rows; the iris of the eye is yellowifh.

This bird is as beautiful as it is rare; and it
is ftill more amiable for its focial temper and
gentle difpofition. It foon grows familiar with
perfons whom it fees frequently, and is pleafed
to receive and repay their careffes. But it has an
averfion to ftrangers, and particularly to children, and flies at them furioufly. Like all other
domefticated Parrots, it clings to the finger
when prefented to it; it alfo clafps wood : but
in winter, and even in fummer, when the
weather is cool and rainy, it prefers the arm or
the fhoulder, efpecially if the perfon has wool-
len clothes, for in general it likes warm. ftuffs.
It is alfo fond of kitchen ftoves when they
are cooled fo much as to retain only a gentle
warmth. For the fame reafon, it avoids fitting
on hard bodies which quickly communicate
cold, fuch as iron, marble, glafs, &c.; and in
cold rainy weather, though in fummer, it fhud-
ders and trembles if water be thrown upon it.
However, in fultry days, it bathes of its own
accord, and often dips its head in the water.

If one ftroke it gently, it fpreads its wings,
and fquats; it then utters its difagreeable cry,
which refembles the chatter of the jay, raifing
its wings during the action, and briftling its
feathers:

feathers: and this habitual cry feems to exprefs
either pleafure, or languor. Sometimes it has
a fhort fhrill cry, which is lefs equivocal than
the former, and denotes joy and fatisfaction; for
it is generally addreffed to perfons whom it loves;
but this cry alfo marks its impatience, fits, and
its pettifh gufts of ill-humour. But it is impoffible to be precife on this fubject; for birds
organized like the Parrots perpetually vary or
modify their voice, as they are prompted by imitation.

The Green Ara is jealous: it is fired at feeing
a young child fharing in its miftrefs's careffes
and favours; it tries to dart at the infant, but,
as its flight is fhort and laborious, it only fhews
its difpleafure by geftures and reftlefs move-
ments, and continues tormented by thefe fits
till its miftrefs is pleafed to leave the child, and
take the bird on her finger. It is then over-
joyed, murmurs fatisfaction, and fometimes
makes a noife exactly like the laugh of an old
perfon. Nor can it bear the company of other
Parrots, and if one be lodged in the fame room,
it will ftrive to deprive it of every comfort. It
would appear, therefore, that the bird can fuffer
no rivals whatever in its miftrefs's favour, and
that its jealoufy is founded on attachment; accordingly it takes no notice when it fees a different perfon fondle a child.

It eats nearly the fame things that we do. It
is particularly fond of bread, beef, fried fifh,
                                            paftry,

paſtry, and ſugar; but it ſeems to prefer roaſted
apples, which it ſwallows greedily. It cracks
nuts with its bill, and picks them dexterouſly
with its claws. It does not chew the ſoft fruits,
but ſucks them, by preſſing its tongue againſt
the upper mandible; and, with reſpect to the
harder ſort of foods, ſuch as bread, paſtry, &c.
it bruiſes or chews them, by preſſing the tip of
the lower mandible upon the moſt hollow part
of the upper. But, whatever be the nature of
its food, its excrements are always green, and
mixed with a ſort of white chalky ſubſtance, as
in moſt other birds, except when it is ſick, and
then they aſſume an orange or deep yellow
caſt.

Like all the other Parrots, the Blue Ara uſes
its claws with great dexterity; it bends forward
the hind toe to lay hold of the fruits and other
crumbs which are given to it, and to carry
them to its bill. The Parrots, therefore, employ their toes nearly as the ſquirrels or monkeys; they alſo cling and hang by them. The
Green Ara almoſt always ſleeps in this way,
hooked to the wires of its cage. There is alſo
another habit common to the Parrots, viz. they
never climb or creep without faſtening by the
bill, with which they begin, and uſe the feet
only as a ſecond point of their motion.

The noſtrils are not viſible in this Ara, as in
moſt of the other Parrots; inſtead of being
placed in the uncovered part of the horn of the
bill,

bill, they are concealed in the firſt ſmall feathers that cover the baſe of the upper mandible,
which riſes and forms a cavity at its root when
the bird makes an effort to imitate difficult
ſounds : in ſuch caſes the tongue folds back at
the tip, and recovers its ſhape when it eats ; a
power not commonly poſſeſſed by birds which
can only move it backwards or forwards in the
direction of the bill. This little Green Ara is
as hardy as moſt of the other Parrots, or even
more ſo. It learns more eaſily to prattle, and
pronounces much more diſtinctly, than the Red
or Blue Aras. It liſtens to the other Parrots,
and improves beſide them. Its cry is like that
of the other Aras, only its voice is not near ſo
ſtrong, and does not articulate ſo diſtinctly the
ſound *ara.*

It is ſaid that bitter almonds will kill Parrots,
but I am not certain of the fact; I know, however, that parſley, of which they are very fond,
if taken even in ſmall quantity, is very pernicious ; as ſoon as they eat it, a thick viſcous liquor runs from the bill, and they die in an hour
or two.

It appears that there is the ſame variety in
the Green Aras as in the Red ; at leaſt Edwards
has deſcribed a *great Green Maccaw*, which is
thirteen inches long, and fifteen to the middle
feather of the tail : the face was red ; the quills
of the wing blue, and alſo the lower part of the
back and the rump. Edwards calls the colour

8   of

of the under furface of the wings and of the tail, *dull orange*, and it is probably the fame with that dull bronze red which we perceived below the wings of our Green Ara; the feathers of the tail, in that of Edwards, were red above, and terminated by blue [A].

The BLACK ARA.

FOURTH SPECIES.

Pfittacus Ater, Gmel.
The *Black Maccaw*, Lath.

THE plumage is black, with reflections of fhining green, and thefe mingled colours are much like thofe of the ani. We can only indicate this fpecies, which is known to the fa-vages of Guiana, but which we have not been able to procure. It differs from the other Aras in fome of its habits; it never approaches the fettlements, but remains on the arid and barren fummits of rocks and mountains. Läet feems to mention this bird by the name of *Ararauna*, or *Machao*, whofe plumage, he fays, is black, but fo well mixed with green that, in the fun beams, it fhines admirably; the legs are yel-

[A] Specific character of the *Pfittacus Severus*: " It is green: its cheeks naked; the quills of its wings and tail blue, below pur-plifh."

low,

low, he fubjoins ; the bill and the eyes reddifh,
and it refides in the interior parts of the country.

Briffon has formed another Ara from a Parrakeet, and called it the *variegated Ara of the
Moluccas.* But, as we have frequently obferved, there are no Aras in India, and we have
fpoken of this bird in treating of the Parrakeets
of the old continent.

[A] Specific chara&er of the *Pfittacus Ater :* " It is black, with
a greenifh fplendour; its bill and eyes are reddifh; its legs yellow."

THE

AMAZONS and CRICKS.

WE fhall apply the name of *Amazon Parrots* to all thofe which are marked with red on the fan of the wing; they have received that appellation in America, becaufe they are brought from the country of the Amazons. We fhall appropriate the term *Crick* to thofe which have no red on the fan, but only on the wing: this too is the name given by the favages of Guiana to thefe Parrots. They are diftinguifhed from Amazons by other properties alfo: 1. the plumage of the Amazons is fhining, and even dazzling, whereas the green colour of the Cricks is dull and yellowifh; 2. in the Amazons the head is covered with a fine and very bright yellow, but, in the Cricks, this yellow is dull and intermixed with other colours; 3. the Cricks are rather fmaller than the Amazons, which are much fmaller than the Aras; 4, the Amazons are exceedingly beautiful and rare, but the Cricks are the moft common of the Parrots, and the moft inferior in point of beauty; they are extenfively fpread, while the Amazons are hardly ever found, except at Para, and in fome other countries bordering on the river of Amazons.

But the Cricks, having red on their wings, ought to be joined with the Amazons, of which this red forms the principal character : their natural habits are likewise the same ; they fly in numerous flocks, perch in multitudes on the same spots, and all scream together so loud that they may be heard at a great distance. They frequent also the woods, both those on the mountains and those which grow in the low grounds, and even the swamps that abound with palms, elastic-gum trees, and bananas, &c. and are fond of the fruit of these trees. They eat, therefore, a greater variety of fruits than the Aras, which commonly subsist on the palmettoes alone ; but these dates are so hard that they can hardly be cut; they are round, and as large as pippins.

Some authors * have said that the flesh of all the American Parrots contracts the odour and colour of the substances on which they feed ; that it smells of garlic, when they eat the fruit of the acajou ; that it has the scent of musk and of cloves, when they eat the fruit of the Indian wood : and that it receives a black tinge, when they live upon the fruit of the *genipa*, whose juice, though at first as limpid as water, becomes as black as ink in the space of a few hours. They subjoin that the Parrots become very fat during the maturity of the mangroves,

* Dutertre, *Hist. des Antilles*, t. II. p. 251.—Labat, *Nouv. Voy. aux Iles de l'Amerique*, t. II. p. 159.

which

which yield excellent food ; and laftly that the feeds of the cotton fhrub intoxicate them to fuch a degree, that they may be caught with the hand.

The Amazons, the Cricks, and all the other Parrots of America, conftruct their nefts in holes formed in decayed trees by the woodpeckers, and only lay two eggs twice a year, which the cock and hen hatch by turns : it is faid that they never forfake their neft, and perfift in hatching, though their eggs be handled and deranged. In the love feafon, they affemble and breed in the fame haunt, and fearch their food in company; when their appetite is fatisfied they make a continual and noify babbling, fhifting their place inceffantly, and fluttering from tree to tree, till the darknefs of night and the fatigue of action invite to repofe. In the morning they are obferved on the naked branches, at fun-rife, and they remain quiet till the dew is dried from their plumage, and their warmth recovered; then they rife in a flock, with a noife like that of gray crows, but louder. They breed in the rainy feafon *.

The favages commonly take the Parrots in the neft, becaufe they are more eafily reared and better tamed. But the Caribbs, according to Labat, catch them alfo after they are old : they obferve the trees on which they perch in

* Note communicated by M. de la Borde, King's Phyfician at Cayenne.

great

great numbers in the evening, and, after dark, they carry near the spot lighted coals, on which they throw gum and green pimento ; the birds are suddenly involved and stifled in thick smoke, and fall to the ground ; the savages then seize them, tie their feet, and recover them from the suffocation, by throwing water on the head *. They also bring down the Parrots without hurting them much, by shooting them with blunt arrows †. But the old ones thus caught are difficult to tame. There is only one method of rendering them tractable ; it is to blow the smoke of tobacco into their bill, which partly intoxicates them, and makes them gentle and pliant. If they grow mutinous again, the dose is repeated, and thus in the course of a few days their disposition is softened. We can hardly form an idea of the envenomed temper of the wild Parrots ; they bite cruelly without provocation, and will not quit their hold. The old birds never learn to prattle in perfection. Tobacco smoke is also used to cure them of their noisy disagreeable cry.

Some authors ‡ alledge that the female Parrakeets never learn to speak ; but this is a mistake : they are more easily taught than the males, and even more docile and gentle. Of

* Labat, *Nouv. Voy. aux Iles de l'Amerique*, t. II. 52.
† " The savages of Brasil have very long arrows, headed with a ball of cotton, for shooting at Parrots." *Belon.*
‡ Frisch, &c.

all

all the American Parrots, the Amazons and the Cricks are the moſt ſuſceptible of education, eſpecially when caught young.

As the ſavages traffic with each other in the feathers of Parrots, they claim a certain number of trees on which theſe birds make their neſts. This is a kind of property from which they derive an income by ſelling the Parrots to ſtrangers, or by bartering the feathers with other ſavages. Theſe trees deſcend from father to ſon, and are often their richeſt inheritance *.

* Fernandez, *Hiſt. Nov. Hiſp. p.* 38.

The AMAZON PARROTS.

WE know five species of thefe, befides.
many varieties; the firft is the Yellow-headed; the fecond, the Tarabé, or Red-
headed; the third, the White-headed; the
fourth, the Yellow Amazon; and the fifth,
the Aourou-Couraou.

———

The YELLOW-HEADED AMAZON,
Buff.

FIRST SPECIES.

Pfittacus Nobilis, Linn. and Gmel.
The *Noble Parrot,* Lath.

THE crown of the head is a fine bright yellow; the throat, the neck, the upper fide
of the back, and the fuperior coverts of the
wings, of a brilliant green; the breaft and the
belly green, with a little yellowifh; the fans of
the wings are of a bright red; the quills of the
wings are variegated with green, black, violet-
blue, and red; the two exterior quills, on each
fide of the tail, have their inner webs red at
their origin, and then deep green, which, at
the extremity, changes into a yellowifh green;
the bill is red at the bafe, all the reft of it ci-

nereous;

nereous; the iris is yellow; the feet gray, and the nails black.

We muſt obſerve that Linnæus commits an error in ſaying that this bird has naked cheeks; which confounds the Amazons with the Aras, to which alone that character belongs. On the contrary, the Amazons are feathered on the cheeks, or between the bill and the eyes, and like all the other Parrots, have only a very ſmall circle of naked ſkin round the eyes [A].

VARIETIES or CONTIGUOUS SPECIES of the
YELLOW-HEADED AMAZON.

THERE are two other ſpecies, or perhaps varieties, related to the preceding.

I. *The Red and Green Parrot of Cayenne,* which has not been mentioned by any naturaliſt, though known in Guiana by the name of *Baſtard Amazon* or *Half-Amazon.* It is ſaid to be a croſs-breed of the Amazon with another Parrot. It is indeed inferior in beauty to the one juſt deſcribed; for it has not the fine yellow on the face near the root of the bill; the green colour of its plumage is not ſo brilliant, but has a yellowiſh caſt; the red on the plumage is the only colour which is ſimilar and diſpoſed in the

[A] Specific character of the *Pſittacus Nobilis:* " It is green; its cheeks naked; its ſhoulders ſcarlet."

N 4　　　　　　ſame

fame way; there is alſo a ſhade of yellowiſh under the tail; the bill is reddiſh, and the feet gray; and as it has the ſame bulk, we can hardly doubt but that it is nearly related to the ſpecies of the Amazon.

II. The ſecond variety was firſt noticed by Aldrovandus, and, according to his deſcription, it appears to differ from this Amazon Parrot only in the colours of its bill, which that author ſays is ochrey on the ſides of the upper mandible, whoſe ridge is bluiſh, with a ſmall white bar near the tip; the lower mandible is alſo yellowiſh in the middle, and lead colour through the reſt of its length. But all the colours of the plumage, the ſize, and ſhape of the body, being the ſame as in the Yellow-headed Amazon, it may be only a variety.

The TARABE, or RED-HEADED A M A Z O N.

SECOND SPECIES.

Pſittacus Taraba, Gmel.
The *Red-headed Amazon's Parrot,* Lath.

THIS Parrot, which is deſcribed by Marcgrave as a native of Braſil, is not found in Peru. The head, the breaſt, the pinions, and tops of the wings, are red; and hence it

ought

THE WHITE-FRONTED PARROT

ought to be ranged with the Amazon Parrots.
All the reft of its plumage is green; the bill
and the feet are dull afh-colour [A].

The WHITE-HEADED AMAZON.

THIRD SPECIES.

Pfittacus Leucocephalus, Linn. Gmel. and Bor.
The *White-headed Parrot*, Edw.
The *White-fronted Parrot*, Lath.

IT would be more accurate to name this bird
the *White-fronted Parrot*; becaufe the white
is generally confined to the face. But fometimes it' furrounds the eye, and extends to the
crown of the head; and often it only borders
the face. The fpecies appears fubject, therefore, to variety. In one fpecimen, the plumage
was alfo of a deeper green, and lefs waved with
black: in another, it was lighter, mixed with
yellowifh, and interfected with black feftoons
all over the body; the throat and the fore-fide
of the neck are of a fine red. That colour is
not fo much fpread in the former, or fo bright,
but there is a fpot of it under the belly. In
both of them, the quills of the wing are blue;
thofe of the tail yellowifh green, tinged with

[A] Specific character of the *Pfittacus Taraba:* " It is green;
its head, its throat, and the leffer coverts of its wings, are red."

red

red in the firſt half; and, on the fan, a red ſpot
is perceived, which is the livery of the Amazons.   Sloane ſays that theſe Parrots are frequently brought from Cuba to Jamaica, and
that they occur alſo in St. Domingo.   They
are found in Mexico, but never in Guiana.
Briſſon divides them into two ſpecies; and this
miſtake was occaſioned by Edwards's White-
headed Parrot being different from his.   The
Martinico Parrot mentioned by Labat, in which
the upper ſide of the head is ſlate colour, with
a little red, is different from our White-headed
Parrot, though Briſſon aſſerts that they are the
ſame [A].

The YELLOW AMAZON.

FOURTH SPECIES.

Pſittacus Aurora, Gmel.
The *Aurora Parrot*, Lath.

THIS bird is probably a native of Brazil, ſince
Salerne ſays that he ſaw one which pronounced Portugueſe words.   We are certain at
leaſt that it comes from the new world, and the
red colour of its vents aſſigns its place among
the Amazons.

All the body and the head are of a very fine

[A] Specific charaĉter of the *Pſittacus Leucocephalus:* " It is
green; its wing-quills blue.; its front white; its orbits ſnowy."

yellow;

THE YELLOW-HEADED AMAZON PARROT.

yellow; the fans are marked with red, and alfo
the great quills of the wings, and the lateral
quills of the tail; the iris is red; the bill and
feet are white [A].

The AOUROU-COURAOU.

FIFTH SPECIES.

Pfittacus Æftivus.
Aiuru-curau, Ray and Johnftone.
The *Common Amazon's Parrot,* Lath.

THE Aourou-Couraou of Marcgrave is a handfome bird, and is found in Guiana and Brazil. Its face is bluifh, with a bar of the fame
colour below the eyes; the reft of the head is
yellow; the feathers of the throat are yellow,
and edged with bluifh green; the reft of the
body is light green, which affumes a yellowifh
tinge on the back and belly; the fan of the
wing is red; the fuperior coverts of the wings
green; the quills of the wing are variegated
with green, black, yellow, blue, and red; the
tail is green, but, when the feathers are fpread,
they appear fringed with black, red, and blue;
the iris is gold colour; the bill is blackifh; and
the feet cinereous [B.]

[A] Specific charaĉter of the *Pfittacus Aurora:* " It is bright
yellow; its *axillæ,* the margins of its wings, and its greater wing-
quills red outwards in the middle."

[B] Specific charaĉter of the *Pfittacus Æftivus:* " It is green,
fomewhat fpotted with yellow; its front blue; its fhoulders blood-
coloured; its orbits carnation."

VARIETIES of the AOUROU-COURAOU.

THERE are several varieties which may be referred to this species.

I. The bird mentioned by Aldrovandus under the appellation of *Pfittacus Viridis Melanorinchos*, which hardly differs at all from the preceding.

II. There is another also described by Aldrovandus, in which the face is beryl blue with a bar of the same colour above the eyes, which is only a shade different from the species of this article. The crown of the head is also of a paler yellow; the upper mandible is red at the base, bluish in the middle, and black at the end; the lower mandible is whitish. In all other properties, the colours are precisely the same as in the *Aourou-Couraou*. It is found in Guiana, Brazil, and Mexico, and also in Jamaica; and it must be very common in Mexico, since the Spaniards give it a proper name, *Catherina* *. From Guiana it has probably been carried into Jamaica, which is at too great distance from the continent to correspond with the excursions of the Parrots. Labat says that they cannot fly from one island to another, and that

* Many beautiful kinds of Parrots are distinguished in New Spain; the *caterinillas* have their plumage entirely green; the *loros* have it green likewise, except the head and the extremity of the wings, which are of a fine yellow; the *pericos* are of the same colour, and are not larger than a thrush." *Hift. Gen. des Voy. t. XII,* *p.* 626.

thofe

thofe of the different iflands may be diftinguifhed. The Parrots of Brazil, Cayenne, and the
reft of the continent of America, which are feen
in the iflands, have been tranfported thither,
and few which are natives of the iflands are
found on the continent, on account of the difficulty of the paffage; for a ftrong current fets
out from the Bay of Mexico, fo that a veffel is
carried in fix or feven days from the continent
to the iflands, though it takes fix weeks or two
months to work back again.

III. Another variety is the *Aiuru-Curuca* of
Marcgrave. There is on the head a blue cap
mixed with a little black, in the midft of which
is a yellow fpot: this indication differs in nothing from the defcription which we have given.
But the bill is afh-coloured at the bafe, and
black at the end; this is the only flight variation.

IV. Marcgrave notices another variety, and'
remarks that it is like the preceding; yet our
nomenclators have ranged them in different fpecies, and even doubled thefe. The only difference is that the yellow extends a little more
on the neck.

V. *The Yellow-fronted Amazon Parrot* of Briffon *(Pfittacus Amazonicus fronte lutea)*. The
only difference is that the face is whitifh, or pale
yellow, but in the other it is bluifh; which is
by no means fufficient to conftitute a diftinct
and feparate fpecies.

The C R I C K S.

THOUGH there is a very great number of birds to which this name is applied, they may be all reduced to feven fpecies, of which the others are varieties. Thefe feven fpecies are: 1. The Yellow-throated Crick; 2. The Meunier or Mealy Crick; 3. The Red and Blue Crick; 4. The Blue-faced Crick; 5. The Crick properly fo called; 6. The Blue-headed Crick; 7. The Violet-headed Crick.

The CRICK with a YELLOW HEAD and THROAT.

FIRST SPECIES.

Pfittacus Ochrapterus, Gmel.
Pfittacus Amazonicus gutture luteo, Briff.
Pfittacus Viridis Alius, capite luteo, Frif. and Klein.
The *Yellow-winged Parrot*, Lath.

THE whole of the head, the throat, and the lower part of the neck, are of a very fine yellow; the under fide of the body is of a fhining green, and the upper fide alfo green, but with a little mixture of yellow; the fan of the wing is yellow, whereas the fame part is red in the Amazons; the firft row of the coverts

3   of

of the wing is red and yellow; the other rows are of a fine green; the quills of the wings and of the tail are variegated with green, black, violet, yellowifh, and red; the iris is yellow; the bill and feet whitifh.

This bird is living at prefent with Father Bougot, who has communicated to us the following account of its difpofition and habits. " It is very fufceptible of attachment to its mafter; it is fond of him, but requires frequent careffes, and feems difconfolate if neglected, and vindictive if provoked. It has fits of obftinacy; it bites during its ill humour, and immediately laughs, exulting in its mifchief. Correction and rigorous treatment only harden it, and make it more ftubborn and wayward; gentle ufage alone fucceeds in mollifying its temper.

" The inclination to gnaw whatever it can reach, is very deftructive in its effects; it cuts the cloth of the furniture, fplits the wood of the chairs, and tears paper, pens, &c. And if it be removed from the fpot, its pronenefs to contradiction will inftantly hurry it back. But this mifchievous bent is counterbalanced by agreeable qualities, for it remembers eafily what it is taught to fay. Before articulating it claps its wings and plays on its rooft; in the cage it grows dejected, and continues filent; never prattles well, except when it enjoys liberty. It chatters lefs in winter than during the fummer months,

months, forgetting its food, when it never ceafes from morning to night.

" In its cheerful days it is affectionate, receives and returns careffes, and liftens and obeys; though a peevifh fit often interrupts the harmony. It feems affected by the change of weather, and becomes filent; the way to re-animate it is to fing befide it; it ftrives by its noify fcreams to furpafs the voice which excites it. It is fond of children; in which refpect it differs from other Parrots. It contracts a pre-dilection for fome of them, and fuffers them to handle and carry it; it careffes them, and if any perfon then touches them, it bites at him fiercely. If its favourite children leave it, it is unhappy, follows them, and calls loudly after them. During moulting it is much reduced, and feems to endure great pain; and that ftate lafts near three months.

" Its ordinary food is hemp-feed, nuts, fruits of all kinds, and bread foaked in wine; it would prefer flefh, but that diet makes it low fpirited and inactive, and, after fome time, occafions its feathers to drop. It is alfo obferved to keep its food in bags under the chin, and to ruminate*."

* Note communicated by the Rev. Father Bougot, Guardian of the Capuchins of Semur, who has long amufed himfelf with rearing Parrots.

[A] Specific character of the *Pfittacus Ochropterus*: " It is green; its front and orbits whitifh; its top, its cheeks, its throat, and the more remote coverts of its wings, fine yellow."

The MEALY CRICK.

Le Meunier, ou Le Crik Poudré, Buff.

SECOND SPECIES.

Pfittacus Pulverulentus. Gmel.
The *Mealy Green Parrot,* Lath.

No naturalift has defcribed this fpecies dif-
tinctly·; only Barrère feems to mention it as
large, whitifh, and powdered with gray. It is
the biggeft of all the Parrots of the new world,
except the Aras. It is called *meunier*, or the
miller, by the fettlers at Cayenne, becaufe its
plumage, whofe ground colour is green, appears fprinkled with meal: there is a yellow
fpot on the head; the feathers on the upper
furface of the neck have a broad edging of
brown; the under fide of the body is of a lighter green than the upper fide, and is not mealy;
the outer quills of the wings are black, except
a part of the outer webs, which are blue; there
is a large red fpot on the wings; the quills of
the tail are of the fame colour with the under
fide of the body, from their origin to three
fourths of their length, and the remaining
fourth yellowifh green.

This Parrot is one of the moft efteemed, as
well for its magnitude and the fingularity of its
colours, as for the facility with which it learns
to fpeak, and the mildnefs of its difpofition.

VOL. VI.   o   There

There is only one flight defect in its appearance,
viz. its bill is like whitish horn [A].

The RED and BLUE CRICK.

THIRD SPECIES.

Psittacus Cæruleocephalus, Linn. and Gmel.
Psittacus Guianensis cæruleus, Briss.
Psittacus Versicolor, Ray.
The *Red and Blue Parrot*, Will. and Lath.

THIS Parrot has been mentioned by Aldrovandus, and all the other naturalists have
copied his account; but they do not agree in
their descriptions According to Linnæus, the
tail is green; Brisson represents it as rose-co-
loured.   As neither of them has seen it, I shall
quote Aldrovandus.

" The epithet *variegated* (ποικιλος) suits it
well, considering the diversity and richness of
its colours; blue and soft red *(roseus)* predominate; the blue appears on the neck, the breast,
and the head, whose crown is marked with a
yellow spot; the rump is of the same colour;
the belly is green; the top of the back light
blue; the quills of the wings and of the tail are

[A] Specific character of the *Psittacus Pulverulentus:* " It is
green, and above is sprinkled with mealy specks; a bright yellow
spot on its head, and a red one on its wings."

all

all rofe colour; the coverts of the former are
mixed with green, yellow, and rofe colour;
thofe of the tail are green; the bill is blackifh;
the feet are reddifh gray." Aldrovandus does
not inform us from what country this bird is
brought; but as there is red on its wings, and
a yellow fpot on the head, we have ranged it
with the American Cricks.

We may obferve that Briffon has confounded
with it the Violet Parrot mentioned by Barrère,
but which is very different, and belongs neither
to the Amazons nor to the Cricks [A.]

The BLUE-FACED CRICK.

FOURTH SPECIES.

Pfittacus Havanenfis, Gmel.
Pfittacus Amazonicus gutturæ cærulco, Briff.
The *Blue-fronted Parrot*, Lath.

THIS Parrot was fent to us from the Havan-
na, and it is probably common in Mexico
and near the ifthmus of Panama; but it is not
found in Guiana. It is much fmaller than the
Mealy Crick, its length being only twelve
inches. Among the quills of the wings, which

[A] Specific charaĉter of the *Pfittacus Cæruleocephalus:* " It is
blue; its belly, its rump, and its tail, are green; its top bright
yellow; the quills of its wings and tail red,"

O 2			are

are indigo colour, there are fome red ones; the
face is blue; the breaft and ftomach are of a
foft red or lilac, and waved with green; all the
reft of the plumage is green, except a yellow
fpot on the lower part of the belly [A].

The C R I C K.

FIFTH SPECIES.

Pfittacus Agilis, Linn. Gmel. and Bor.
Pfittacus Cayanenfis, Briff.
The *Little Green Parrot,* Edw.
The *Agile Parrot,* Lath.

THE name *Crick* is beftowed on this bird at
Cayenne, where it is fo common that the
fame appellation is extended to a confiderable
tribe of Parrots. It is fmaller than the Amazons; but we ought not, with the nomenclators, to range it among the Parrakeets *: they
have miftaken it for the Guadeloupe Parrakeet,
becaufe it is entirely green. They would have
avoided this error, if they had confulted Marcgrave, who fays expreſsly that it is large as a
hen; and this character is alone fufficient to exclude it from the Parrakeets.

This Crick has alfo been confounded with

[A] Specific character of the *Pfittacus Havanenfis:* "It is green;
its front and throat-afh-blue; a large red fpot on its breaft; its orbits cinereous."
* Willughby, Ray, Linnæus, and Briffon.

the

the *Tahua*, or *Tavoua**, which is widely different; for the Tavoua has no red on its wings,
and is therefore neither an Amazon nor a Crick,
but rather a Popinjay, of which we fhall fpeak
in the following article.

The Crick is near a foot long from the tip of
the bill to the extremity of the tail, and its
wings, when clofed, extend a little beyond the
middle of the tail; both the upper and under furface are of a pretty light handfome green, particularly on the belly and the neck, where the
green is very brilliant; the front and the crown
of the head are alfo of a pleafant green; the
cheeks are greenifh-yellow; there is a red
fpot on the wings, and their quills are black,
terminated with blue; the two middle quills of
the tail are of the fame green with the back,
and the outer quills, being five on either fide,
have each an oblong red fpot on the inner webs,
and which fpread more and more from the inner
quill to the outer one; the iris is red; the bill
and feet whitifh.

Marcgrave notices a variety in this fpecies,
which differs only in point of fize, being rather
fmaller than the preceding. The former he
calls *Aiuru-catinga*, and the latter *Aiuru-apara*.

* Barrere and Briffon.
[A] Specific character of the *Pfittacus Agilis:* " It is green;
the coverts of its bluifh primary wing-quills are fulvous; its tail
fcarcely elongated, red below; its orbits cinereous."

The BLUE-HEADED CRICK.

SIXTH SPECIES.

Pſittacus Autumnalis, 1ſt Var. Gmel.
The *Blue-headed Creature,* Baner.
The *Blue-faced Green Parrot,* Edw. and Lath.

THIS is deſcribed by Edwards; it is found alſo in Guiana. All the fore-ſide of the head and the throat are blue, which colour is terminated, on the breaſt, by a red ſpot; the reſt of the body is green, which is deeper on the back than beneath; the ſuperior coverts of the wings are green; their great quills blue, thoſe adjacent red, and the upper part blue at the extremity; the quills near the body are green; the quills of the tail are green on their upper ſurface as far as the middle, and yellowiſh green below; the lateral quills are red on their exterior webs; the iris is orange coloured; the bill is blackiſh cinereous, with a reddiſh ſpot on the ſides of the upper mandible; the feet are fleſh coloured, and the nails black.

VARIETIES of the BLUE-HEADED CRICK.

To this ſixth ſpecies we ſhall refer the following varieties.

I. The

I. The *Cocho Parrot*, mentioned by Fernandez, which differs in fo far only as it is variegated with red and whitifh inftead of red and bluifh; in every other refpect it is the fame with the Blue-headed Crick. The Spaniards call it *Catherina*, which name they apply alfo to the fecond variety of the *Aouarou-couraou*, and Fernandez fays that it prattles well.

II. The *Leffer Green Parrot* of Edwards, which is diftinguifhed only by its red face and orange cheeks; its other colours, and its fize, are the fame with thofe of the Blue-headed Crick.

III. The *Brafilian Green Parrot* of Edwards is alfo another variety. Its face, and the top of its neck, are of a fine red, whereas thefe parts are bluifh in the Blue-headed Crick; but, in other refpects, the refemblance is exact.—We cannot conceive why Briffon ranges this bird with the Dominica Parrot, mentioned by Labat; for that author fays only that there are a few red feathers in the wings, in the tail, and under the throat, and that all the reft of its plumage is green. But thefe characters are too general, and will apply equally to many other Amazons and Cricks.

The VIOLET-HEADED CRICK.

SEVENTH SPECIES.

Pſittacus Violaceus, Linn. and Gmel.
Pſittacus Aquarum-Lupiarum Inſulæ, Briſſ.
The *Ruff necked Parrot,* Lath.

THIS Parrot is found in Guadaloupe, and was
firſt deſcribed by Father Dutertre. " Its
colours are ſo beautiful," he ſays, " and ſo ſingular, that it deſerves to be ſelected from all the
reſt for deſcription. It is almoſt as large as a
hen; its bill and eyes are edged with carnation;
all the feathers of its head, of its neck, and of
its belly, are violet, mixed with a tincture of
green and black, changing like the neck of a
pigeon; all the upper ſide of the back is of a very
brown green; the great quills of the wings are
black, all the others yellow, green, and red;
on the coverts of the wings are two roſe-ſhaped
ſpots of the ſame colours. When it briſtles the
feathers of its neck, it makes a fine ruff round
the head, on which it ſeems to pride itſelf, as
the peacock does on its tail; it has a ſtrong
voice, ſpeaks very diſtinctly, and is eaſily taught,
if taken young *."

We have not ſeen this Parrot; it is not found
at Cayenne, and it muſt now be very rare in
Guadeloupe, ſince none of the inhabitants of

* *Hiſt. des Antilles,* t. II. p. 251.

this

this ifland could give us any account of it. But this is not extraordinary; for as the iflands advance in population, the number of Parrots gradually decreafes, and Dutertre remarks in particular, that the French colonifts commit great havock among the Violet-headed Cricks in the feafon of the maturity of the guavas, cachimans, &c. when their flefh is exceffively fat and juicy. He adds that they are of a gentle difpofition, and eafily tamed: " We have two," fays he, " which build their neft in a large tree a hundred paces from our hut; the male and female fit alternately, and come one after another to the hut for food, and bring their young ones with them as foon as thefe can leave the neft."

We may obferve that, as the Cricks are the moft common kind of Parrots, and at the fame time fpeak the beft, the favages have amufed themfelves in rearing thefe, and in trying to vary their plumage. For that operation they ufe the blood of a fmall frog, which is very different from thofe of Europe; it is only half the fize, and of a fine azure colour, with longitudinal bars of gold: it is the handfomeft of all the frogs, and feldom frequents marfhes, but inhabits the fequeftered forefts. The favages take a young Crick from the neft, and pluck the fcapular feathers and fome of thofe on the back; then they rub it with the frog's blood, and the new feathers which grow are no longer green, but fine yellow or beautiful red. Thefe

birds

birds thus altered are called *Tapired Parrots* in France. The operation muſt have been an-ciently in uſe among the ſavages, for it is no-ticed by Marcgrave; thoſe which inhabit Gui-ana and the banks of the Amazons equally prac-tiſe it *. The plucking of the feathers hurts the birds greatly, and ſo many die of it, that thoſe which ſurvive are very rare, and are ſold much dearer than the other Parrots.

The Parrot mentioned by Klein and Friſch is one of theſe artificial birds; it would therefore be idle to copy their deſcription.

* Voy. de M. de Gennes au detroit de Magellan. *Paris,* 1698, *p.* 163.

[A] Specific charaĉter of the *Pſittacus Violaceus:* " It is violet, variegated with green, and a mixture of black; its back partly duſky green; its greater wing-quills black, the reſt variegated with yellow, green, and red; a roſy ſpot on the coverts."

THE ARTIFICIAL PARROT.

The POPINJAYS.

Les Papegais, Buff.

THESE are in general fmaller than the Amazons, from which and from the Cricks
they are diftinguifhed by having no red on the
wings. They are all peculiar to the new world.
We are acquainted with eleven fpecies of Popinjays, to which we fhall fubjoin fuch as are
flightly mentioned by authors without defcribing the colours of the wings, and of which we
cannot therefore decide to what genus they belong.

The PARADISE POPINJAY.

FIRST SPECIES.

Pfittacus-Paradifi, Linn. Gmel. and Klein.
Pfittacus Luteus infulæ Cubæ, Briff.
The *Cuba Parrot,* Brown and Catefby.
The *Paradife Parrot,* Lath.

THIS Parrot is very handfome. Its body is
yellow, and all the feathers edged with
dark gloffy red; the great quills of the wings
are white, and all the others yellow, like the
feathers on the body; the two quills in the
middle of the tail are alfo yellow, and all the
lateral ones red, from their origin as far as two
thirds

thirds of their length; the reft is yellow; the iris is red; the bill and feet white.

It would feem that this fpecies admits of fome variety; for in the fpecimen defcribed by Catefby, the throat and belly were entirely red, though there are others in which thefe parts were yellow, and the feathers only edged with white. Perhaps the breadth of the red borders differs according to age or fex, which would account for the diverfity.—The bird is found in the ifland of Cuba [A].

The MAILED POPINJAY, *Buff.*

SECOND SPECIES.

Pfittacus Accipitrinus, Var. Gmel.

THIS American Parrot appears to be the fame with the Variegated Parrot of the old continent, and we prefume that thofe imported into France had been carried from the Eaft Indies to America; and if fome are found in the interior parts of Guiana, they have been naturalized, like the canaries, and feveral other birds and quadrupeds, introduced by navigators. No naturalift or traveller who has vifited the new

[A] Specific charaƈter of the *Pfittacus Paradifi:* " It is yellow; its throat, its belly, and the bafe of its tail-quills, are red."

world

THE MAILED PARROT.

world takes notice of it, though it is well known to our bird-fanciers. Its voice is different from that of the other Parrots of America, and its cry is ſharp and ſhrill. All theſe circumſtances conſpire to prove that it is not indigenous in the new world.

The top of the head and the face are ſurrounded with narrow long feathers, white and radiated with blackiſh, and which it briſtles when irritated, and diſpoſes into a fine ruff. The nape and ſides of the neck are of a fine brown red, and edged with lively blue; the feathers on the breaſt and ſtomach are clouded with the ſame colours, only more dilute, and with a mixture of green; a more beautiful ſilky ſhining green covers the upper ſide of the body and of the tail, except that ſome of the lateral feathers on each ſide appear blue exteriorly, and the primaries of the wing are brown, and alſo the under ſurface of thoſe of the tail.

The T A V O U A.

THIRD SPECIES.

Pſittacus Feſtivus, Linn. and Gmel.
The *Feſtive Parrot,* Lath.

THIS is a new ſpecies, of which M. Duval ſent two ſpecimens for the King's Cabinet.

It

It is rare in Guiana; yet it sometimes approaches the dwellings. Bird-fanciers are eager to obtain it, for of all the Parrots it speaks the best, and even excels the Red-tailed Gray Parrot of Guinea; and yet it was not known till lately, which is somewhat singular. But its talents are attended with an essential defect; it is faithless and mischievous, and bites cruelly when it pretends to caress: it would even seem to lay plans of malice, and its physiognomy, though sprightly, is dubious. It is an exceedingly beautiful bird, and more nimble and agile than any other Parrot.

Its back and its rump are of a very beautiful red; it has also some red on the front, and the upper side of its head is light blue; the rest of the upper side of the body is a fine full green, and the under side of a lighter green; the quills of the wings are of a fine black, with deep blue reflections; so that in some positions they appear entirely of a very deep blue: the coverts of the wings are variegated with deep blue and green.

We have observed that Brisson and Brown have confounded this Popinjay with our fifth species of Cricks.

[A] Specific character of the *Psittacus Festivus:* " It is green; its front purplish; its eye-brows and throat blue; its back blood-coloured."

THE

RED-BANDED POPINJAY, *Buff.*

FOURTH SPECIES.

Pſittacus Dominicenſis, Gmel.
The *Red-banded Parrot*, Lath.

THIS Parrot is found in St. Domingo. On the front a ſmall red band extends between the eyes. This and the blue tinge of the primaries of the wings are almoſt the only interruptions in the colour of the plumage, which is all green and dark complectioned, and ſcaled with blackiſh on the back, and with reddiſh on the ſtomach. It is nine inches and a half long [A].

THE

PURPLE BELLIED POPINJAY, *Buff.*

FIFTH SPECIES.

Pſittacus Leucocephalus, Var. 3, Gmel.

THIS Parrot is found in Martinico. It is not ſo beautiful as the preceding ones : the face is white ; the crown and ſides of the head blue cinereous ; the belly variegated with purple and

[A] Specific character of the *Pſittacus Dominicenſis*: " It is green ; a red band on its front ; black creſcents on its neck and back ; its wing quills blue."

green,

green, but the purple predominates; all the reft of the body, both above and below, is green; the fan of the wing white; the quills variegated with green, blue, and black; the two middle quills of the tail are green, the others variegated with green, red, and yellow; the bill is white; the feet are gray, and the nails brown.

The POPINJAY with a BLUE HEAD and THROAT.

SIXTH SPECIES.

Pfittacus Menfiruus, Linn. Gmel. and Scop.
The *Blue headed Parrot,* Edw. and Lath.

THIS Popinjay is found in Guiana, though rare; and it is befides little fought after, for it cannot be taught to fpeak. The head, neck, throat, and breaft, are of a fine blue, which receives a tinge of purple on the breaft; the eyes are furrounded by a flefh-coloured membrane, whereas in all the other Parrots this membrane is white; on each fide of the head is a black fpot; the back, the belly, and the quills of the wing are of a handfome green; the fuperior coverts of the wings are yellowifh green; the lower coverts of the tail are of a fine red; the quills of the middle of the tail are en-

tirely

tirely green; the lateral ones are of the fame
green colour, but they have a blue fpot, which
extends the more the nearer the quills are to the
edges; the bill is black, with a red fpot on both
fides of the fuperior mandible; the feet are
gray [A].

We have obferved that Briffon has confounded this bird with Edwards's *Blue-faced Green
Parrot,* which is our *Blue-headed Crick.*

The VIOLET POPINJAY.

SEVENTH SPECIES.

Pfittacus Purpureus, Gmel.
The *Little Dufky Parrot,* Edw. and Lath.

THIS is called, both in America and in France,
the *Violet Parrot.* It is common in Guiana; and, though handfome, is not much
efteemed, becaufe it never learns to fpeak.

We have already remarked that Briffon confounds this with the Red and Blue Parrot of
Aldrovandus, which is a variety of our Crick.
The wings and tail are of a fine violet; the
head and the borders of the face are of the fame
colour, which is waved on the throat, and
melted into the white and lilac; a fmall red

[A] Specific charaƈter of the *Pfittacus Menfruus:* " It is green;
its head bluifh; its vent black."

ftreak edges the front; all the upper fide of the body is brown, obfcurely tinged with violet; the under fide of the body is richly clouded with blue-violet, and purple-violet; the lower coverts of the tail are rofe colour, which alfo tinges exteriorly the edges of the outer quills of the tail, through their firft half [A].

The S A S S E B E.

EIGHTH SPECIES.

Pfittacus Collarius, Linn. and Gmel.
Pfittacus Jamaicenfis gutture rubro, Briff.
Pfittacus Minor collo miniaceo, Ray.
The *Common Parrot of Jamaica,* Sloane.
The *Red-throated Parrot,* Lath.

OVIEDO is the firft who has mentioned this Popinjay under the name of *Xaxebès,* or *Saffebe.* Sloane make it a native of Jamaica. The head, and both the upper and under furface of the body, are green; the throat and the lower part of the neck are of a fine red; the quills of the wings are fome green and others blackifh. It is a pity that Oviedo and Sloane, who faw this bird, did not defcribe it more fully [B].

[A] Specific charaĉter of the *Pfittacus Purpureus:* " Above it is dark brown, below purple; its top and its cheeks black; its orbits blue; a collar with dirty points; and the quills of its wings and tail blue."

[B] Specific charaĉter of the *Pfittacus Collarius:* " It is green, with a reddifh throat."

The BROWN POPINJAY.

NINTH SPECIES.

Pſittacus Sordidus, Linn. and Gmel.
Pſittacus Novæ Hiſpaniæ, Briſſ.
The *Duſky Parrot,* Edw. and Lath.

THIS bird is deſcribed, figured, and coloured,
    by Edwards : it is one of the rareſt, and
of the leaſt beautiful in the whole genus of Parrots. It is found in New Spain. It is nearly
as large as a common pigeon; the cheeks and
the upper ſide of the neck are greeniſh ; the
back is dull brown; the rump is greeniſh ; the
tail is green above and blue below ; the throat
is of a beautiful blue, which is about an inch
broad; the breaſt, belly, and legs, are brown,
with a little cinereous ; the wings are green,
but the quills next the body are edged with yellow ; the under coverts of the tail are of a fine
red; the bill is black above, its baſe yellow,
and the ſides of the two mandibles are of a fine
red; the iris is brown nut colour.

[A] Specific character of the *Pſittacus Sordidus:* " It is browniſh ; its throat blue ; its wings and tail green ; its bill and vent
red."

THE
AURORA-HEADED POPINJAY.

TENTH SPECIES.

Pſittacus Ludovicianus, Gmel.
Pſittacus Viridis, capite luteo, fronte rubra, Friſ.
The *Orange-headed Parrot,* Lath.

DUPRATZ is the only perſon who has deſcribed this bird. " It is not," ſays he,
" ſo large as the Parrots which are commonly
brought into France; its plumage is of a beautiful celadine-green; its head is enveloped in
orange, which receives a red tinge near the bill,
and melts into the green on the ſide of the body;
it learns with difficulty to ſpeak, and when it
has made that acquiſition, it ſeldom diſplays it.
Theſe Parrots always appear in flocks, and if
they are ſilent when tamed, they are very noiſy
in the air, and their ſhrill ſcreams are heard at a
diſtance. They live on walnuts, the kernels
of pine tops, the ſeeds of the tulip tree, and
other ſmall ſeeds *.

* Voyage à la Louiſiane, *par le Page Dupratz,* t. II. p. 128.
[A] Specific character of the *Pſittacus Ludovicianus:* " It is ſea-
green; its head fulvous, inclining to reddiſh near the bill."

The P A R A G U A.

ELEVENTH SPECIES.

Pfittacus Paraguanus, Gmel.
Lorius Brafilienfis, Brifl.
The *Paraguan Lory,* Lath.

THIS bird, which is defcribed by Marcgrave, appears to be found in Brafil. It is partly black, and larger than the Amazon; the breaft, and the upper part of the belly, and alfo the back, are of a very beautiful red; the iris is likewife of a fine red; the bill, the legs, and the feet, are deep afh colour.

The beautiful red colours would indicate a relation to the Lory; but as that bird occurs only in India, while the other is probably indigenous in Brafil, I fhall not venture to pronounce whether they are of the fame, or of different fpecies; efpecially as Marcgrave, who faw the Parrot, only gives it the name *Paragua,* without faying that it is a native of Brafil. It is perhaps a Lory, as Briffon conceives. The conjecture derives force from another circumftance: Marcgrave fpeaks alfo of a gray Parrot * as brought from Brazil, which we fufpect to be originally from Guinea; becaufe none of thefe

* *Pfittacus Cinereus,* Linn. and Gmel.
Maracana Prima, Marc. Johnft. Will. and Ray.
Pfittacus Brafilienfis Cinereus, Brifl.
Specific character: " It is entirely bluifh afh-colour."

gray

gray Parrots are found in America, though they are frequent in Guinea, from whence they are often carried with the negroes. Indeed the manner in which Marcgrave expreſſes himſelf ſhews that he did conſider it as an American Parrot ; *A Bird evidently like the Parrot**.

* Avis pſittaco planè ſimilis.

[A] Specific character of the *Pſittacus Paraguanus:* " It is ſcarlet ; its head, its neck, its vent, its tail, its ſhoulders, and its wings, black."

The PARROQUETS.

Les Perriches, Buff.

BEFORE e confider the great tribe of Parroquets, we fhall furvey feparately a little genus that appears to belong neither to the Parroquets nor to the Popinjays, and which is intermediate in regard to fize. It contains only two fpecies, the *Maipouri* and the *Caica*; which laft was unknown till very lately.

The MAIPOURI, *Buff.*

FIRST SPECIES.

Pfittacus Melanocephalus, Linn. and Gmel.
Pfittacus Mexicanus pectore albo, Briff.
Pfittacus Africapillus, Miller.
The *White-breafted Parrot,* Edw. and Lath.

THE name is very applicable ; for this Parrot whiftles like the tapir, which is called *maipouri* in Cayenne ; and though there is a vaft difference between that huge quadruped and this little bird, they utter founds fo exactly fimilar, as not to be diftinguifhable. It is found in Guiana, in Mexico, and as far as the Caraccas ; it never comes nigh the fettlements, but commonly lives in woods furrounded with wa-

P 4   ter,

ter, or even among the trees which grow in
the deluged favannas. It has no other note
than the fharp whiftle, which it repeats often
while on the wing, and it never learns to
fpeak.

These birds commonly affociate in fmall bodies, but often without any tie of affection; for
they fight frequently, and with rancourous obftinacy. When any are caught, they reject
every kind of food, fo that it is impoffible to
keep them alive; and their temper is fo ftubborn that it cannot be foftened by the fmoke of
tobacco, which calms the moft froward of the
Parrots. The Maipouris require to be bred
when young, and they would not repay the
trouble of educating them, were not their plumage fo beautiful, and their figure fo fingular:
for their fhape is very different from that of the
Parrots, or even of the Parroquets; their body
is thicker and fhorter, their head much larger,
their neck and tail extremely fhort; fo that
they have an heavy unwieldy air. All their
motions are fuitable to their figure; even their
feathers are entirely different from thofe of other
Parrots and Parrakeets, being fhort, clofe, and
cohering to the body; fo that they feem compreffed and glued artificially on the breaft and
on all the lower parts of the body —The Maipouri is as large as a fmall Popinjay, and, for
this reafon perhaps, have Edwards, Briffon, and
Linnæus, claffed it with the Parrots; but the
difference

difference is so great as to require a distinct genus.

The upper side of the head is black ; there is a green spot below the eyes ; the sides of the head, the throat, and the lower part of the neck, are of a fine yellow ; the upper side of the neck, the belly, and the legs, are orange; the back, the rump, and the superior coverts of the wings, and the quills of the tail, are of a fine green ; the breast and belly are whitish when the bird is young, and yellowish after it is grown up ; the great quills of the wings are exteriorly blue on the upper side, and blackish below ; the following ones are green, and edged exteriorly with yellow ; the iris is of a deep chesnut ; the bill flesh coloured; the feet ash brown, and the nails blackish [A].

The CAICA.

SECOND SPECIES.

Psittacus Pileatus, Gmel.
The *Hooded Parrot,* Lath.

C AICA, in the Galibi language, is the name of the largest Parroquet, and hence we have

[A] Specific character of the *Psittacus Melanocephalus:* " It is green, below yellow ; its cap black ; its breast white ; its orbits carnation."

applied

applied it to the prefent bird. It is of the fame
genus with the preceding; for it has all the peculiarities of the form, and alfo the black hood.
Its fpecies is not only new in Europe, but even
in Cayenne. M. Sonini de Mononcour tells us,
that he faw it the firft in 1773. Prior to that
date none ever appeared in Cayenne, and it is
ftill uncertain from what country they come.
But they have fince continued to arrive annually in fmall flocks, about the months of September and October, and halt only a fhort time
during the fine weather, fo that they are only
birds of paffage.

The hood which envelopes the Caica is pierced with a hole, in which the eye is placed; the
hood extends very low, and fpreads into two
chin pieces of the fame colour; the circuit of
the neck is fulvous and yellowifh; the beautiful
green which covers the reft of the body is interrupted by an azure tinge, that marks the
edges of the wing from the fhoulder, borders
the great quills on a darker ground, and tips
thofe of the tail, except the two middle ones,
which are entirely green, and appear rather
fhorter than the lateral ones.

[A] Specific character of the *Pfittacus Pileatus:* " It is green;
its head black; its orbits white; a fky-blue fpot on its fhoulders;
the tail tipt with blue."

PARROQUETS

OF THE NEW CONTINENT.

THE diftinction of long and fhort tailed Parrakeets obtains both in the new and in the old continent. Of the long-tailed ones, fome have the tail equally tapered, others unequally. We fhall therefore purfue the former plan; we fhall begin with fuch as have long and equal tails, then confider fuch as have long but unequal tails, and conclude with the fhort-tailed ones.

PARROQUETS

WITH LONG AND EQUALLY TAPERED TAILS.

The PAVOUANE PARROQUET.

FIRST SPECIES,

With a long and equal Tail.

Pfittacus Guianenfis, Gmel. and Briff.
The *Pavouane Parrot*, Lath.

THIS is one of the handfomeft of the Parroquets. It is pretty common in Cayenne, and is alfo found in the Antilles, as M. de la Borde affures us. It learns more eafily to fpeak than any of the Parroquets of the new continent;

8   nent;

nent; but, in other refpects, it is indocile, for
it always retains its wild favage character. Its
afpect is angry and turbulent, but as it has a
quick eye and a flender active fhape, its figure
is pleafing. Our bird-fanciers have adopted the
name *Pavouane*, which it has in Guiana. Thefe
Parroquets fly in flocks *, perpetually fcreaming
and fqualling; and they range through the woods
and favannas, and prefer the fruit of a large tree,
called in that country the *immortal*, and which
Tournefort denominates the *corallo dendron*.

It is a foot long; its tail is near fix inches,
and regularly tapered; the upper fide of the
wings and tail of a very fine green. In proportion as the bird grows older, the fides of the
head and neck are covered with fmall fpots of a
bright red, which become more and more numerous; fo that, in fuch as are aged, thefe parts
are almoft entirely covered with beautiful red
fpots. Thefe never begin to appear till the fecond
or third year. The fmall inferior coverts of the
wings are of the fame bright red, in every period of its age, only the colour is not quite fo
bright when the bird is young. The great inferior wings are of a fine yellow; the quills of
the wings and tail are of a dull yellow below;
the wing is whitifh, and the feet are gray.

---

* " It is remarked that the Parrakeets never affociate with the
Parrots, but always keep together in great flocks." *Wafer*, in
Dampier's Voyage.

The BROWN-THROATED PARROQUET.

SECOND SPECIES,
With a long and equal Tail.

Pfittacus Æruginofis, Linn. and Gmel.
Pfittacu Martinicana, Briff.
The *Brown-throated Parrakeet,* Edw. and Lath.

EDWARDS is the firft who defcribed this Parrakeet. It is found in the new world;
Briffon received a fpecimen from Martinico.

The front, the fides of the head, the throat,
and the lower part of the neck, are of a brown
gray; the crown of the head is bluifh green;
all the upper part of the body yellowifh green;
the great fuperior coverts of the wings blue;
all the quills of the wings blackifh below, but
the primaries are blue above, with a broad
blackifh border on the under fide; the middle
ones are of the fame green with the upper fide
of the body; the tail is green above and yellowifh below; the iris is chefnut; the bill and feet
afh coloured [A].

[A] Specific charaéter of the *Pfittacus Æruginofa:* It is green;
its top and its primary wing-quills blue; its orbits cinereous."

The PARROQUET with a VARIE-GATED THROAT.

THIRD SPECIES,
With a long and equal Tail.

THIS Parroquet is very rare and handſome; it is not frequently ſeen in Cayenne, nor do we know whether it can be taught to ſpeak. It is not ſo large as a blackbird; the greateſt part of its plumage is of a fine green, but the throat and the fore-ſide of the neck are brown, with ſcales and mails of ruſty gray; the great quills of the wings are tinged with blue; the front is water-green; behind the neck and a little below the back, is a ſmall zone of the ſame colour; on the fold of the wing are ſome feathers of a light vivid red; the tail is partly green above and partly dun-red, with copper reflections, and below it is entirely copper co-loured; the ſame tinge appears under the belly.

The PARROQUET with VARIE-GATED WINGS.

FOURTH SPECIES,
With a long and equal Tail.

Pſittacus Vireſcens, Gmel.
Pſittaca Cayanenſis, Briſſ.
The *Yellow-winged Parrakeet*, Lath.

THIS ſpecies is called the *Common Parrakeet* in Cayenne. It is not ſo large as a black-bird, being only eight inches and four lines long, including the tail, which is three inches and a half. Theſe Parroquets keep in numer-ous flocks, prefer the cleared grounds, and even reſort to the ſettled ſpots. They are very fond of the buds of the *immortal* tree, and when in bloſſom they perch on it in crowds. One of theſe large trees planted in the new town of Cayenne draws the viſits of theſe birds; they are frightened away by firing upon them, but they ſoon return. It is difficult to teach them to ſpeak.

In this Parroquet the head, the whole body, the tail, and the ſuperior coverts of the wings, are of a fine green; the quills of the wings are variegated with yellow, bluiſh green, white, and green; the quills of the tail are edged with yellowiſh on the inſide; the bill, the feet, and the nails, are gray.

In

In the female the colours are not fo bright, which is the only difference.

Barrere confounds this bird with the *Anaca* of Marcgrave; but thefe two birds, though of the fame genus, are of different fpecies.

The A N A C A.

FIFTH SPECIES,
With a long and equal Tail.

Pfittacus Ana, Gmel.
Pfittacula Brafilienfis Fufca, Briff.
The *Chefnut Crowned Parrakeet,* Lath.

THE Anaca is a very handfome Parroquet, which is found in Brazil. It is only of the fize of a lark; the crown of the head is chefnut; the fides of the head brown; the throat cinereous; the upper fide of the neck and the flanks green; the belly is rufty brown; the back green with a brown fpot; the tail light brown; the quills of the wings green, terminated with blue, and there is a fpot or ra- ther a fringe of blood colour on the top of the wings; the bill is brown; the feet cinereous.

Briffon has ranged this Parrakeet among thofe which have a fhort tail, but Marcgrave never mentions that property; and as that author ne- ver omits, in his defcriptions, to note when
they

they have a fhort tail, and yet ranges the prefent between two long-tailed ones, we prefume that belongs to that tribe. We have
drawn the fame inference with regard to the
following, which Marcgrave names *Jendaya*,
without faying that it has a fhort tail [A].

The JENDAYA.

SIXTH SPECIES,
With a long and equal Tail.

Pfittacus-Jandaya, Linn. and Gmel.
Pfittacula Brafilienfis Lutea, Briff.
The *Yellow-headed Parrot*, Lath.

THIS Parroquet is equal in bulk to the blackbird. The back, the wings, the tail, and
the rump, are of a bluifh green, inclining to
that of beryl; the head, the neck, and the
breaft, are orange yellow; the extremities of
the wings blackifh; the iris of a fine gold colour; the bill and feet black. It is found in
Brazil, but no perfon has feen it except Marcgrave, and all the other writers have copied his
account [B].

[A] Specific character of the *Pfittacus Anaca*: " It is green, below brown rufous; its top bay; a fpot on its back, and its tail
pale brown; the margin of its wings red."

[B] Specific character of the *Pfittacus Jandaya*: " Above it is
green, below bright yellow; its head and neck bright yellow."

The EMERALD PARROQUET.

SEVENTH SPECIES,
With a long and equal Tail.

Pſittacus Smaragdinus, Gmel.
The *Emerald Parrot,* Lath.

THE rich and brilliant green that covers the whole of the body, except the tail, which is cheſnut with a green point, ſeems to entitle this bird to the name of *Emerald Parroquet.* The appellation of *Magellanic Parrakeet,* which is given in the *Planches Enluminées,* ought to be rejected ; for no Parrot or Parroquet inhabits ſo high a latitude. It is not likely that theſe would paſs the tropic of Capricorn in queſt of regions which are colder than thoſe at an equal diſtance on the northern hemiſphere. Farther, is it credible that birds which live upon tender and juicy fruits would wing their courſe to frozen tracts, which yield nothing but a few ſtarved berries? Yet ſuch are the lands which border on the Straits of Magellan, where ſome travellers are ſuppoſed to have ſeen Parrots. This aſſertion, which is preſerved in the work of a reſpectable author *, would have appeared extraordinary, had we not found, in tracing it to its ſource, that it reſts on an evidence which de-

* Hiſt. des Navig. aux terres Auſtrales, *t. I. p.* 347.

ſtroys

ftroys itfelf; it is that of the navigator Spilberg, who places the Parrots in the Straits of Magellan, near the fame place where, a little before, he fancied that he faw Oftriches *. For a fimilar reafon, perhaps, we ought to reject the relation that Parrots are found in New Zealand and in Diemen's Land †, in the 43d degree of fouth latitude.

We fhall now proceed to enumerate and defcribe the Parroquets of the new continent, which have a long tail unequally tapered.

* Hift. Gen. des Voy. t. XI. pp. 18 & 19.

† Captain Cook's fecond Voyage.

[A] Specific character of the *Pfittacus Smaragdinus*: " It is brilliant green; the hind part of its belly, its rump, and its tail, ferruginous chefnut."

PARROQUETS

The SINCIALO.

FIRST SPECIES,
With a long and unequal Tail.

Pſittaeus Ruſiroſtris, Linn. Gmel. and Gerin.
The *Long-tailed Green Parrakeet,* Edw. and Lath.

THIS bird is called *Sincialo* at St. Domingo. It is not larger than a blackbird, but is twice as long, its tail being ſeven inches, and its body five. It is diſpoſed to chatter, and eaſily learns to ſpeak, to whiſtle, and to mimic the cries of all the animals which it hears : theſe Parroquets fly in flocks, and perch on the cloſeſt and moſt verdant trees ; and as they are green themſelves, they can hardly be perceived. They make a great noiſe among the trees, many at once ſcreaming, ſqualling, and chattering ; and if they overhear the voice of men or other animals, they cry the louder *. This habit is not peculiar to the Sincialos, for almoſt all Parrots that are kept in the houſe babble with more vociferation when a perſon ſpeaks high. They feed like the other Parrots, but are more lively

* Dutertre, *t. II. p.* 252.

and

and cheerful: they are foon tamed; they feem fond of being taken notice of, and they feldom are filent, for whenever a perfon talks, they fcream and chatter likewife. They grow fat and delicate to eat, during the maturity of the feeds of Indian wood, which principally fupports them.

The whole plumage of this Parroquet is yellowifh green; the inferior coverts of its wings and tail are almoft yellow; the two quills in the middle of the tail are longer, by an inch and nine lines, than thofe contiguous on either fide, and the other lateral quills contract gradually, fo that the outermoft are five inches fhorter than the mid-ones. The eyes are encircled by a flefh coloured fkin; the iris is fine orange; the bill is black, with a little red at the bafe of the upper mandible; the feet and nails are flefh coloured. This fpecies is fcattered through almoft all the warm parts of America.

The Parroquet mentioned by Labat is a variety of this*; the only difference being that there are fome fmall red feathers on the head, and the bill is white.—We muft obferve that Briffon has confounded this laft bird with the *Aiuru-catinga* of Marcgrave, which is one of our Cricks.

* *Perrique de la Guadaloupe*, Labat.
Pfittaca Aquarum Lupiarum, Briff.

[A] Specific character of the *Pfittacus Rufiroftris*: " It is green; its bill and feet are red; its tail-quills tipt with bluifh; its orbits carnation."

THE
RED-FRONTED PARROQUET.

SECOND SPECIES,
With a long and unequal Tail.

Pſittacus Canicularis, Linn. and Gmel.
Pſittacus Braſilienſis fronte rubro, Briſſ.
The *Red and Blue-headed Parrakeet,* Edw. and Lath.

THIS bird is found, like the preceding, in almoſt all the warm parts of America. It
was firſt deſcribed by Edwards. The front is
of a bright red; the crown of the head of a fine
blue; the back of the head, the upper ſide of
the neck, the ſuperior coverts of the wings and
thoſe of the tail are deep green; the throat and
all the under ſide of the body are a little yellowiſh; ſome of the great coverts of the wings are
blue; the primaries are dull aſh colour on the
inſide, and blue on the outſide, and at the extremity; the iris is orange; the bill cinereous;
and the feet reddiſh.

We muſt obſerve that Edwards, and Linnæus,
who has copied him, confound this Parroquet
with the *Tui-apute-juba* of Marcgrave, which
conſtitutes a different ſpecies, as will appear
from the following deſcription.

[A] Specific character of the *Pſittacus Canicularis :* " It is
green, with a red front; the back of its head and the outermoſt
quills of its wings, are blue; its orbits fulvous."

THE ILLINOIS PARROT.

The A P U T E J U B A.

THIRD SPECIES,
With a long and unequal Tail.

Pſittacus Pertinax, Linn. Gmel. and Bor.
Pſittaca Illiniaca, Briſſ.
Pſittacus Viridis malis croceis, Klein.
The *Yellow-faced Parrot,* Edw.
The *Illinois Parrot,* Penn. and Lath.

THE front, the ſides of the head, and the top of the throat, are of a fine yellow ; the crown and back of the head, the upper ſurface of the neck and of the body, the wings and the tail, are of a fine green. Some of the ſuperior coverts of the wings, and the great quills, are edged exteriorly with blue ; the two quills in the middle of the tail are longer than the lateral ones, which continually ſhorten, infomuch that the mid-ones exceed the outermoſt by an inch and nine lines ; the lower belly is yellow ; the iris deep orange ; the bill and feet cinereous.

From this defcription alone it is manifeſt that this ſpecies is not the ſame with the preceding, and is even widely different. Befides, it is very common in Guiana, where the former is never found. It is vulgarly called at Cayenne the *Wood-lice Parrakeet,* becauſe it generally lodges in the holes where theſe infects neſtle. It remains the whole year in Guiana, and frequents the favannas and the cleared lands. It is very

Q 4   improbable

improbable that this fpecies extends to the country of the Illinois, or roams fo far north, as
Briffon afferts ; efpecially as no fpecies of Parrot is found beyond Carolina, and only one fpecies in Louifiana, which we have before defcribed [A].

The GOLDEN-CROWNED PARROQUET.

FOURTH SPECIES,
With a long and unequal Tail.

Pfittacus Aureus, Linn. and Gmel.
Pfittaca Brafilienfis, Briff.

THIS name was beftowed by Edwards, who
took the bird for a female of the preceding
fpecies. What he defcribed was really a female,
fince it layed five or fix fmall white eggs in England, and lived fourteen years in that climate.
But the fpecies is different from the foregoing,
for though both are common in Cayenne, they
never affociate together, but keep in great feparate flocks ; and the males refemble the females.
The Golden-crowned Parroquet is called in
Guiana the *Parrakeet of the Savannas* ; it fpeaks
extremely well, is very fondling and intelligent ;

[A] Specific character of the *Pfittacus Pertinax*: " It is green ;
its cheeks fulvous ; the quills of its wings and tail fomewhat
hoary."

whereas

whereas the preceding is not efteemed, and articulates with difficulty.

This handfome Parroquet has a large orange fpot on the fore part of the head; the reft of the head, all the upper fide of the body, the wings, and the tail, are of a deep green; the throat and the lower part of the neck, are of a yellowifh green, with a flight tinge of dull red; the reft of the under fide of the body is pale green; fome of the great fuperior coverts of the wings are edged exteriorly with blue; the outer fide of the feathers of the middle of the wings is alfo of a fine blue, which forms on each wing a broad longitudinal band of that beautiful colour; the iris is vivid orange; the bill and feet blackifh [A].

The GUAROUBA, or YELLOW PARROQUET.

FIFTH SPECIES,
With a long and unequal Tail.

Pfittacus-Guarouba, Linn. and Gmel.
Pfittaca Brafilienfis Lutea, Briff.
Qui Jubo Tui, Marc. Ray, &c.
The *Brafilian Yellow Parrot*, Lath.

MARCGRAVE and De Laët are the firft who take notice of this bird, which is found in

[A] Specific chara&er of the *Pfittacus Aureus*: " It is green; its cere and its orbits bluifh carnation; its top golden; an oblique blue ftripe on the coverts of its wings."

Brazil,

Brafil, and fometimes in the country of the
Amazons, where however it is rare*, nor is
it ever feen near Cayenne. This Parroquet,
which the Brazilians call *Guiaruba*, that is,
Yellow Bird, does not learn to fpeak at all;
and it is melancholy and folitary. Yet the favages hold it in great eftimation for the fake of
its plumage, which is very different from that
of the other Parrots, and on account of its being eafily tamed. It is almoft entirely yellow;
only there are fome green fpots on the wings,
whofe fmall quills are green, fringed with yellow; the primaries are violet fringed with blue;
and the fame mixture of colours appears on the
tail, whofe extremity is blue-violet; its middle
and rump are green, edged with yellow; all
the reft of the body is pure yellow, and vivid faffron, or orange. The tail is five inches
in length, which is that of the body; it is much
tapered, fo that the laft lateral feathers are one
half fhorter than thofe of the middle. The
Yellow Mexican Parrakeet, given by Briffon
from Seba, appears to be a variety of this; and
the little pale red which Seba reprefents on the
head of his bird *Cocho*, and which was perhaps
only an orange tint, does not form a fpecific
character.

* " The rareft of the Parrots are thofe which are entirely yellow, with a little green at the extremity of the wings: I never faw
any of this fort but at Para." La Condamine, *Voyage a Riviere
des Amazones*, p. 173.

THE
YELLOW-HEADED PARROQUET.

SIXTH SPECIES,
With a long and unequal Tail.

Pſittacus Carolinenſis, Linn. and Gmel.
Pſittaca Carolinenſis, Briſſ.
The *Carolina Parrot,* Cateſby, Penn. and Lath.

THIS Parroquet appears to be one of thoſe which travel from Guiana to Carolina, to Louiſiana*, and even to Virginia. The front is of a beautiful orange; all the reſt of the head, the throat, the half of the neck, and the fan of the wing, are of a fine yellow; the reſt of the body, and the ſuperior coverts of the wings, are light green; the great quills of the wings are brown on the inner ſide; the outer ſide is yellow, as far as one third of its length, it then grows green and blue near the extremity; the middle quills of the wings, and thoſe of the tail, are green; the two middle ones of the tail are an inch and half longer than thoſe adjacent on either ſide; the iris is yellow; the bill is yellowiſh white; and the feet are gray.

* "I ſaw alſo that day, for the firſt time, Parrots (in Louiſiana); they appear along the Teakiki. but in ſummer only: theſe were ſtray-birds, which repaired to the Miſſiſſippi, where they occur in all ſeaſons. They are ſcarcely larger than black-birds; their head is yellow, with a red ſpot on the middle; on the reſt of their plumage green predominates." *Hiſt. de la Nouv. France,* par Charlevoix. *Paris,* 1744, t. III. p. 384.

Theſe

Thefe birds, fays Catefby, feed upon the
feeds and kernels of fruits, particularly apples,
and the grains contained in cyprefs cones. In
autumn they refort to the orchards in great
flocks, and as they tear and mangle fruits to
obtain the kernel, which is the only part that
they eat, they do much injury. They penetrate as far as Virginia, which is the moft
northern colony, fubjoins Catefby, where I
heard of Parrots being feen. This is alfo the
only fpecies found in Carolina, where a few
breed; but moft of them retire fouthwards in
the love feafon, and appear again during the
harveft; being enticed by the fruit trees, and
rice crops. The colonies between the tropics
fuffer greatly from the influx of Parrots on their
plantations. In the months of Auguft and September of 1750 and 1751, a prodigious number
of Parrots of all kinds arrived in Surinam, and
fpread in flocks among the ripe coffee; they ate
the red hufks, without touching the beans,
which they fuffered to fall to the ground. In
1760, about the fame feafon, new fwarms of
thefe birds appeared, and, extending along the
coaft, did much injury, though it could not
be conjectured whence they came *. In general, the ripenefs of fruits, the plenty or
fcarcity of food in different countries, compel

* Piftorius. *Befchriving van colonie van Surinaamen*, Amfterdam,
1768.

certain

certain fpecies of Parrots to *flit* from one tract
to another *.

The ARA PARROQUET.

SEVENTH SPECIES,
With a long and unequal Tail.

Pfittacus-Makawuanna, Gmel.
The *Parrot Maccaw,* Lath.

BARRERE is the firft who has noticed this
bird. It is however frequently feen in Cayenne, where it is reckoned migratory. It haunts
the overflowed favannas, like the Aras, and alfo
fubfifts on the fruits of the palmetto. It is
called the *Ara Parroquet,* becaufe it is larger
than the other Parroquets; its tail very long,
being nine inches, and its body the fame; like
the Aras alfo, it has a naked fkin from the corners of the bill to the eyes, and pronounces diftinctly the word *Ara,* though with a raucous
voice, and lower and fhriller. The natives at
Cayenne call it *Makavouanne.*

* " In the *Antis* are found Parrots of all fizes and colours. Thefe
birds iffue from the country of the Antis, when the cara or maize
is fown, of which they are very fond ; and accordingly they make
great havock. The *Guacamayas* alone, on account of their unwieldi-
nefs, never fally from the country of the Antis; they all fly in
flocks, yet one fpecies intermingles not with another." *Garcilaffo,*
Hift. des Incas. *Paris,* 1744, t. II. p. 83.

[A] Specific character of the *Pfittacus Carolinenfis:* " It is
green; its head, its neck, and its knees, yellow."

The

The quills of the tail are unequally tapered; all the upper fide of the body, of the wings, and of the tail, deep green, with a dark caft, except the great quills of the wings, which are blue, edged with green, and terminated with brown on the outfide ; the upper part and the fides of the head are green mixed with deep blue, fo as, in certain pofitions, to appear entirely blue; the throat, the lower part of the neck, and the top of the breaft, have a deep rufty caft ; the reft of the breaft, the belly, and the fides of the body, are of a paler green than that of the back; laftly, on the lower belly there is fome brown-red, which extends over fome of the lower coverts of the tail ; the quills of the wings and of the tail are yellowifh-green below.

We have only to defcribe the fhort-tailed Parroquets of the new continent, to which we have given the generic name of *Toui*, by which they are known in Brazil.

The TOUIS, or SHORT-TAILED PARROQUETS.

THESE are the fmalleft of all the Parrots which inhabit the new continent : their tail is fhort, and their bulk exceeds not that of the fparrow, and moft of them are incapable of being taught to fpeak ; for of the five fpecies with which we are acquainted, there are only two which can acquire that talent. The Tuis appear to be found in both continents, and, though not exactly of the fame fpecies, they are analogous and related, becaufe they have been tranfported, as I formerly mentioned. Yet I am inclined to think that they are all originally natives of Brazil, whence they have been introduced into Guinea and the Philippine iflands.

The YELLOW-THROATED TOUI.

FIRST SPECIES
Of fhort-tailed Toui.

Pfittacus-Tovi, Gmel.
Pfittacula gutture luteo, Briff.
The *Yellow-throated Parrakeet*, Lath.

THE head and all the upper fide of the body are of a fine green; the throat is of a fine orange

orange colour; all the under fide of the body
yellowifh green; the fuperior coverts of the
wings are variegated with green, brown, and
yellowifh; the inferior coverts are fine yellow;
the quills of the wings are variegated with
green, yellowifh, and deep cinereous; thofe of
the tail are green and edged internally with yellowifh; the bill, the feet, and the nails, are
gray [A].

The S O S O V E.

SECOND SPECIES
Of Toui or fhort-tailed Parroquet.

Pfittacus-Sofove, Gmel.
The *Cayenne Parrakeet*, Lath.

SOSOVE is the Galibi name of this charming
little bird, which is eafily defcribed, fince
it is entirely of a brilliant green, except a fpot
of light yellow on the quills of the wings, and
on the fuperior coverts of the tail; the bill is
white, and the feet gray.

This fpecies is common in Guiana, efpecially
near Oyapoe, and the river Amazons. It can
eafily be tamed, and taught to fpeak. Its voice
is like that of Punch in the puppet-fhews; and
when well trained it chatters perpetually [B].

[A] Specific charaƈter of the *Pfittacus Tovi*: " It is green; a
pale orange fpot on its throat; a broad chefnut bar on its wings,
with a green gold luftre."
[B] Specific charaƈter of the *Pfittacus Sofove*: " It is green, with
a dilute yellow fpot on its wings and the covert of its tail."

The T I R I C A.

THIRD SPECIES
Of Toui, or ſhort-tailed Parroquets.

Pſittacus-Tirica, Gmel.
Pſittacula Braſilienſis, Briſſ.
The *Green Parrakeet,* Lath.

MARCGRAVE firſt deſcribed this bird. Its plumage is entirely green; the eyes are black; the bill carnation; and the feet bluiſh. It is ſoon tamed and taught to ſpeak, and is very gentle, and eaſily managed.

The *ſmall Chatterer* of the *Planches Enlumineés* ſeems to be of the ſame ſpecies: it is alſo entirely green; its bill fleſh coloured, and of the uſual ſize of a Toui.

The *Tuin* of Jean de Laët * does not mean any particular ſpecies, but comprehends all the Parroquets in general; and therefore we ought not, with Briſſon, to refer it to the *Tui-tirica* of Marcgrave.

Sonnerat mentions a bird which he ſaw in the iſland of Luçon, and which much reſembles the *Tui-tirica* of Marcgrave. It is of the ſame bulk, and its plumage wholly dyed green, though deeper above, and lighter below. But it is diſtinguiſhed by the gray colour of its bill, which is carnation in the other, and by the gray caſt

* *Deſcription des Indes Occidentalis, p.* 490.
VOL. VI.   R   of

of its feet, which are bluish in the former: these differences would be insufficient to constitute a species, if the climates were not so distant. It is possible, and even probable, that this bird was carried from America to the Philippines, where it might undergo those small changes [A].

The ETE, or TOUI-ETE.

FOURTH SPECIES
Of Toui, or short-tailed Parroquet.

Pſittacus Paſſerinus, Linn. Gmel. and Bor.
Pſittaca Braſilienſis uropygio cyano, Briſſ.
Tuite, Ray and Will.
The *Short-tailed Green Parrakeet*, Bancroft.
The *Leaſt Blue and Green Parrakeet*, Edw. and Lath.

WE are likewiſe indebted to Marcgrave for the account of this bird. It is found in Brazil; its plumage is in general light green; but the rump, and the top of the wings, are of a fine blue; all the quills of the wings are edged with blue on the outſide, which forms a long blue band when the wings are cloſed; the bill is fleſh-coloured, and the feet cinereous.

To the ſame ſpecies we may refer the bird denominated by Edwards the *Leaſt Green and Blue Parrakeet*, the only difference being that

[A] Specific character of the *Pſittacus Tirica*: "It is green; its bill carnation; its feet and nails bluiſh."

its

its wing-quills are not edged with blue, but
with yellowifh-green, and that the bill and feet
are fine yellow [A].

The GOLDEN-HEADED TOUI.

FIFTH SPECIES
Of fhort-tailed Parroquet.

Pfittacus-Tui, Gmel.
Pfittacula Brafilienfis Iɛterocephalos, Briff.
The *Gold-head Parrakeet,* Lath.

Tʜɪs bird is alfo found in Brazil. All its
plumage is green, except the head, which
is of a fine yellow ; and, as its tail is very fhort,
we muft not confound it with another Parroquet which has alfo a gold colour, but, at the
fame time, a long tail [B].

A variety, or at leaft a contiguous fpecies, is
delineated in the *Planches Enluminées,* where it
is denominated the *Little Parrakeet of the ifland
of St. Thomas,* becaufe the Abbe Aubry, Reɛtor
of St. Louis, in whofe cabinet the fpecimen was
lodged, faid that it came from that ifland. But
the only difference between it and the Gold-

[A] Specific charaɛter of the *Pfittacus Pafferinus:* " It is yellow greenifh; a fpot on its wings, and their under furface blue.''
[B] Specific charaɛter of the *Pfittacus-Tui:* " It is green; its
front orange ; its orbits bright yellow."

head

head Toui is that the yellow tinge is much paler.

These five species are all the Touis of the new world that we are acquainted with ; nor are we certain whether the two small short-tailed Parrakeets, the first noticed by Aldrovandus, the second by Seba, ought to be classed with the rest, for the descriptions are very imperfect. That of Aldrovandus seems rather to be a *Cockatoo*, by reason of the tuft on its head, and that of Seba appears to be a *Lory*, because its plumage is almost entirely red. But we know none of the Cockatoos or Loris that resemble them closely, or with which we could venture to class them.

THE CURUCULI.

The CURUCUIS.

Les Couroucous, ou Couroucoais, Buff.

SUCH is the name which thefe birds bear in their native climate of Brazil. This word imitates their cry fo exactly, that the natives of Guiana have omitted only the firft letter, and call them *Urucoos.* Their characters are thefe: Their bill fhort, hooked, indented, broader than it is thick, and much like that of the Parrots; it is furrounded at its bafe by ragged feathers, projecting forwards, but not fo long as in the bearded birds, which we fhall afterwards defcribe; the legs alfo are very fhort, and feathered within a little of the infertion of the toes, which are placed two behind and two before. We know only three fpecies, and thefe may perhaps be reduced to two, though nomenclators reckon fix, fome of which are varieties, and others belong to a different genus.

The RED-BELLIED CURUCUI.

Le Couroucou à Ventre Rouge, Buff.

FIRST SPECIES.

Trogon-Curucui, Edw. and Lath.
Trogon Brafilienfis Viridis, Briff.
Tzinitzian, Fernand. Johnft. Will. and Ray.

THIS bird is ten inches and a half long; the head, the whole of the neck and the rife of the breaft, the back, the rump, and the coverts of the upper fide of the tail, are of a fine brilliant green, but changing, and, in a certain pofition, blue; the coverts of the wings are blue gray, variegated with fmall black zig-zag lines; and the great quills of the tail are black, except their fhafts, which are partly white; the quills of the tail are of a fine green, like the back, except the two outer ones, which are blackifh, and have fmall tranfverfe gray lines; a part of the breaft, the belly, and the coverts of the under fide of the tail, are of a fine red; the bill is yellowifh, and the legs are brown.

Another fubject, which appears to have been a female, differed in no refpect, except that all the parts, which were of a fine brilliant green in the firft, are blackifh-gray in this, and without any reflections; the fmall zig-zag lines are much more indiftinct, becaufe the dark brown predominates, and the three outer quills of the tail

have,

have, on their exterior webs, alternate black
and white bars ; the upper mandible is entirely
brown, and the lower yellowifh ; laftly, the
red colour is much lefs fpread, occupying only
the lower belly, and the coverts of the under
furface of the tail.

There is a third fubject in the King's Cabinet, which differs chiefly from the two preceding in thefe refpects : the tail is longer, and
the three outer quills on each fide have their outer
webs and their tips white ; the three exterior
quills of the wing are marked with tranfverfe
fpots, that are alternately white and black at
their margin ; there is alfo a gold-green fhade,
waving on the back and on the quills in the
middle of the tail, which has not place in the
preceding. But the red tinge is difpofed in the
fame way, and begins only at the lower belly,
and the bill is fimilar in its fhape and colour.

The Chevalier Le Febvre Defhayes, corref-
pondent of the Cabinet, whom we have often
had occafion to quote as an excellent obferver,
has fent us a coloured drawing of this bird, with
excellent obfervations. He fays that it is called at
St. Domingo *the red drawers* *, and in many of
the other iflands it is termed *the Englifh lady* †.
" This bird retires," he adds, " into the depths
of the forefts during the feafon of its amours ;
its melancholy and even difmal accents feem to

* *Caleçon rouge.*   † *Demoifelle ou Dame Angloife.*

R 4   exprefs

exprefs that profound fenfibility which carries it
into the defert, to enjoy in folitude the tendernefs of love, and that languor, which is more
delicious perhaps than its tranfports. This cry
alone reveals its retreat, which is often inacceffible, and difficult to difcover.

" Their loves commence in April : they
choofe the hole of a tree, and line it with the
duft of worm-eaten wood ; and this bed is as
foft as cotton or down. If they cannot find
fuch duft, they break frefh wood with their
bill, and reduce it to powder ; and their bill,
which is indented near the point, is fufficiently
ftrong for that purpofe : it alfo ferves to enlarge
the hole, when not fufficiently wide. They
lay three or four eggs, which are white, and
fomewhat fmaller than thofe of a pigeon.

" While the female hatches, the employ-
ment of the male is to bring fupplies of food, to
keep watch on a neighbouring bough, and to
fing. At other times he is filent and referved ;
but during incubation, he fatigues the echos
with thofe languifhing founds, which how infipid foever they may appear to us, undoubtedly footh the tedious occupation of his dear
companion,

" The young, at the moment of their exclufion, are entirely naked, without any trace of
feathers, but which begin to fprout two or
three days after. Their head and bill appear
uncommonly thick, compared with the reft of
their

their body; their legs too feem exceffively long, though they are very fhort when the bird is grown. The male becomes filent the inftant that the brood are hatched; but he again re-fumes his fong, with his loves, in the months of Auguft and September.

" They feed their young with worms, ca-terpillars, and infects. Their enemies are the rats, the ferpents, and both the nocturnal and diurnal birds of prey; fo that the fpecies of the *Ooroocoais* is not numerous, moft of them fall-ing a facrifice to depredation.

" After the young ones are flown, they re-main not long together; they yield to their fo-litary inftinct and difperfe.

" In fome individuals the legs are reddifh, in others they are flaty blue. It has not been ob-ferved whether this diverfity is occafioned by age, or refults from the difference of fex."

The Chevalier Defhayes tried to raife fome of the preceding year, but his attempts were fruit-lefs; and, either from a languid or a lofty tem-per, they obftinately refufed to eat. " Per-haps," fays he, " I fhould have fucceeded bet-ter, if I had taken them juft after hatching; but a bird, which lives fo remote from us, and which feeks felicity in the freedom and filence of the defert, feems not adapted for flavery, and muft continue a ftranger to all the habits of the domeftic ftate."

[A] Specific character of the *Trogon-Curucui*; " It is gold-green, below fulvous; its throat black."

THE

YELLOW-BELLIED CURUCUI.

Le Couroucou à Ventre Jaune, Buff.

SECOND SPECIES.

Pfittacus Viridis, Linn. and Gmel.
Trogon Cayanenfis Viridis, Briff.

THIS bird is about eleven inches long; the
wings when clofed do not reach quite to
the tail; the head and the upper fide of the
neck are blackifh, with reflections of handfome
green in fome parts; the back, the rump, and
the coverts of the upper fide of the tail, are brilliant green, like the thighs; the great coverts
of the wings are blackifh, with fmall white
fpots; the great quills of the wings are blackifh, and the four or five outer ones have a white
fhaft; the quills of the tail are of the fame colour with thofe of the wings, except that they
have fome reflections of a brilliant green; the
three outer ones on each fide are radiated tranfverfely with black and white; the throat and
the under fide of the neck are dark brown; the
breaft, the belly, and the coverts below the
tail, are of a fine yellow; the bill is indented, and appears dark brown, as well as the
legs; the nails are black; the tail is tapered,
the feather on each fide being two inches

4   fhorter

shorter than the two middle ones, which are the longest [A].

Between the Red-bellied Curucui and the Yellow-bellied Curucui lie some varieties, which our nomenclators have taken for different species. Such, for instance, is the one denominated, in the *Planches Enluminées*, the *Guiana Curucui*, which is only a variety of the Yellow-bellied Curucui, occasioned by age; the sole difference being, that the upper side of the back, which in the adult is fine azure, is ash-coloured in the young one.

Further, the bird represented in the *Planches Enluminées* by the name of the *Rufous-tailed Curucui of Cayenne*, is a variety of the same Yellow-bellied Curucui, produced by moulting; since the only difference is that the feathers of the back and tail are rufous instead of blue *.

There is also a variety of this Yellow-bellied Curucui: it is the bird termed by Brisson the *White-bellied green Curucui of Cayenne*. The only difference lies in the colour of the tail, which may be owing to age, for the feathers were not completely formed. It might also be an acci-

[A] Specific character of the *Trogon Viridis*: " It is gold-green, below yellow; its throat black; a gold-green bar on its breast."

* *Trogon Rufus*, Gmel.
The *Rufous Curucui*, Lath.
Specific character: " It is rufous; its belly, its vent, and its thighs, yellow; the coverts of its wings streaked with black and gray; its wing-quills and the middle quills of the tail tipt with black."

dental

dental diverfity; but certainly none of thefe three birds can be regarded as a diftinct and feparate fpecies *

We have feen another individual whofe breaft and belly were whitifh, with a tinge of citron in many parts; which made us fufpect that the White-bellied Curucui, juft mentioned, was only a variety of the Yellow-bellied Curucui.

The VIOLET-HOODED CURUCUI.

Le Couroucou à Chaperon Violet, Buff.

THIRD SPECIES.

Trogon Violaceus, Gmel.
The *Violet-headed Curucui*, Lath.

THE throat, neck, and breaft, are of a very dufky violet; the head is of the fame colour, except that of the front, and of the fpace round the eyes and ears, which is blackifh; the eye-brows yellow; the back and rump of a deep green, with gold reflections; the fuperior coverts of the tail are bluifh-green, with the fame gold reflections: the wings are brown, and their coverts, as well as the middle quills, are dotted with white; the two central quills of the tail are green, verging on bluifh, and terminated with black; the two adjacent pairs are of the

* *Trogon Viridis*, Var. Linn. and Gmel.

fame

same colour in the uncovered part, and blackiſh
in the reſt; the three lateral pairs are black,
ſtriped and terminated with white; the bill is
lead colour at the baſe, and whitiſh near the
point; the tail exceeds the wings when cloſed,
by two inches and nine lines, and the total
length of the bird is nine inches and a half.

M. Koelreuter calls this bird *Lanius* *; but it
is of a genus very different from that of a ſhrike,
a lanner, or another bird of prey. A broad ſhort
bill, and briſtles around the lower mandible;
ſuch are the characters which it has in common with the Curucuis. But the properties
wherein it reſembles the cuckoos, that the legs
are very ſhort and feathered to the nails, which
are ſlender and diſpoſed in pairs, the one before
and the other behind; that the nails are ſhort,
and ſlightly hooked; and laſtly, the want of a
membrane around the baſe of the bill: all theſe
differ from the characters of the rapacious tribe.

The Curucuis are ſolitary birds which live in
the heart of damp foreſts, where they ſubſiſt on
inſects; they are never obſerved to conſort in
flocks; they generally ſit on the middle branches,
the cock and hen on ſeparate but adjacent.trees,
and call each other alternately, by repeating their
hollow monotonous cry, *ooroocoais*. They never
fly far, but only from tree to tree, and ſeldom even
do that; for they remain during the greateſt part

* *Comment. Petropol.* 1763.

of

of the day in the fame fpot, concealed beneath
the thickeft boughs; where, though their voice
is continually heard, yet, as they are motion-
lefs, they can hardly be difcovered. They are
clothed fo thickly with plumage, that they appear larger than in reality; they would feem to
equal the bulk of a pigeon, though they have
not more flefh than a thrufh. But if their
feathers be numerous and clofe, they are
weakly rooted, for they drop with the leaft
rubbing; fo that it is difficult to prepare fpeci-
mens for the cabinet. Thefe birds are among
the moft beautiful of South America. Fernandez fays that the fine feathers of the Red-bellied Curucui were ufed by the Mexicans in
making portraits, a gaudy kind of paintings,
and other ornaments which they wore at fefti-
vals, or in battle [A].

There are two other birds mentioned by Fernandez, which Briffon fuppofes to be Curucuis;
but they undoubtedly belong not to that genus.

The firft is what Fernandez compares to the
ftare, and which we have formerly noticed *. I
am aftonifhed that Briffon could fancy that it
was a Curucui, fince Fernandez himfelf refers
it to the genus of the ftares, and their figures
are fimilar. But the fhape of the bill, the dif-

[A] Specific chara&ter of the *Trogon Violaceus:* " It is violet; its
eye-brows bright yellow; its back and rump gold-green; its wings
brown; its intermediate tail-quills bluifh-green, tipt with black."

* *Trogon Mexicanus,* Briff.

pofition of the toes, the form of the body, every
property of the bird in fhort, is fo widely different from thofe of the Curucuis, that they never with propriety can be affociated.

The fecond bird which Briffon has taken for
a Curucui is one which Fernandez * fays is exceedingly beautiful, and of the fize of a pigeon;
that it frequents the fea fhore ; and that its bill
is long, broad, black, and a little hooked : this
form of the bill is very different from what obtains
in the Curucuis, a circumftance alone fufficient
to exclude it from this genus. Fernandez fubjoins
that it does not fing, and that its flefh is unfit
for eating; that its head is blue, and the reft of
its plumage blue, variegated with green, black,
and whitifh. But thefe indications are not pre
cife enough to determine the fpecies.

* *Trogon Mexicanus Varius*, Briff.

The CURUCUCKOO.

Cuculus Brafilienfis, Linn. and Gmel.
Cuculus Brafilienfis Criftatus Ruber, Briff.
The *Red-cheeked Cuckoo,* Lath.

BETWEEN the extenfive family of the
Cuckoo and that of the Curucui, we fhall
place a bird which feems to participate of both ;
fuppofing that the indication of Seba is lefs faulty
than moft of thofe inferted in his bulky work :
his account is as follows—" The head is of a
pale red, bearing a fine tuft of brighter red, variegated with black. The bill is pale red ; the
upper fide of the body is bright red ; the coverts
of the wings and the under fide of the body are
pale red ; the quills of the wings and thofe of
the tail are yellow, fhaded with a blackifh tint."

This bird is not fo large as the magpie, its
total length being about ten inches.

We muft obferve that Seba takes no notice of
the difpofition of its toes, and in his figure they
are difpofed by three and one, not by two and
two. But the afferting the bird to be a *Cuckoo,*
implied the latter.

THE TOURACO CUCKOO

The T O U R A C O.

Cuculus-Perfa, Linn. and Gmel.
Cuculus Guineenfis Criftatus Viridis, Briff.
The *Crown Bird from Mexico,* Alb.

THIS is one of the moft beautiful of the
African birds; for befides that its plumage is brilliant and its eyes fparkle with fire, it
has a fort of creft on the head, or rather a
crown, which confers an air of diftinction. I
cannot conceive, therefore, why our nomenclators range it with the cuckoos, which, every
body knows, are ugly birds; it is alfo difcriminated by its tuft, and by the fhape of its bill, of
which the upper part is more arched than in the
cuckoos. Indeed the only common character is
that it has two toes before and two behind; a
property which belongs to many birds.

The Touraco is as large as a jay; but its long
broad tail feems to increafe its bulk : its wings
are however very fhort, not reaching to the origin of the tail. The upper mandible is convex,
and covered with feathers reflected from the
forehead, and in which the noftrils are concealed. Its eye, which is lively and full of fire, is
encircled by a fcarlet eye-lid, which has a great
number of protuberant *papillæ* of the fame colour. The beautiful tuft, or rather *mitre,* that

VOL. VI.   S   crowns

crowns its head is a bunch of briftled feathers, which are fine and filky, and confift of fuch delicate fibres that the whole is tranfparent. The beautiful green which covers all the neck, the breaft, and the fhoulders, is alfo compofed of fibres of the fame kind, and equally fine and filky.

We know two fpecies, or rather two varieties, of the genus; the one termed the *Abyffinian Touraco*, and the fecond the *Touraco from the Cape of Good Hope*.

The only difference lies in the tints, for the bulk of the colours is the fame. The Abyffinian Touraco has a blackifh tuft, compact like a lock, and reflected backwards: the feathers on the forehead, the throat, and compafs of the neck, are meadow green; the breaft, and top of the back, are of the fame colour, but with an olive tinge, which melts into a purple brown, heightened by a fine green glofs; all the back, the coverts of the wings, and their quills next the body, and all thofe of the tail, are coloured in the fame way; all the primaries are of a fine crimfon, with a black indenting on the fmall webs, near the tip: we cannot conceive how Briffon faw only four of thefe red feathers. The under fide is dun gray, flightly fhaded with light gray.

The Touraco of the Cape of Good Hope differs not from the Abyffinian one, except that
the

the tuft is of a light green, and sometimes fringed with white. The neck is of the same green, which melts on the shoulders into a darkish tint, with glossy green reflections.

We had a Touraco alive from the Cape, and were assured that it lived upon rice. No other food was offered to it at first, and this it would not touch, but grew famished, and, in that extremity, it ate its own excrements. During three days, it subsisted only on water and a bit of sugar. But observing grapes brought to the table, it shewed a strong appetite for them; some were given to it, which it swallowed greedily. It discovered in the same way a fondness for apples, and afterwards for oranges. From that time it was fed on fruit for several months. This seemed to be the natural food, its curved bill not being in the least adapted for collecting grain. The bill is wide, and cleft as far as under the eyes; the bird hops, but does not walk; its nails are sharp and strong, its hold firm; its toes are stout, and invested with thick scales. It is lively, and bustles much; it continually utters a weak, low, and hoarse cry, *crêu, crêu*, from the bottom of its gizzard, without opening its bill. But sometimes it has a very loud scream, *cō, cō, cŏ, cŏ, cŏ, cŏ, cŏ*; the first notes low, the others higher, rapid, and noisy, with a shrill and harsh voice. It vents this cry of its own accord, when it is hungry;

but it may be made to repeat it at pleasure, if one prompts it by imitation.

This beautiful bird was given to me by the Princess of Tingri, to whom my most respectful thanks are due. It is even handsomer than at first, for it was in moult when I made the foregoing description. At present, which is four months since that time, the plumage is restored with fresh beauties; there are two white streaks of small feathers, or short silky hairs, the one pretty near the inner corner of the eye, the other before the eye, and extending backwards to the outer corner; between these two is another streak of the same down, but of a deep violet cast; the upper surface of the body and of the tail shines with a rich purplish blue, and the crest is green and not fringed. These new characters dispose me to think that it does not resemble the Touraco from the Cape of Good Hope so much as I at first supposed; it seems also to differ in the same properties from the Abyssinian Touraco. We have therefore three varieties; but we cannot determine whether the diversity belongs to the species or to the individual, whether it is periodical or constant, or only sexual.

It does not appear that this bird is found in America, though Albin gives it as brought from Mexico. Edwards affirms that it is indigenous in Guinea, from whence the individual
mentioned

mentioned by Albin was poffibly tranfported into the new world. We are unacquainted with the habits of the bird, when it enjoys its native freedom ; but as it is exceedingly beautiful, we may hope that travellers will obferve them, and communicate their remarks,

[A] Specific chara&ter of the *Cuculus Perfa:* " Its tail is equal ; its head crefted ; its body bluifh green ; its wing-quills blood coloured."

The CUCKOO*.

Le Coucou, Buff.
Cuculus Canorus, Linn. Gmel. and Muller.
Cuculus, Gefner, Will. Johnft. Briff. &c.

IN the age of Ariftotle it was generally faid
that no one had ever feen the hatch of the
Cuckoo: it was known that this bird lays like
the reft, but makes no neft; that it drops its
eggs, or its egg (for it feldom depofits two in
the fame place) in the nefts of other birds, whe-

* In Hebrew, according to the different authors, *Kaath*, *Kik*,
Kakik, *Kakata*, *Schalac*, *Schafchaph*, *Kore*, *Banchem*, *Euchem*: In
Syriac, *Coco*: In Greek, Κοκκυξ: In Latin, *Cuculus*: In Italian,
Cucculo, *Cucco*, *Cuco*, *Cucho*: In Spanifh, *Cuclillo*: In German,
Kukkuk, *Gucker*, *Guggauch*, *Gugckufer*: In Flemifh, *Kockok*, *Kockuut*,
Kockuunt: In Swedifh, *Giock*: In Norwegian, *Gouk*: In Danifh,
Gioeg-Kukert, *Kuk*, *Kukmanden*: In Lapponic, *Geecka*: In old Englifh and in Scotch, *Gowk*.
A paffage from an Italian author, Gerini, will illuftrate the mif-
application of the name Cuckoo. " It lay its eggs in the neft of
the *curruca* (pettychaps); and hence a fottifh hufband, indifferent
to the difhonour of his marriage-bed and the impofition of fpuri-
ous children, has been called *curruca*: and afterwards that name
was corrupted, from ignorance, into *cornuto* (horned). Formerly,
and even at prefent, this word, as well as Cuckoo (cuckold), is
beftowed on a fot, who is infenfible to fhame."
The Latins applied the word *cuculus* to a hufband who was un-
faithful to his bed; and among the Greeks, it was beftowed on
thofe caught in any difgraceful action, or on perfons lazy and floth-
ful. In general, the term conveyed an imputation of indolence
and ftupidity; in which fenfe it is ftill ufed among fome nations in
Europe. (*Silly Gowk* is an expreffion of reproach among the po-
pulace in Scotland.)

ther

ther larger or fmaller than itfelf, fuch as the
warblers, the green-finches the larks, the wood
pigeons, &c.; that it often fucks the eggs which
it finds, and leaves its own in their ftead, to
be hatched by the ftranger ; that this ftranger,
particularly the pettychaps, acts the part of a
tender mother to her fuppofititious brood, fo that
the young ones become very fat and plump * ;
that their plumage changes much when they
arrive at maturity; and laftly, that the Cuckoos
begin to appear and are heard early in the fpring;
that they are feeble on their arrival ; that they
are filent during the dog-days ; and that a certain fpecies of them build in craggy rocks †.
Such are the principal facts in the hiftory of the
Cuckoo: they were known two thoufand years
ago, and fucceeding ages have added nothing to
the ftock. Some circumftances had even fallen
into oblivion, particularly their breeding in holes

* It is faid even that the adults are not bad eating in autumn ;
but there are countries where, at no period of their age, in no condition of their flefh, at no feafon of the year, they are ever eaten,
being regarded as birds unclean and unlucky : in others they are
held propitious, and venerated as oracles: and fome countries there
are, where it is imagined that the foil under the perfon's right
foot, who firft hears the Cuckoo's note, is a certain prefervative
againft fleas and vermin.

† May not this be the Andalufian Cuckoo of Briffon, and the
Great Spotted Cuckoo of Edwards? The fubject mentioned by the
latter was killed on the rock of Gibraltar, and its fellows might
have been bred in Greece, whofe climate is fo nearly the fame:
laftly, might not thefe have been fparrow-hawks, miftaken for
Cuckoos by reafon of the refemblance of their plumage ; and it is
known that fparrow hawks breed in the holes of craggy rocks.

of

of precipices. Nor have even the fabulous stories related of this singular bird undergone any alteration : error has its limits as well as truth, and, on a subject of so great celebrity, both have been exhausted.

Twenty centuries ago it was asserted, as at present, that the Cuckoo is nothing else than a little sparrow-hawk metamorphosed; that this change is effected every year at a certain stated season ; that when it appears in the spring, it is conveyed on the shoulders of the kite, which, to assist the weakness of its wings, is so obliging as to carry it (remarkable complaisance in a bird of prey like the kite); that it discharges upon plants a saliva which proves pernicious to them by engendering insects ; that the female Cuckoo takes care to lay into each nest she can discover, an egg like those contained in it *, the better to deceive the mother ; that the mother nurses the young Cuckoo, and sacrifices her own brood to it, because they appear not so handsome † ; that, like a true step-mother, she neglects them, or kills them, and directs the intruder to eat them : some supposed that the

* See Ælian, Salerne, &c. The true egg of the Cuckoo is larger than that of the nightingale; of a longer shape, of a gray colour almost whitish, spotted near the large end with violet-brown, very obscure, and with deeper and more apparent brown; and lastly, marked in the middle with some irregular streaks of chesnut.

† Observe that the Cuckoos are frightful when first hatched, and even many days after.

female

female Cuckoo returned to the neft where fhe had depofited her egg, and expelled or devoured the other young, that her own might fare the better: others fancied that the little pretender deftroyed its fofter-brothers, or rendered them victims to its voracity, by feizing exclufively all the food provided by their common nurfe. Elian relates that the young Cuckoo, fenfible that it is a baftard, or rather an intruder, and afraid of being betrayed by its plumage and treated as fuch, flies away as foon as it can ufe its wings, and joins its real mother *. Others pretend that the nurfe difcovers the fraud from the colours of the plumage, and abandons the intruder. Laftly, others imagine that the young bird, before it flies, devours even its fecond parent †, which had given it every thing but life; and the Cuckoo has been made the great fymbol of ingratitude ‡.—But it is abfurd to impute crimes that are phyfically impoffible. How could the young Cuckoo, which can hardly feed without affiftance, have ftrength fufficient to devour a wood-pigeon, a lark, a yellow bunting, or a pettychaps? It is true, that, in fupport of the poffibility of the fact, the evidence of a grave author, Klein, may be

* *Nat. Anim. Lib. III.* 30. It is alfo faid, by running into an oppofite extreme, that the Hen-cuckoo, neglecting her own eggs, hatches thofe of others. *See* Acron, *in Sat. VII. Horat. Lib. I.*

† Linnæus, and others.

‡ " Ungrateful as a Cuckoo," fay the Germans. Melancthon has left a fine harangue on the ingratitude of this bird.

adduced,

adduced, who made the obfervation at the age
of fixteen. Having difcovered in his father's
garden a pettychaps' neft with a fingle egg,
which was fufpected to belong to a Cuckoo, he
fuffered the incubation to proceed, and even
waited till the bird was feathered. He then
fhut both it and the neft in a cage, which he
placed on the fame fpot. A few days after he
found the hen-pettychaps entangled in the wires
of the cage, and its head fticking in the throat
of the young Cuckoo, which had fwallowed it
through miftake, while catching greedily at a
caterpillar that was probably too near. To
fome accident of this kind the Cuckoo owes its
bad name. But it is not true that it devours its
nurfe, or its fofter brothers: for, in the firft
place, its bill, though large, is too weak; the
one mentioned by Klein could not crufh the
head of the pettychaps, and was choked by it;
in the fecond place, to remove all objections
and fcruples, I have decided the point by experiment. On the 27th of June, I put a young
Cuckoo, which had been hatched in the fpring,
and was already nine inches long, in an open
cage, with three young pettychaps, which
were not one quarter feathered, and could not
eat without affiftance. The Cuckoo, far from
devouring them, or even threatening them,
feemed eager to repay its obligations to the fpecies. It fuffered the little birds, which were
not in the leaft afraid, to warm themfelves under

der its wings. On the other hand, a young
owl, which had as yet only been fed, began
of itſelf to eat by devouring a pettychaps,
which was lodged with it. I know that
ſome qualify the account by ſaying that the
Cuckoo ſwallows the chicks juſt as they burſt
from the ſhell; and as theſe little embryos
might be regarded as beings intermediate between
eggs and birds, they might therefore be eaten
by an animal which habitually feeds on eggs,
whether hatched or not. But though this ſtate-
ment is leſs improbable, it ought not to be admitted till it is evinced by obſervation.

With reſpeſt to the ſaliva of the Cuckoo, it
is nothing elſe than a frothy exudation from the
larva of a certain kind of graſsſhopper*. Perhaps the Cuckoo was obſerved to ſeek the larva
under this froth, which might give occaſion to
its being ſuppoſed to depoſit its ſaliva; and as
an inſeſt was perceived to emerge, it would be
imagined, that the ſaliva of the Cuckoo engen-
dered vermin.

* This inſeſt is the *Cicada Spumaria* of Linnæus. It inhabits
Europe, and is frequent on brambles, withies, and graſs ſtalks; it
ſettles in the forking of the ſtalks, and evacuates numerous veſi-
cles, reſembling froth, under which the larva lies concealed. This
ſpittle, ſo frequent in the fields, is termed in French *ecume print-
annière*, or ſpring froth, and the inſeſt which emerges is denominated *ſauterelle-puce* (graſshopper flea), or *cigale bédaude*. This inſeſt, it is ſaid, kills the Cuckoo by pricking it beneath the wing;
which at beſt is only ſome miſrepreſented faſt.—This frothy ſubſtance is well known in England by the name of *Cuckoo-ſpittle*, or
woodſare.

I will

I will not ferioufly combat the notion, that the Cuckoo is annually metamorphofed into a fparrow-hawk *. It is an abfurdity which never was believed by the real naturalifts, and fome of them have confuted it. I fhall only obferve that the opinion feems to have taken rife from the following circumftances: the two birds are feldom found in our climates at the fame time; they refemble each other in their plumage †, in the colour of their eyes and legs, in the length of their tail, in having a membranous ftomach, and a long tail, in their fize, in their flight, and in their little fecundity; both live folitary, and have long feathers that defcend from the legs on the tarfus, &c.; their plumage is alfo fubject to vary, fo that a bird which was taken for a beautiful merlin from its colours, was found on diffection to be a female Cuckoo ‡. But thefe qualities are not what

* 1 have juft witneffed an odd enough fcene. A fparrow-hawk alighted in a pretty populous court yard; a young cock of this year's hatching inftantly darted at him, and threw him on his back; in this fituation the hawk, fhielding himfelf with his talons and his bill, intimidated the hens and turkies, which fcreamed tumultuoufly round him: when he had a little recovered himfelf, he rofe and was taking wing, when the cock rufhed upon him a fecond time, overturned him, and held him down fo long that he was caught.

† Efpecially feen from below when they fly. The Cuckoo ruftles with its wings in rifing, and then fhoots along like the tiercel falcon.

‡ See Salerne, Hift. des Oifeaux, p. 40. M. Heriffant faw many Cuckoos which, by their plumage, refembled different kinds of male hawks, and one that refembled a wood pigeon. Mem. de l'Acad. des Sciences, 1752, p. 417.

conftitute

conftitute a bird of prey; there are wanting the
proper bill and talons, and the requifite courage
and ftrength, in which the Cuckoo, confidering its bulk, is very deficient*. M. Lottinger
has obferved that Cuckoos of five or fix months
old are as helplefs as young pigeons; that they
remain for hours in the fame fpot, and have fo
little appetite, that they muft be affifted in fwal-
lowing. It is true that when they grow up,
they affume a little more refolution, and may
fometimes pafs for birds of rapine. The Vifcount
de Querhoent, whofe teftimony has the greateft
weight, faw one which, being apprehenfive of
an attack from another bird, briftled its feathers,
and raifed and depreffed its head flowly and repeatedly, and then fcreamed out; fo that in this
manner it often put to flight a keftril, which
was kept in the fame houfe †.

The Cuckoo, far from being ungrateful,
feems confcious and mindful of its obligations.
On its return from its winter retreat, it eagerly

* Ariftotle juftly obferves that it is a timid bird; but I know
not why he cites, as a proof of this timidity, its laying in the neft
of another. *De Generatione*, Lib. III. 1.

† An adult Cuckoo, raifed by M. Lottinger, charged all other
birds, the ftrongeft equally as the weakeft, thofe of its own kind
or thofe of another, aiming preferably at the head and eyes. It
rufhed even upon ftuffed birds, and, though roughly repelled, it
would never defift from the attack. For my own part, I know
from experience that the Cuckoos menace the hand extended to
catch them, that they rife and fink alternately, briftling their feathers, and that they even bite in a fort of anger, though with little effect.

haftens,

haftens, it is faid, to the place of its birth, and
if it finds its nurfe or fofter-brothers, they all
join in mutual gratulations, each venting its joy
in its own manner*. Thefe different expreffions, thefe reciprocal careffes, thefe falutations
of gladnefs, and thefe fportive frolics, are what
have no doubt been miftaken for battles between
the fmall birds and the Cuckoo. A real combat, however, may fometimes take place, as
when the birds furprife a Cuckoo about to deftroy their eggs, in order to depofit its own †.
This well afcertained fact, that it lays in another's neft, is the chief fingularity in its hiftory, though not altogether unexampled. Gefner fpeaks of a certain bird of prey, which is
much like the gofs-hawk, that lays in the neft
of the jackdaw : and though this unknown
bird fhould be fuppofed to be nothing but the
Cuckoo, efpecially as this is often taken for one
of the rapacious tribe ; it at leaft cannot be denied that the wry-necks fometimes raife their
numerous progeny in the neft of the nuthatch,
as I myfelf have afcertained ; that the fparrows
fometimes occupy the fwallows' nefts, &c.
Thefe inftances, however, are very rare, and

* Frifch.

† Ariftotle, Pliny, and thofe who have copied or amplified from
them, agree that the Cuckoo is timid ; that all the fmall birds an-
noy them, and that it can put none of thefe to flight : others add,
that this perfecution originates from its refemblance to a bird of
prey ; but when did the fmall birds ever purfue the birds of
prey ?

the

the conduct of the Cuckoo muſt be regarded as an extraordinary phenomenon.

Another ſingularity in its hiſtory is, that it drops only one egg, at leaſt in the ſame neſt. It may indeed lay two eggs, as Ariſtotle ſuppoſes, and which appears poſſible from the diſſection of females, of which the ovarium frequently contained two eggs, well formed and of equal ſize *.

Theſe two ſingularities ſeem to imply a third: it is that their moulting is ſlower and more complete than in moſt birds. Sometimes in the winter ſeaſon we find, in the hollows of trees, one or two Cuckoos entirely naked, infomuch that they may be taken for real toads. Father Bougaud, whom we have often quoted with that confidence which he merits, avers that he ſaw one in that ſtate, which was taken out of a hollow tree about the end of December. Of four other Cuckoos raiſed, the one by Johnſon, as mentioned by Willughby, the other by the Count de Buffon, the third by Hebert, and the fourth by myſelf; the firſt languiſhed on the approach of winter, grew ſcabby, and died; the ſecond and third caſt the whole of their feathers in November, and the fourth, which died towards the end of October, had loſt more than half. The ſecond and third alſo ſoon died; but, previous to their death, they fell into a

* Linnæus and Salerne.

kind

kind of numbneſs and torpor. Many other ſimilar facts are adduced; and though it has been erroneouſly concluded that all the Cuckoos which made their appearance in ſummer remain torpid during the winter*, concealed in hollow trees or under ground, diſrobed of their plumage, and, according to ſome, with an ample proviſion of corn (which this ſpecies never eats): if theſe concluſions ought not to be admitted, we may at leaſt ſafely infer that thoſe which, on the moment of their departure, are ſick or wounded, or too young, or in ſhort too weak, from whatever cauſe, to perform their diſtant retreat, remain behind, and paſs the winter ſheltered in the firſt hole they meet with which has a good aſpect, as do the quails †: 2. That, in general, theſe birds are very late in moulting, and conſequently ſlow in reſuming their plumage, which is hardly reſtored on their appearance in the be-

* Thoſe who ſpeak of theſe Cuckoos found in winter lodged in holes, agree that they are abſolutely naked, and reſemble toads. This account makes me ſuſpect that the ſuppoſed Cuckoos were often toads or frogs, which really paſs the winter without food, their mouth being then ſhut, and their jaws, as it were, glued together. —Ariſtotle poſitively aſſerts that the Cuckoos never appear during winter in Greece.

† In winter, ſportſmen ſometimes meet with quails ſquatted under a large root, or in ſome other hole facing the ſouth, with a little proviſion of grain and heads of different ſorts of corn. I muſt own that the Marquis de Piolenc and another perſon aſſured me that two Cuckoos which they reared and kept ſeveral years did not drop all their feathers in the winter: but as they remarked not the time, nor the duration, nor the quantity of the moult, we can draw no concluſion from theſe two obſervations.

ginning

ginning of spring; accordingly, their wings are then very weak, and they seldom perch on lofty trees, but struggle from bush to bush, and sometimes alight on the ground, where they hop like the thrushes. We may therefore say that, during the love season, the surplus food is almost entirely spent on the growth of the feathers, and can furnish very little towards the reproduction of the species ; that, for this reason, the female Cuckoo never lays above one egg, or at most two; and that, as the bird has little abilities for generation, it has also less ardour for all the subordinate functions, which have the preservation of the species as their object, such as nestling, hatching, and rearing their young, &c. which all originate from the same source, and are proportioned to it. Besides, as the male instinctively devours birds' eggs, the female must be careful to conceal hers; she must not return to the spot where she has deposited one, left the male discover it ; she must therefore choose the most concealed nest, and which is also the most remote from his usual haunts; and if she has two eggs, she must entrust them to different nurses : and thus she takes all the precautions suggested by concern for her progeny, and yet carefully avoids betraying it through indiscretion. Viewed in this way, the conduct of the Cuckoo will coincide with the general rule, and imply in the mother an affection for her young, and even a

rational kind of concern, which prefers their
intereſt to the tender ſatisfaction of fondling and
aſſiſting them by her offices. The diſperſion
too of the eggs in different neſts, whatever be
the cauſe, whether the neceſſity of concealing
them from the male, or the ſmallneſs of the
neſt *, would alone render it impoſſible for the
female to hatch them. This fact is the more
probable, as two eggs are often found completely
formed in the ovarium, but very ſeldom two
eggs in the ſame neſt. Beſides, the Cuckoo is
not the only bird which never builds ; many
ſpecies of titmice, wood-peckers, king-fiſhers,
&c. come under the ſame deſcription. We have
already ſeen that it is not the only one that lays in
other's neſts ; there is alſo another example of a
bird which never hatches its eggs ; the oſtrich,
in the torrid zone, depoſits its eggs in the ſand,
and the heat of the ſun accompliſhes the developement of the embryos. It never loſes ſight
of them indeed, and guards them aſſiduouſly ;
but it has not the ſame motives as the Cuckoo
to conceal its attachment, and therefore does
not take all the precautions which might ex-
empt it from farther ſolicitude. The conduct
of the Cuckoo is not then an abſurd irregularity,
a monſtrous anomaly, a deviation from the laws

* Perſons of veracity have told me that they twice ſaw two
Cuckoos in a ſingle neſt, but both times in the neſt of a throſtle:
but the throſtle's neſt is larger than that of the pettychaps, of the
willow-wren, or of the red-breaſt.

of

of nature, as Willughby expreffes it ; it is the
neceffary confequence of eftablifhed principles,
and the want of it would occafion a void in the
general fyftem, and interrupt the chain of phæ-
nomena.

What feems to have aftonifhed fome naturalifts the moft, is that attention which they
term unnatural in the nurfe of the Cuckoo,
which neglects its own eggs to cherifh thofe of a
foreign, and even hoftile bird. One of thefe, an
excellent ornithologift, ftruck with the appearance, has made a feries of obfervations on this
fubject : he took the eggs of feveral fmall birds
out of their nefts, and in their place fubftituted
a fingle egg of a bird of a different kind, and
not a Cuckoo; and he inferred from his experiments that, in fimilar circumftances, birds will
hatch no fingle egg but the Cuckoo's, which is
therefore favoured by a fpecial law of the Creator.

But this conclufion will appear rafh and pre-
carious, if we attend to the following confider-
ations : 1. the affertion being general, one contrary fact is fufficient to overturn it; and for
this reafon, forty-fix experiments made on twenty fpecies are too few : 2. It would require
many more, and thofe performed with greater
nicety, to eftablifh a propofition which is an ex-
emption to the general laws of nature : 3. Ad-
mitting the experiments to be fufficiently numerous and accurate, they would be inconclufive,
if not made precifely in the fame manner,

in

in like circumſtances. For inſtance, the caſe is not ſimilar when the egg is left by a bird or dropped by a man, eſpecially by one who is bi-aſſed to a favourite hypotheſis; nay the frequent appearance of a perſon will diſturb the moſt eager brooder, and even cauſe her to abandon the education of a Cuckoo, though far advanced *, as I have myſelf experienced: 4. The fundamental aſſertions of this author are not quite accurate; for, though it ſeldom happens, the Cuckoo ſometimes lays two eggs in the ſame neſt. Further, he ſuppoſes that the Cuckoo ſucks all the eggs in the neſt, or deſtroys them ſomehow, leaving only its own; but this is hardly ſuſceptible of proof, and is improbable. But I have often received neſts, in which were ſeveral beſides the Cuckoo's egg which properly belonged to theſe neſts †, and even many of theſe eggs hatched, as well

---

* A meadow green-finch, whoſe neſt was on the ground under a thick root, abandoned the education of a young Cuckoo, merely from the inquietude occaſioned by the repeated viſits of ſome cu-rious perſons.

† 16 May, 1774, five eggs of a titmouſe with one of the Cuckoo; the eggs of the titmouſe diſappeared by degrees.

19 May, 1776, five eggs of the redbreaſt with one egg of the Cuckoo.

10 May, 1777, four eggs of the nightingale with one egg of the Cuckoo.

17 May, 1777, two eggs of the titmouſe under a young Cuckoo, but which did not ſucceed, Some incident of this ſort might have given occaſion to ſay that the young Cuckoo charges itſelf with hatching the eggs of its nurſe. (*See* Geſner, p. 365.)

as

as that of the Cuckoo * : 5. But, what is no lefs decifive, there are inconteftible facts obferved by perfons attached to no hypothefis †, which are directly oppofite to thofe related by the author, and entirely overturn his inconclufive inductions.

FIRST EXPERIMENT.

A hen canary, which fat on her eggs and hatched them, continued to fit when two blackbirds' eggs, brought from the woods, were put under her, though eight days afterwards ; and the incubation would have fucceeded if they had not been removed.

SECOND EXPERIMENT.

Another hen canary fat four days on feven eggs, five of which were her own, and two thofe of pettychaps ; but, the cage being car-

* 14 June, 1777, a Cuckoo newly hatched in a throftle's neft, with two young throftles, began to fly.

8 June, 1788, a young Cuckoo in the neft of a nightingale, with two young nightingales, and an addle egg.

16 June, 1778, a young Cuckoo in the neft of a red-breaft, with a little red-breaft that feemed to have been hatched before it.

M. Lottinger, in a letter dated 17 October, 1776, has related to me a fact, which he proved himfelf: in the month of June, a Cuckoo newly hatched in the neft of a blackcap, with a young blackcap that already flew, and an addle egg. I could cite many other fimilar examples.

† I owe the greateft part of thefe facts to one of my relations, Madame Potot de Montbeillard, who has many years ufefully amufed herfelf with birds, has ftudied their habits, and traced their purfuits ; and fometimes has been fo obliging as to make obfervations and try experiments relative to the fubjects in which I was engaged.

ried

ried to the lower ſtory, ſhe forſook them all.
Afterwards ſhe laid two eggs, but did not ſit.

THIRD EXPERIMENT.

Another hen canary, whoſe mate had eaten
her ſeven firſt eggs, ſat on the two laſt, along
with three others, the one a canary's, the ſecond a linnet's, and the third a bulfinch's; but
all theſe happened to be addle.

FOURTH EXPERIMENT.

A hen wren hatched a blackbird's egg; and
a hen tree-ſparrow hatched a magpye's egg.

FIFTH EXPERIMENT.

A hen tree-ſparrow ſat on ſix eggs which ſhe
had laid; five were added, and ſhe ſtill ſat;
five more were added, and finding the number
too large, ſhe ate ſeven of them, and continued
to ſit on the reſt; two were taken away, and a
magpye's egg put in their place, and the ſparrow
hatched it, along with the ſeven others.

SIXTH EXPERIMENT.

There is a well-known method of hatching
canaries' eggs, by putting them under a hen
goldfinch, taking care that they are previouſly
as far advanced in their incubation as thoſe of
the goldfinch.

SEVENTH EXPERIMENT.

A hen canary having ſitten nine or ten days
on three of her own eggs and two of thoſe of
the blackcap, one of the latter was removed,
in which the embryo was not only formed,
but

but living; two young yellow buntings, juft hatched, were entrufted to her, and fhe treated them with the fame attention as fhe would do her own, and ftill continued to fit on the four eggs that were left, but they turned out to be addle.

About the end of April 1776, another hen canary having laid an egg, it was taken away; and three or four days after, it being replaced, the bird ate it. Two or three days afterwards, fhe laid another egg, and fat on it; two chaffinch's eggs were then put under her, and fhe continued to fit, though fhe had broken her own eggs; at the end of ten days the chaffinch's eggs were removed, being tainted. Two newly hatched yellow buntings were given, which fhe reared very well. After which fhe laid two eggs, ate one, and though the other was taken away, fhe continued to brood as if fhe had eggs; a fingle egg of the redbreaft was put under her, which fhe hatched fuccefsfully.

Another hen canary, having laid three eggs, broke them almoft immediately; two chaffinch's eggs, and one of the blackcap, were fubftituted, on which fhe fat, and on three others, which fhe laid fuccefsively. In four or five days, the cage having been carried to a room in the lower ftory, the bird forfook them. A fhort time afterwards, fhe laid an egg, to which

was

was joined one of the nuthatch, and then two others, to which a linnet's egg was added. She fat on them all feven days, but preferring the two ftrangers, fhe threw out her own fuc-ceffively on the three following days, and on the eleventh fhe alfo toffed out that of the nuthatch, fo that the linnet's was the only one that fuc-ceeded. If this laft egg had been that of a Cuckoo, what falfe inferences might have been drawn.

TENTH EXPERIMENT.

On the 5th of June, a Cuckoo's egg was placed under the hen canary mentioned in the feventh experiment, which fhe hatched, along with three of her own. On the 7th, one of thefe eggs difappeared; another on the 8th, and the third and laft on the 10th; on the 11th, fhe alfo ate the Cuckoo's egg.

Laftly, a hen red-breaft, ardently bent on brooding, has been feen to unite with her mate in repelling a female Cuckoo from the neft; they fcreamed, attacked furioufly, and hotly purfued her *.

It

* See Obfervations fur l'inftinét des Animaux, t. I. p. 167, note 32. The author of that note adds fome details relative to the hif-tory of our bird: " While one of the red-breafts was ftriking with its bill the lower belly of the Cuckoo, this bird fhivered its wings with an almoft infenfible quiver, opened its bill fo wide that an-other red-breaft, which affailed it in front, drove its head feveral times into the cavity, without receiving any injury; for the Cuckoo was no way irritated, but feemed to be in the condition of a fe-male under the neceffity of laying. In a little while the Cuckoo, being

It follows from thefe experiments, 1. That the females of many fpecies of fmall birds which hatch the Cuckoo's egg, hatch likewife other eggs along with their own : 2. That they often fit on thefe eggs in preference to their own, which they fometimes entirely deftroy : 3. That they will hatch a fingle egg, though it be not a Cuckoo's : 4. That they boldly drive off the female Cuckoo, when they furprife her dropping the egg in their neft : 5. Laftly, that they fometimes eat this favoured egg, even in cafes where it is fingle and alone. But a more general and important confequence is, that the inftinct of hatching which fometimes appears fo powerful in birds, is not determined by the kind or quality of the eggs; fince they often eat or break them, or fit on addle ones ; they fit even on balls of chalk or wood, and fometimes brood in the empty neft. When a bird hatches the egg of a Cuckoo, or of any other bird, fhe follows therefore the general inftinct; and it is unneceffary to recur to any fpecial appointment of

being exhaufted, began to totter, loft its balance, and turned on the branch, from which it hung by the feet, its eyes half-fhut, its bill open, and its wings expanded. Having remained about two minutes in this attitude, conftantly haraffed by the two red-breafts, it quitted the branch, flew to perch at a diftance, and appeared no more. The female red-breaft refumed her incubation, and all her eggs were hatched, and formed a little family, that long lived attached to this diftrict." M. le Marquis de Piolenc alfo tells me in his letters of a Cuckoo being repelled by buntings.

the

the Author of Nature in accounting for the
conduct of the female Cuckoo *.

I afk my reader's pardon for this long difcuffion, of the importance of which he may not be
convinced. The bird which is the fubject of
this article has given rife to fo many errors, that
I have thought it neceffary, not only to extir-
pate thefe from natural hiftory, but to oppofe
the attempts of thofe who endeavour to convert
them into metaphyfical principles. Nothing
is more inconfiftent with found philofophy, than
to multiply the laws of the univerfe; a phænomenon appears fingle and unconnected, becaufe it is not accurately known; and it requires an attentive comparifon of the works of
nature, a clofe inveftigation of the relations
which fubfift, to enable us to penetrate into
her views.

I know more than twenty fpecies of birds, in
the nefts of which the Cuckoo depofits her eggs;
the pettychaps, the blackcap, the babbling
warbler, the wagtail, the red-breaft, the common wren, the yellow wren, the titmoufe,

* Frifch fuppofes another particular law, to explain why the
prefent Cuckoos never hatch their eggs; it is, he fays, becaufe a
bird never hatches unlefs itfelf has been hatched by a female of its
own fpecies. He admits indeed that the firft female Cuckoo emitted from Noah's ark muft have laid in its own neft, and muft itfelf
have taken the trouble of hatching its eggs. He might have fpared
this exception, for there are many inftances of fmall birds fucceeding with their own eggs along with that of the Cuckoo.

the

the nightingale, the red-tail, the fky-lark, the wood-lark, the tit-lark, the linnet, the green-finch, the bulfinch, the throftle, the jay, the black-bird, and the fhrike. The Cuckoo's eggs are never found in the nefts of partridges or quails, at leaft they never fucceed in them, be-caufe the young of thefe birds run almoft the inftant they are hatched. It is even fingular that the young Cuckoos, which, when bred in the cage, require feveral months before they eat without affiftance, can ever be raifed in the nefts of larks, which, as we have feen in their hif-tory, beftow only fifteen days on their educa-tion. But in the ftate of nature, neceffity, li-berty, and the proper choice of food, will con-fpire to unfold their inftinct, and haften their growth *; and may not the attention of the nurfe be proportioned to the wants of her adopt-ed child ?

We fhall perhaps be furprifed to find ma-ny granivorous birds, fuch as the linnet, the greenfinch, and the bulfinch, in the lift of the Cuckoo's nurfes. But many of thefe, it fhould be remembered, feed their brood with in-fects; and even the vegetable fubftances mace-rated in their craw, may fuit the Cuckoo for a certain time, till it can pick up caterpillars,

---

* I muft own that Salerne fays, that this bird is fed whole months by its adoptive mother, which it follows crying inceffantly for food. But this fact would be difficult to obferve.

fpiders,

fpiders, and beetles, &c. which fwarm about its manfion.

When the neft, where it is lodged, belongs to a fmall bird, and confequently is conftructed on a narrow fcale, it is ufually found fo much flattened that it can hardly be recognifed ; the natural effect of the bulk and weight of the young Cuckoo. Another confequence is, that the eggs or young birds are frequently thruft out of the neft; but, though expelled from their paternal abode, they fometimes furvive ; for if they be fomewhat grown, if the neft be near the ground, and if the afpect is favourable and the feafon mild, they find fhelter under the mofs or foliage, and the parents, without forfaking the intruder, continue to feed and watch them.

All the inhabitants of forefts affert, that when the female Cuckoo has once depofited her egg in the neft which fhe has felected, fhe retires to a diftance, and feems to forget her progeny; and that the male never difcovers the fmalleft concern in the matter. But M. Lottinger has obferved that, though the parent Cuckoos do not vifit their offspring, they approach within a certain diftance of the fpot, calling, and feem to liften and reply to each other. He adds that the young Cuckoo conftantly anfwers to the call, whether in the woods or in a volery, provided it be not difturbed by the fight of a perfon. It is certain that the old

ones

ones can be enticed by imitating their call, and that they fometimes chant in the vicinity of the neft ; but there is no proof that thefe are the parents of the young bird. They never render thofe tender offices which mark parental attachment, and their calls proceed only from the fympathy common between birds of the fame fpecies.

Every body knows the ordinary fong of the Cuckoo * ; it is fo diftinctly formed, and fo often repeated, that, in almoft all languages, it has given name to the bird. It belongs exclufively to the male, and is heard only during the fpring, the feafon of love, and either when he fits on a dry branch, or while he moves on the wing : fometimes it is interrupted by a dull rattling found, *crou, crou*, uttered with a hoarfe lifping voice. There is alfo another occafional cry, which is loud but rather quavering, and compofed of feveral notes, like that of the little diver : it is heard when the male and female purfue each other in amorous frolic †. Some have alfo fufpected it to be the cry of the female. When fhe is animated fhe has befides a

---

* *Cou cou, cou cou, cou cou cou, tou cou cou (ou,* in French, pronounced *oo)*. This frequent repetition has given rife to two modes of proverbial expreffions: when a perfon dwells upon the fame fubject, he is faid in German *to fing the Cuckoo's fong*. The fame phrafe is applied to a fmall body of people, who, by their tumultuary vociferation, feem to form a numerous affembly.

† Thofe who have heard it exprefs it thus; *go, go, guet, guet. guet.*

fort

fort of clucking, *glou, glou*, which is repeated
five or fix times with a ftrong clear voice, while
fhe flies from tree to tree. This would feem
intended to incite the male; for as foon as he
hears the call he haftens to her with ardour,
uttering *tou, cou, cou* *. But notwithftanding this
variety of inflection, the fong of the Cuckoo
ought never to be compared with that of the
nightingale, except in the fable †. It is very
uncertain whether thefe birds ever pair; they
are ftimulated by appetite, but they fhew nothing like fentiment or attachment. The males
are much more numerous than the females, and
often contend for them ‡; yet the object of the
ftruggle is a female in general, without any
fymptom of choice or predilection; and when
their paffion is fatisfied, they defert her with
the coldeft indifference. They difcover no folicitude, and make no provifion, for their off-
fpring. The mutual attachment between pa-

* Note communicated by the Count de Riollet, who makes a
laudable amufement of obferving what fo many others only look at.

† It is faid that the nightingale and the Cuckoo difputed the
merits of the fong in prefence of the afs, which adjudged the prize
to the Cuckoo; but that the nightingale appealed to man, who
pronounced in its favour, fince which time the nightingale fings as
foon as it fees a man, in gratitude for his decifion, or in juftifica-
tion of it.

‡ Seldom or never do perfons kill or take any but the finging
Cuckoos, and, by confequence, the males. I have feen three or
four killed in a fingle excurfion, and not one female among them.
" In a trap, which we placed on a tree frequented by Cuckoos,"
fays the author of the Britifh Zoology, " we caught not fewer
than five male birds in one feafon."

rents

rents is founded on the common tendernefs to their young.

The young Cuckoos, foon after their exclufion, have alfo a call not fhriller than that of the pettychaps and red breafts, their nurfes, whofe tone they affume from the force of imitation * . and as if fenfible of the neceffity of foliciting and importuning an adoptive mother, who cannot have the compaffion of a real parent, they continually repeat their entreaty ; and, to remove ambiguity, their broad bill is opened to its utmoft width, and the expreffion is rendered ftill mcre fignificant by the clapping of their wings. After their wings have acquired fome ftrength, they purfue their nurfe among the neighbouring branches, or meet her when fhe brings food. The young Cuckoos are voracious †, and·can hardly be maintained by little birds, fuch as the red-breaft, the pettychaps, the common and yellow wrens, which have befides a family to fupport. They retain their call, ac-

---

* " The fingular ftruflure of their noftrils contributes perhaps," fays M. Frifch, " to produce this fharp cry." It is true that the noftrils of the Cuckoo have, with regard to their exterior, a pretty fingular ftruflure ; but I am convinced that they contribute not in the leaft to modify this cry, which continued the fame after I had ftopt the noftrils with wax. I have difcovered, by repeating this experiment upon other birds, and particularly upon the wren, that the cry remains unvaried, whether the noftrils be fhut or left open. It is befides known that the voice of birds is formed not in the noftrils, or at the *glottis,* but at the lower part of the *trachea arteria,* near its forking.

† Hence the proverb *to fwallow like a Cuckoo.*

cording

cording to Frifch, till the fifteenth or twentieth
of September; it then begins to grow flat, and
is foon loft entirely.

Moft ornithologifts agree, that infects are the
chief part of the Cuckoo's food, and that, as I
have already remarked, it is peculiarly fond of
birds' eggs. Ray found caterpillars in its ftomach; I have alfo perceived veftiges of vegetable fubftances, fmall beetles, &c. and fometimes pebbles. Frifch afferts that the young
Cuckoos ought, in every feafon, to be fed in
the morning and evening at the fame time as
in the longeft days of fummer. That author
has alfo obferved the way in which they eat the
infects alive: they lay hold of the caterpillar
by the head, and, drawing it into their bill,
they fqueeze the juicy matter through the anus,
and then fhake it feveral times before they fwallow it. They alfo feize butterflies by the head,
and, preffing with their bill, they crufh the
breaft, and fwallow the whole together with
the wings. They likewife eat worms; but they
prefer fuch as are alive. When infects could
not be had, Frifch gave a young Cuckoo which
he raifed, fheep's liver, and efpecially kid-
neys, cutting them into fmall ftrips like the infects for which they were fubftituted. When
thefe were too dry, he foaked them a little,
that they might be eafily fwallowed. The bird
never drank unlefs its food was too dry, and
then it drank awkwardly and with reluctance.

In

In every other cafe it rejected water, and fhook off the drops which were forcibly or artfully introduced into its bill*; in fhort, it is habitually under the impreffion of a hydrophobia.

The young Cuckoos never fing during their firft year, and the old ones ceafe towards the end of June, at leaft their fong is then unfrequent. But this filence does not announce their departure; they are found in the open country until the end of September, and even later †. It is the fcarcity of infects, no doubt, which determines them to retire to warmer climates : they migrate for the moft part into Africa, fince the Commanders of Godeheu and des Mazys reckon them among thofe birds which are feen twice a year paffing and repaffing the ifland of Malta ‡. On their arrival in our climates they approach neareft our dwellings; during the reft of their ftay they fly about among the woods, the meadows, &c. and wherever they can difcover nefts to plunder or depofit their egg, or find infects and fruits. Towards autumn the adults, and efpecially the females, are excellent food, and as fat as they were lean in the

---

* I have remarked this fact, as well as the Carthufian of Salerne. May not this be the reafon why a decoction of the Cuckoo's dung in wine has been recommended as a fure remedy for the hydrophobia ?

† Querhoent and Hebert.

‡ Salerne relates, from Voyagers, that the Cuckoos fometimes alight in great numbers on fhips.

spring *. Their fat is collected chiefly under
the neck +, and is the moſt delicate morſel of
the bird. They are commonly ſingle ‡, reſt-
leſs, and perpetually changing their place ; and
though they never fly to any great diſtance,
they range over a conſiderable tract during the
courſe of the day. The ancients watched the
times of the appearance and diſappearance of the
Cuckoo in Italy. The vine-dreſſers who had
not lopped their vines before its arrival were
reckoned lazy, and held the object of public
deriſion. The by-paſſers who ſaw the back-
ward ſtate of the vineyard, mocked the owner's
indolence by repeating the call of that bird,
which was itſelf, and with much propriety, regarded as the emblem of ſloth, ſince it diſpenſes
with the ſacred obligations of nature. It was
an uſual expreſſion alſo, *crafty as a Cuckoo* (for
cunning and indolence may ſubſiſt together);

* It is in this ſeaſon only that the proverb, *lean as a Cuckoo*, is
juſt.

+ I obſerved the ſame thing in a young rock ouzel, which I
brought up, and which died in the month of October.

‡ In the month of July were a dozen Cuckoos on a large oak ;
ſome ſcreamed with all their might, others were ſilent ; a fowling-
piece was diſcharged at them, and one dropt, which was a young
one. This would give room to ſuppoſe that the old and young
gather together in ſmall bodies, to migrate.

Note communicated by M. le Comte de Riollet.

* *Inde natam exprobationem fœdam putantium vites per imitationem*
cantus alitis temporarii quem cuculum vocant ; dedecus enim habetur
. . . falcem ab illâ volucre in vite deprehendi, ut ob id petulantiæ ſales
etiam cum primo vere ludantur. Pliny, *Lib. XVII.* 26.

either

either becaufe, declining the tafk itfelf, it con-
trives to make other birds hatch its eggs ; or
for another reafon furnifhed by the ancient
mythology *.

Though fubtle, though folitary, the Cuckoos
are fufceptible of a fort of education. Many perfons of my acquaintance have reared and tamed
them ; they are fed with minced meat, either
cooked or raw, with infects, eggs, foaked bread,
fruits, &c. One of the Cuckoos thus bred
knew its mafter, anfwered his call, followed
him to the fport, perched on his fowling-piece,
and if it found a black cherry tree on the road,
it flew to it, and returned not till fatiated with
the fruit ; fometimes it did not join its mafter
again the whole day, but followed him at a diftance, fhifting from tree to tree : when at home
it was permitted to run about, and to rooft at
night. The dung of this bird is white and
abundant, which is a great inconvenience in its
education ; it muft be carefully guarded againft
the cold during the tranfition of autumn
into winter : this is a critical period for the
Cuckoo ; at leaft I loft at this time all thofe

* Jupiter having perceived that his fifter Juno was alone on the
Dictean mount, or Thronax, raifed a violent ftorm, and came in
fhape of a Cuckoo, and alighted on the knees of the goddefs, who
feeing it drenched and beaten by the tempeft, pitied it, and cherifhed it under her robe : the god then refumed his proper form,
and became the hufband of his fifter. From that moment the
Dictean mount was called *Coccygian*, or *Cuckoo-mountain*; and hence
the origin of the *Jupiter-cuculus*. *See* Gefner, p. 368.

which

which I tried to rear, and indeed many of other kinds.

Olina fays that the Cuckoo may be trained for the chafe like the fparrow-hawks and the falcons; but he is the only perfon who makes that affertion, which, perhaps, as well as many other errors in the hiftory of this bird, has its fource in the refemblance of its plumage to that of the fparrow-hawk.

The Cuckoos are generally diffufed in the ancient continent; and though thofe of America have different habits, many of them ftill retain a certain family likenefs. The common kind appear only during fummer in the colder temperate climates, fuch as thofe of Europe; and they refide in the winter only in the warmer countries, fuch as thofe of the north of Africa: they feem to fhun both extremes alike.

When the Cuckoos alight on the ground, they hop, as I have remarked; but this feldom happens: and even though it were not afcertained by experience, we might infer it from the fhortnefs of their legs and thighs. A young Cuckoo, which I had occafion to obferve in the month of June, made no ufe of its legs in walking, but crept on its belly, affifting its motion by means of its bill, like the parrot in climbing. When it clambered in its cage, I perceived that the largeft of the hind toes was turned forwards, but was lefs ufed than the two

other

other anterior ones *. It aſſiſted its progreſſive motion by flapping its wings.

I have already ſaid that the plumage of the Cuckoo is very ſubject to vary in different individuals; and conſequently, in deſcribing the bird, we can do nothing more than give an idea of the colours and their diſtribution, ſuch as they are moſt commonly obſerved. The greater number of the full-grown males, which were brought to me, reſembled much the one deſcribed by Briſſon; in all of them, the upper-ſide of the head and body, including the coverts of the tail, the ſmall coverts of the wings, the great ones next the back, and the three quills covered by them, are of a fine aſh-colour; the great coverts of the middle of the wing are brown, ſpotted with rufous, and terminated with white; the moſt remote on the back, and the ten firſt quills of the wing are deep cinereous, the inſide of the latter being ſpotted with ruſty white; the ſix contiguous quills brown, and marked, on both ſurfaces, with rufous ſpots, terminated with white; the throat and the fore-part of the neck are light cinereous; the reſt of the underſide of the body is radiated tranſverſely with brown, on a dirty white ground; the fea-

---

* If this habit be common to the ſpecies, what becomes of the expreſſion, *digiti ſcanſorii*, applied by many naturaliſts to the toes diſpoſed, as in the Cuckoo, two before and two behind? Beſides, do not the nut-hatches, the titmice, and the creepers, excel in climbing, though their toes are placed in the ordinary way, three before and one behind?

thers

thers of the thighs the fame, and falling on each fide on the tarfus, like ruffles; the tarfus clothed exteriorly with afh coloured feathers as far as the middle; the quills of the tail blackifh, and terminated with white; the eight intermediate ones are fpotted with white near the fhaft on the infide; the two middle ones are fpotted in the fame manner on the outer edge, and the laft of the lateral quills is barred tranfverfely with the fame colour; the iris is chefnut, fometimes yellow; the internal membrane of the eye is very tranfparent; the bill is black without, and yellow within; the corners of the bill are orange; the legs yellow, and a little yellow appears alfo at the bafe of the lower mandible.

I have feen feveral females which were very like the males; and in fome I perceived, on the fides of the neck, traces of thofe brown ftreaks mentioned by Linnæus.

Dr. Derham fays, that, in females, the neck is variegated with rufty, and the upperfide of the body is of a darker caft*; that the wings alfo are of a rufty hue, and the eyes not fo yellow. According to other obfervers, the plum-

* A perfon of veracity affured me that he faw fome of thefe individuals of a browner caft, which were alfo larger: if they were females, this would be another point of analogy to the rapacious birds. On the other hand, Frifch remarks, that of two young Cuckoos of different fexes, which he raifed, the male was the browneft.

age

age of the male is always darker than that of
the female; but the variation is fo great, that
no certain difcrimination can be drawn.

In young Cuckoos, the bill, the legs, the
tail, and the underfide of the body, are nearly
the fame as in the adults, except that the quills
are more or lefs retained in their tubes: the
throat, the forepart of the neck, and the underfide of the body, are barred with white and
blackifh; in fuch manner, however, that the
blackifh predominates on the anterior parts more
than on the pofterior (in fome fubjects there is
hardly any white under the throat); the upperfide of the head and body is beautifully variegated with blackifh, white, and rufty, diftri-
buted fo that the rufty appears more on the
middle of the body, and the white on the extremities; there is a white fpot behind the head,
and fometimes above the face; all the quills of
the wings are brown tipped with white, and
fpotted more or lefs with rufty and white; the
iris is greenifh gray, and the ground of the
plumage is a very light afh colour. It is probable that the female fo handfomely *mottled*, of
which Salerne fpeaks, was hatched the fame
year. Frifch tell us, that young Cuckoos reared in the woods have a plumage lefs variegated,
and more like the plumage of thofe which are
bred in the houfe. If this be not the cafe, we
fhould at leaft expect it; for domeftication in
general alters the colours of animals, and we

might

might prefume that thofe fpecies of birds which participate of that ftate, undergo a proportional change of colours. At the fame time I muft own, that I could perceive no difference between the young wild Cuckoos which I have feen (and I have feen many of them) and thofe which I reared. Perhaps what Frifch examined were older than the domeftic ones with which he compared them. The fame author adds that the young males have a darker plumage than the females, and that the infide of their mouth is redder, and their neck thicker *.

The weight of a full grown Cuckoo, weighed on the 12th of April, was four ounces two gros and a half; that of another, on the 17th of Auguft, was about five ounces. But they are heavier in the autumn, being then much fatter, and the difference is not inconfiderable. I weighed a young one on the 22d of July, whofe total length was near nine inches, and found it two ounces two gros ; another almoft as large, though much leaner, was only one ounce four gros, that is near one third lefs than the firft.

In the full grown male the inteftinal tube is about twenty inches; there are two *cæca* of unequal length, the one fourteen lines (fometimes

---

* Frifch fufpeéts that the thicknefs of the throat, which is peculiar to the male, may have fome relation to the cry of the males. Yet I muft obferve that, in the numerous diffeétions which I have made, I could not perceive that the organs of voice were larger in the males than in the females.

twenty-

twenty-four), the other ten (fometimes eighteen , both directed forwards, and adhering all
along to the great inteftine, by means of a thin
tranfparent membrane ; there is a gall bladder ;
the kidnies are placed on each fide of the fpine,
divided into three principal lobes, and thefe fubdivided into others fmaller, fecreting a whitifh
liquor ; the two tefticles are of an oval fhape,
and of unequal fize, attached to the upper part
of the kidnies, and feparated by a membrane.

The œfophagus dilates, at its lower part, into
a fort of glandular fac, parted from the ftomach
by a contraction ; the ftomach is flightly mufcular on its circumference, membranous on its
middle, and adhering by means of fibrous net-
work to the mufcles of the lower belly, and to
the different parts which furround it.   It is befides much fmaller and better proportioned in a
young bird reared by a red-breaft or pettychaps
than in one bred and tamed by man ; in the latter, the ftomach, being diftended by exceffive
feeding, is as large as a common hen's egg, and
occupies all the anterior part of the cavity of the
belly, from the fternum to the anus, and fometimes ftretches under the fternum five or fix
lines *, and, at other times, it conceals the

* *See* Mémoires de l'Academie Royale des Sciences, *annѐe*, 1752,
p. 420.   The Cuckoo of M. Heriffant was domeftic, to judge by
the quantity of flefh with which its ftomach was filled.   In the
nutcracker the ftomach is alfo very bulky, fituated likewife in the
middle of the belly, and not covered by the inteftines.

whole

whole of the inteſtines; whereas in the wild
Cuckoo, which I directed to be killed the inſtant they were brought to me, it never extended quite to the ſternum, but left uncovered,
between its lower part and the anus, two circumvolutions of the inteſtines, and three on the
right ſide of the abdomen. I ſhould add, that
in moſt of the birds which I diſſected, nothing
was required to be compreſſed or diſplaced in order to perceive one or two circumvolutions of
inteſtines in the cavity of the belly on the right
ſide of the ſtomach, and one between the lower
part of the ſtomach and the anus. This difference of conformation, therefore, conſiſts but in
degree, ſince, in the greater number of the
winged tribes, not only is the poſterior ſurface
of the ſtomach ſeparated from the ſpine by a portion of the inteſtinal tube which is interpoſed,
but the left ſide is never covered by any portion
of theſe inteſtines; and this ſingle difference is
far from proving that the Cuckoo is incapable
of hatching, as an ornithologiſt alleges. It is not,
probably, becauſe the ſtomach is too hard, ſince
the ſides being membranous, its tenſeneſs pro-
ceeds from accident or repletion; nor is the reaſon becauſe the bird avoids cooling its ſtomach,
which is leſs ſhielded than that of other birds;
for it is evident, that it would be much leſs liable to be affected by cold, if employed hatching than in fluttering and perching on trees.
The nut-hatch has the ſame ſtructure, and yet
it

it fits. Befides, not only the ftomach, but the
whole of the lower part of the body rests upon
the eggs, otherwise moft birds which, like the
Cuckoo, have a long fternum, could not cover
above three or four at once, which falls much
fhort of the ufual number.

I found in the ftomach of a young Cuckoo,
which I reared, a lump of raw flefh almoft dry,
and which had not paffed through the *pylorus* ;
it was decompofed, or rather divided into exceedingly fine *fibrillæ*. In another, which was picked
up dead in the woods about the beginning of
Auguft, the internal membrane of the ventricle was hairy, the briftles being about a line in
length, and directed towards the orifice of the
œfophagus : in general, one meets with very few
pebbles in the ftomach of young Cuckoos, and
there are hardly any which have not fome vef-
tige of vegetable matter in their ftomach. Perhaps thefe have been bred by the greenfinches,
the larks, and other birds, which neftle on the
ground ; the fternum forms a re-entrant angle.

Total length thirteen or fourteen inches ; the
bill thirteen lines and a half ; the edges of the
upper mandible fcolloped near the point (but
not in all the young ones) ; the noftrils ellipti-
cal, encircled by a projecting margin, and in
the centre there is a fmall whitifh fpeck which
rifes almoft as high as the margin ; the tongue
is white at the point, and not forked ; the tarfus ten lines ; the thigh lefs than twelve ; the
inner

inner one of the hind nails is the weakeſt and the leaſt hooked; the two fore toes are connected together at their baſe by a membrane; the under ſide is of a very fine texture, like ſhagreen; the alar extent about two feet; the tail is ſeven inches and a half, conſiſting of ten tapered quills*, and exceeding the wings by two inches [A].

VARIETIES of the CUCKOO.

It might be regarded as ſingular that the figure and aſpect of a bird, which lives in the ſtate of nature, ſhould be ſo inconſtant and variable. But as the Cuckoos never pair, they are ſtimulated only by vague and general appetite unreſtrained by perſonal attachment, and contract irregular alliances; which has given riſe to diverſity in the bulk, in the ſhape, and in the plumage, and which, in the opinion of many, has converted them into falcons, merlins, gos-hawks, ſparrow-hawks, &c. But, not to enter into a detail of theſe exhauſtleſs varieties, which appear to fluctuate, I ſhall only

* Ray reckons only eight in one which he obſerved in 1693; but aſſuredly two were wanting.

[A] Specific character of the *Cuculus Canorus:* " Its tail is rounded, blackiſh, and dotted with white." In England it arrives about the middle of April, and retires in the end of July; its appearance is a month later in the cold climate of Sweden. Its retreat is abſolutely unknown.

observe

obferve that in Europe Cuckoos are found differing greatly in fize; that the colours, afh-
gray, rufous, brown, and whitifh. are diftri-
buted differently in different individuals, fo that
each of thefe predominates more or lefs, and,
by the multiplicity of their fhades, increafes the
variations of the plumage. With regard to the
foreign Cuckoos, I find two which feem to be
varieties of the European kind, occafioned by
the influence of climate; and perhaps I fhould
add feveral others, if I had an opportunity of
furveying them more clofely.

I. The Cuckoo from the Cape of Good
Hope * refembles that of Europe both in its
proportions and in the tranfverfe bars below
the belly, and in its fize, which is not much
fmaller.

The upper fide of the body is brown green;
the throat, the cheeks, the fore part of the
neck, and the fuperior coverts of the wings,
are deep rufous; the quills of the tail are of a
lighter rufous, terminated with white; the
breaft, and all the reft of the under fide of the
body, are barred tranfverfely with black on a
white ground; the iris is yellow; the bill deep
brown; and the legs reddifh brown. The total
length is fomewhat lefs than twelve inches.

May not this be the bird known at the Cape
of Good Hope under the name of *edolio*, be-

* *Cvculus Capenfis,* Gmel.
The *Cape Cuckoo,* Lath.

caufe

caufe it repeats that word in a low melancholy
tone? It has no other fong, and many inhabitants of that country, not Hottentots but
Europeans, believe that the foul of a certain
fhip-mafter, who often pronounced the fame
word, entered into the body of this bird; for
modern ages have alfo their metamorphofes.
This is no doubt as true as the ftory of *Jupiter
cuculus*, and yet we probably owe to it our
knowledge of the bird's cry. It were fortunate
if every error conveyed fome truth.

II. Travellers fpeak of a Cuckoo of the kingdom of Loango in Africa, which is rather
larger than ours, but having the fame colours,
and differing chiefly in its fong; this muft be
underftood of its tone, for it pronounces *cuckoo*
as ours does. The cock begins, it is faid, with
founding the gamut, and fings alone the three
firft notes; then the hen ftrikes in, and accom-
panies him through the reft of the octave. She
differs from our female Cuckoo, which calls
much lefs than the male, and in a different
manner. This is ftill another reafon for dif-
tinguifhing the Loango Cuckoo from our fpecies, and confidering it as a variety.

The FOREIGN CUCKOOS.

THE principal characters of the European Cuckoo are, as we have feen, a large head, a broad wide bill, the toes placed two before and two behind, the legs feathered and fhort, the thighs ftill fhorter, the nails flender and flightly hooked, the tail long and confifting of ten tapered quills. It is difcriminated from the curucuis by the number of tail-quills, thefe being twelve, and particularly by the greater length of its bill, and the greater convexity of its upper mandible: it is alfo diftinguifhed from the barbus, by having no briftles round the bafe of the bill. But the whole muft be received with a degree of latitude, nor muft we expect to find all the characters exactly combined in each bird that ought to be claffed with the Cuckoo of Europe. The various tribes of animals melt into each other, and no abfolute limits can be affigned. It is enough if the birds which we fhall refer to the genus of the Cuckoos be more clofely related to it than to any other genus. Our object is to trace nature as fhe actually exifts, not to form artificial arrangements; and to facilitate the progrefs of knowledge, by condenfing and abridging the detail of facts, which ferves as the foundation. Among the foreign Cuckoos,
therefore,

therefore, we fhall meet with fome fpecies that
have even tails, as the fpotted Cuckoo of China, that of the ifland of Pannay, the vourou-
driou of Madagafcar, and a variety of the
brown Cuckoo fpotted with rufous, from the
Eaſt Indies; with fome, that are in fome degree forked, fuch as the Cuckoo which has
two long fhafts inſtead of the two outer quills;
with others, which have the tail exceedingly
tapered, like the widow birds, fuch as the fanhia of China, and the crefted collared Cuckoo;
with others, wherein it is only partly tapered,
as the old-man with rufous wings from Carolina, which has only two pairs of quills tapered, and as a variety of the crefted jacobine from
Coromandel, which has only the outer pair tapered, that is, fhorter than the four other pairs,
which are equal to each other in length; with
others, which have twelve quills in the tail,
fuch as the honey-guide of the Cape; with
others, which have only eight, as the guira-
cantara of Brafil, if Marcgrave was not deceived
in counting them; with others, which fpread
their tail even when in a ſtate of repofe, as the
Madagafcar coua, the gold-green and white
Cuckoo of the Cape of Good Hope, and the fecond coukeel of Mindanao; with others, which
have all the quills clofe and imbricated, both the
middle and lateral ones; with others, which
have fome hairs about the bill, as the fanhia,
the honey-guide, and a variety of the greenifh

8

Cuckoo

Cuckoo of Madagafcar; with others, which
have a proportionally longer and flenderer bill,
as the tocco of Cayenne; with others, wherein
the internal and pofterior toe is armed with a
long fpur, like that of our larks, as the houhou
of Egypt, the Cuckoo of the Philippines, the
green Cuckoo of Antigua, the toulou and the
rufty-white; and laftly, with others, wherein
the legs are more or lefs fhort, more or lefs
feathered, or even without any feather or down.
Even what is regarded as the moft fixed and
certain charaéter, I mean the difpofition of the
toes, two placed before and two behind, is alfo
liable to vary; fince I have obferved, in the
Cuckoo, that one of the hind toes fometimes
turns forwards, and others have found, in the
owls, that one of the anterior toes fometimes
turns backwards; but thefe flight differences,
far from confounding the genus of the Cuckoos,
difplay the true order of nature; as they fhew
the richnefs of her plans, and the eafe with
which they are executed; exhibiting the endlefs variety of fhades, the infinite diverfity of
features which diftinguifh the individual, and
yet without obliterating the general family likenefs.

It is a remarkable circumftance with regard
to the tribe of Cuckoos, that the branch fettled
in the new world feems lefs fubjeét to the variations which I have mentioned, and retains a
greater refemblance to the European fpecies

confidered

considered as the common trunk, and from
which it was perhaps later separated. In fact,
the European Cuckoo visits the northern countries, and pushes its excursions as far as Den-
mark and Norway, and consequently might
cross the narrow straits which divide the continents at these high latitudes : but it could
much more easily pass the isthmus of Suez, or
some narrow inlets, to spread through Africa ;
and nothing could oppose or retard its progress
into Asia. In these countries, therefore, the
settlement must have been more early, and a
greater deviation from the primitive stock may
be expected. Accordingly, though there are
scarcely two or three exceptions or exterior ano-
malies in fifteen species or varieties found in
America, there are fifteen or twenty in thirty-
four species that occur in Asia and in Africa ;
and no doubt there are more, if all the Cuckoos
were known, which is so far from being the
case, that it is still undecided whether, among
so many, there be one that drops its eggs in the
nests of other birds ; we know only that many
of these foreign kinds build their nest, and hatch
their own eggs. But though we are commonly acquainted with the superficial differences
only, some general and radical ones must obtain, especially between the two branches settled in the old and in the new world, which
must undoubtedly receive an impression from
the climate. I have noticed that the American

species

fpecies are in general fmaller than thofe of the old continent, probably owing to the fame caufes which check the growth and expanfion of quadrupeds in the new world, whether thofe be indigenous, or introduced. There are at moft two fpecies of Cuckoos in America of nearly the fame fize with ours ; but the others exceed not our blackbirds and thrufhes. In Africa and Afia, there are a dozen fpecies as large as the European, or larger, and fome of them equal to hens in bulk.

This ftatement will I hope juftify the plan which I have adopted, of feparating the Cuckoos of America from thofe of Afia and Africa. Time and obfervation, the two great fources of inform- ation, are ftill wanting to illuftrate the habits and economy of thefe birds, and to point out their true differences, external or internal, ge- neral or particular.

BIRDS

OF THE OLD CONTINENT,

WHICH ARE RELATED TO THE CUCKOO.

I.

The Great SPOTTED CUCKOO.

Cuculus Glandarius, Linn. Gmel. and Bor.
Cuculus Andalusiæ, Briff.

I BEGIN with this Cuckoo, which is not
entirely foreign to Europe, since one was
killed on the rock of Gibraltar. It would seem
to be migratory, wintering in Asia or in Africa,
and appearing sometimes in the south of Europe.
Both this and the following species may be considered as intermediate between the common
and the foreign kinds: it differs, however, from
the common not only by its size and plumage,
but by its proportions.

The most remarkable ornament of this Cuckoo is a silky tuft of a bluish gray, which can
be erected at pleasure, but in the ordinary state
lies flat on the head; there is a black bar on the
eyes, which gives it a marked aspect; brown
predominates on all the upper parts, including
the wings and the tail, but the middle quills,
and almost all the coverts of the wings, the
four

four lateral pairs of the tail, and their fuperior coverts, are terminated with white, which forms a pleafant enamel ; all the under furface of the body is brown orange, which is pretty bright on the anterior parts, and duller on the pofterior ; the bill and legs are black.

It is as large as a magpie ; the bill fifteen or fixteen lines ; the legs fhort ; the wings not fo long as in the ordinary Cuckoo ; the tail about eight inches, confifting of ten tapered quills, and exceeding the wings about four inches and a half[A].

II.

The BLACK and WHITE CRESTED CUCKOO.

Cuculus Pifanus, Gmel.
Cuculus ex nigro et albo mixtus, Gerini.
The *Pifan Cuckoo*, Lath.

THIS Cuckoo likewife muft be regarded as only half foreign, fince it was feen once in Europe. The authors of the Italian Ornithology inform us that in 1759 a male and female of this fpecies neftled in the vicinity of Pifa, and that the hen laid four eggs, which

[A] Specific charaéter of the *Cuculus Glandarius*: " Its tail is wedge-fhaped, its head fomewhat crefted, with a black bar on its eyes."

ſhe hatched, &c *. It muſt, therefore, be very different from the ordinary Cuckoo, which undoubtedly never recur to incubation in our climates.

The head is black, and ornamented with a tuft of the ſame colour, which lies backwards; all the upper ſide of the body, including the ſuperior coverts, is black and white; the great quills of the wings rufous, terminated with white; the quills of the tail are blackiſh, terminated with light rufous; the throat and breaſt are rufous; the inferior coverts of the tail are ruſty; the reſt of the under ſide of the body white, and even the feathers on the lower part of the thigh, which deſcend to the tarſus; the bill is greeniſh-brown, and the legs green.

This Cuckoo is rather larger than the ordinary kind, and its tail is proportionally longer; its wings are alſo longer, and its tail more tapered than in the Great Spotted Cuckoo, which it reſembles in other reſpects [A].

* Theſe authors ſay expreſsly that neither before nor ſince that time have theſe birds been ſeen in the neighbourhood of Piſa.

[A] Specific character of the *Cuculus Piſanus:* " Its tail is wedge-ſhaped; its body variegated with black and white above, and white below; its head creſted with black; its throat and breaſt, rufous."

III.

The GREENISH CUCKOO of MADAGASCAR.

Cuculus Madagafcarienfis, Gmel.

The *Great Madagafcar Cuckow*, Lath.

THIS Cuckoo is chiefly diftinguifhed by its magnitude: all the upper fide of the body is deep olive, variegated with waves of a darker brown ; fome of the lateral quills of the tail are terminated with white; the throat is light olive, fhaded with yellow ; the breaft, and the top of the belly, fulvous ; the lower belly brown, and alfo the inferior coverts of the tail ; the thighs are of a gray wine colour; the iris orange ; the bill black ; the legs yellowifh brown; the tarfus not feathered.

Total length twenty-one inches and a half ; the bill twenty-one or twenty-two lines; the tail ten inches, and compofed of ten tapered quills, and extending more than eight inches beyond the wings, which are not long [A]:

I find a note of Commerfon on a Cuckoo from the fame country, and very fimilar to the prefent:—I fhall only remark the differences.

It is nearly as large as a hen, and weighs thir-

[A] Specific character of the *Cuculus Madagofcarienfis:* " It is olive, waved with brown; below fulvous; the throat olive, diluted with yellow."

teen

teen ounces and a half; on the head there is a
naked fpace, furrowed flightly, tinged with
blue, and encircled by fine black feathers; thofe
of the head and neck are foft and filky; there
are fome briftles around the bafe of the bill,
whofe infide is black; the tongue is alfo black
and forked; the iris reddifh; the thighs and
the infide of the wing-quills blackifh; the legs
black.

Total length twenty-one inches and three
quarters; the bill nineteen lines, its edges
fharp; the noftrils like thofe of the gallinaceous
tribe; the outer of the two hind toes can turn
backwards or forwards (as I have obferved in
the European Cuckoo); the alar extent twenty-
one inches, and each wing contains eighteen
quills.

All that Commerfon fays with regard to the
habits of this bird is, that it affociates with
the other Cuckoos. It would appear to be a
variety of the greenifh Cuckoo; perhaps it differs only in fex, in which cafe I fhould reckon
it the male.

THE MADAGASCAR CRESTED CUCKOO.

IV.
The COUA.

Cuculus Criftatus, Linn. and Gmel.
Cuculus Madagafcarienfis Criftatus, Briff. and Ger.
The *Madagafcar Crefted Cuckoo,* Lath.

I RETAIN the name given by the inhabitants
of Madagaſcar, becauſe it certainly denotes
the bird's call. It has a tuft which reclines
back, whoſe feathers, and alſo thoſe of the reſt
of the head and of all the upper ſide of the body,
are greeniſh cinereous; the throat and fore-
part of the neck are cinereous; the breaſt red
wine colour; the reſt of the under ſide of the
body whitiſh; the legs barred almoſt imper-
ceptibly with cinereous; what appears of the
quills of the tail and of the wings is light green,
changing into blue and ſhining violet, but the
lateral quills of the tail are terminated with
white; the iris is orange; the bill and legs
black; it is rather larger than the ordinary
Cuckoo, and is differently proportioned.

Total length fourteen inches; the bill thirteen lines; the tarſus nineteen lines; and the
toes are alſo longer than in our Cuckoo; the
alar extent is ſeventeen inches; the tail is ſeven inches, conſiſting of quills that are a little
tapered, and exceeds the wings ſix inches.

Commerſon made this deſcription in the
month

month of November, from the living fubject, and in its native climate; he adds that its tail diverges, or rather expands, that its throat is fhort, its noftrils oblique and uncovered, that its tongue terminates in a cartailginous tip; its cheeks naked, wrinkled, and bluifh.

The flefh is excellent. The bird is found in the woods near Fort Dauphin [A].

V.

The HOUHOU of EGYPT.

Cuculus Ægyptius, Gmel.
The *Egyptian Cuckoo,* Lath.

THIS Cuckoo has named itfelf; for its cry, *hou, hou,* is repeated feveral times fuccef-fively, in a hollow tone. It is feen frequently in the Delta; the male and female feldom feparate from each other; but it is more rare to find fe-veral pairs collected together. They are *acri-dophagites* * in the full fenfe of the word; for they feem to fubfift moftly, if not entirely, on grafs-hoppers. They never alight on large trees, and far lefs on the ground; but perch on bufhes near fome brook. They have two un-

---

[A] Specific character of the *Cuculus Criftatus:* " Its tail is rounded; its head crefted; its body of a gloffy greenifh afh-co-lour."

* i. e. *locuft-eaters,* from αχριδες, *locuftæ.*

common

common characters: the firft is, that all the
feathers which cover the head and neck are
thick and haid, while thofe on the belly are foft
and loofe; the fecond is, that the nail of the
inner hind toe is long and ftraight, like that of
the lark.

In the female (for I have no certain account
of the male) the head and upper fide of the neck
are dull green, with polifhed fteel reflections;
the fuperior coverts of the wings are greenifh
rufous; the quills of the wings rufous, terminated with fhining green, except the two laft,
which are entirely of that colour, and the two
or three preceding thefe, which are mixed with
it; the back is brown, with greenifh reflections; the rump brown, and alfo the fuperior
coverts of the tail, whofe quills are fhining
green, with the glofs of burnifhed fteel; the
throat and all the under fide of the body are rufty
white, which is lighter under the belly than on
the anterior parts, and on the flanks; the iris is
bright red; the bill black, and the legs blackifh.

Total length from fourteen inches and a half
to fixteen and a half; the bill fixteen or feventeen lines; the noftrils three lines, and very
narrow; the tarfus twenty-one lines; the inner
hind nail nine or ten lines; the wings fix or feven inches; the tail eight lines, confifting of
ten tapered quills, and exceeding the wings five
inches.

M. de Sonini, to whom I am indebted for the

account of this bird, adds that the tongue is
broad, and flightly cut at the tip; the ftomach
is like that of the European Cuckoo; the inteftines are twenty inches, and there are two *cæca*,
of which the fhorteft is an inch.

After attentively comparing the defcription
of this female, in all its details, with the bird
reprefented No. 824, *Planches Enluminées*, under the name of *Philippine Cuckoo*, I am of
opinion that it is the male, or at leaft a variety,
of that fpecies: the fame fize, the fame proportions, the fame lark's fpur, the fame ftiffnefs in
the feathers of the head and neck, and the fame
tapered tail; only the colours are duller, for, except thofe of the wings, which are rufous, as
in the Houhou, all the reft of the plumage is
gloffy black. The bird defcribed by Sonnerat
in his Voyage to New Guinea, under the name
of the *green Antigua Cuckoo*, refembles the preceding fo much, that what I have faid equally
applies to it: the head, the neck, the breaft,
and the belly, are of a dull green verging on
black; the wings are of a deep brown rufous;
the nail of the inner to is more flender, and
perhaps rather longer; all its feathers are in general hard and ftiff; the webs are ragged, and
each fibre is a new fhaft, to which fhorter fibres
are attached. The tail does not indeed appear
tapered in the figure, but this was perhaps an
overfight. This Cuckoo is hardly larger than
that of Europe.

8

Laftly,

THE LONG HEELED CUCKOO.

Laftly, the Madagafcar bird, called *Toulou*, refembles the female of the Egyptian Houhou in the fame properties as the Cuckoo of the Philippines; its plumage is not fo dark, efpecially on the anterior part, where the black is foftened by fpots of light rufous; in fome individuals the olive affumes the place of the black on the body, and it is fprinkled with whitifh longitudinal fpots, which appear alfo on the wings. I fhould therefore fuppofe that they are young ones of the year's hatch, efpecially as, in the birds of this genus, the plumage undergoes confiderable change, it is well known, at the firft moulting [A].

VI.

The RUFOUS-WHITE CUCKOO.

Le Rufalbin, Buff.
Cuculus Senegalenfis, Linn. Gmel. Briff. and Bor.
The *Straight-heeled Cuckoo*, Lath.

THE two prevailing colours of its plumage are rufous and white. When it perches, the tail fpreads, as that of the *Coua*, like a fan; it is almoft perpetually in motion; its cry is precifely the found made by two or three jerks of

[A] Specific character of the *Cuculus Ægyptius*: " It is brown, below rufous white; its head and neck dull green; its tail wedge-fhaped, and bright green; its wing-quills rufous."

the

the tongue upon the palate. Like the two preceding, it has a ftraight long nail on the inner hind claw, fimilar in form to that of the lark's; the upper furface of the head and neck blackifh; the fides of each feather deeper coloured, and yet more brilliant; the wings rufous, and their coverts rather darker near the tip: the back is of a very brown rufous; the rump and the fuperior coverts of the tail barred tranfverfely with light brown on a deeper brown ground; the throat, the fore part of the neck, and all the under fide of the body, are dirty white, only the feathers of the throat and neck have a more brilliant fhaft, and the reft of the under fide of the body is radiated tranfverfely with delicate ftripes of a lighter colour; the tail is blackifh; the bill black, and the legs gray brown; the body is fcarcely larger than that of a blackbird, but it has a much longer tail.

Total length fifteen or fixteen inches; the bill fifteen lines; the tarfus nineteen; the nail of the inner hind toe is above five lines; the alar extent a foot and feven or eight inches; the tail eight inches, confifting of ten tapered quills, and exceeding the wings about four inches [A].

[A] Specific charaćter of the *Cuculus Senegalenfis:* " Its tail is wedge-fhaped; its body gray, white below; its cap and its tail-quills blackifh."

VII.

The B O U T S A L L I C K, *Buff.*

Cuculus Scolopaceus, Linn. and Gmel.
Cuculus Bengalenſis Nævius, Briſſ. and Klein.
The *Brown and Spotted Indian Cuckoo*, Edw.
The *Indian Spotted Cuckoo*, Lath.

EDWARDS perceived ſo much reſemblance between this Bengal Cuckoo and the European, that he particularly marked the points
of diſparity, which, in his opinion, were ſufficient to conſtitute a diſtinct ſpecies. I ſhall
here ſtate theſe differences.

It is above one third ſmaller, though it has a
longer ſhape, and though its body meaſures,
between the bill and the tail, half an inch more
than that of the ordinary Cuckoo. Its head is
larger, its wings ſhorter, and its tail longer in
proportion.

The prevailing colour is brown, which is
deeper and ſpotted with a lighter brown on the
upper part, more dilute and ſpotted with white,
orange, and black, on the lower part; the light
brown or ruſty ſpots form, by their arrangement on the quills of the tail and wings, a tranſ-
verſe barring, which is a little inclined towards
the point of the quills; the bill and legs are
yellowiſh.

Total length thirteen or fourteen inches; the
bill twelve or thirteen lines; the tarſus eleven

or

or twelve; the tail about feven inches, compofed of ten tapered quills, and exceeding the
wings by near five inches [A].

VIII.

The VARIEGATED CUCKOO of
MINDANAO.

Cuculus Mindanenfis, Gmel.
The *Mindanao Cuckoo*, Lath.

THIS bird is fo much variegated, that, at firft
fight, we might take its coloured figure
as defigned for the young Cuckoo of Europe.
Its throat, head, neck, and all the upper part
of the body, are fpotted with white or rufous,
which is more or lefs dilute on a brown ground,
which is itfelf variable, and verges on a gold
green more or lefs brilliant on all the upper part
of the body, including the wings and tail; but
the difpofition of thefe fpots is changed on the
quills of the wings, where they form tranfverfe
bars of pure white externally, and tinged internally with rufous, and on the quills of the tail,
where they form tranfverfe bars of a rufty co-

[A] Specific charaƈter of the *Cuculus Scolopaceus:* " Its tail is
wedge-fhaped; its body clouded on both fides with gray and
brown."

lour;

lour; the breaſt and all the under ſide of the
body, as far as the extremity of the lower coverts of the tail, are white, barred tranſverſely
with blackiſh; the bill is alſo blackiſh above,
but ruſty below, and the legs are brown gray.

This Cuckoo is found in the Philippines;
and it is much larger than the European kind.

Total length fourteen inches and a half; the
bill fifteen lines; the longeſt toe ſeventeen lines;
the ſhorteſt ſeven lines; the alar extent nineteen lines and a half; the tail ſeven inches, conſiſting of ten quills, which are nearly equal, and
exceeds the wings four inches and a half [A].

IX.

The CUIL.

Cuculus Honoratus, Linn. and Gmel.
Cuculus Malabaricus Nævius, Briſſ.
The *Sacred Cuckoo,* Lath.

THE name *Cuil* is applied to this bird by the
inhabitants of Malabar. It is a new ſpecies introduced by M. Poivre, and differs from
the preceding by its ſmallneſs, by the ſhortneſs
of its bill, and by the unequal length of its tail
quills.

[A] Specific character of the *Cuculus Mindanenſis:* " Its tail is
rounded; its body green gold, ſpotted with brown white; below
waved with white and blackiſh."

The head and all the upper fide of the body are blackifh cinereous, fpotted regularly with white; the throat and all the under fide of the body white, barred tranfverfely with cinereous; the quills of the wings blackifh, thofe of the tail cinereous; and both are barred with white; the iris light orange; the bill and legs of a dilute afh colour.

The Cuil is rather fmaller than the ordinary Cuckoo; it is held facred on the Malabar coaft, becaufe, no doubt, it deftroys the pernicious infects. General fuperftition originates from error; but fome particular ceremonies and obfervances may have their foundation in reafon.

Total length eleven inches and a half; the bill eleven lines; the tarfus ten; the tail five and a half, confifting of ten tapered quills, the outer pair being fcarcely half the length of the inner, and it exceeds the wings three inches and a half [A].

[A] Specific character of the *Cuculus Honoratus:* " Its tail is wedge-fhaped; its body blackifh, fpotted with white; below ftriped with white and cinereous."

X.

The BROWN CUCKOO, variegated with BLACK.

Cuculus Tahitius, Gmel.
Cuculus Taitenfis, Lath. Ind.
The *Society Cuckoo,* Lath. Syn.

ALL that we know of this bird befide what is contained in the appellation, is that it has a long tail, and that it is found in the Society Ifles, and there called the *ara wereroa.* The only account given of it is contained in Capt. Cook's fecond voyage, from which we have extracted this fhort indication with the view of inducing travellers addicted to natural hiftory to procure fuller information concerning this new fpecies [A].

XI.

The RUFOUS SPOTTED CUCKOO.

Le Coucou Brun Piquete de Roux, Buff.
Cuculus Punctatus, Linn. and Gmel.
Cuculus Indicus Nævius, Briff.

THIS is found in the Eaft Indies, and as far as the Philippines. The head and all the

[A] Specific character of the *Cuculus Tahitius:* "It is brown, below white, ftriped with brown; its eye-brows white; its wing-quills fpotted with ferruginous."

upper

upper part of the body are dotted with rufous on
a brown ground, but the quills of the wings and
tail, and its fuperior coverts, are barred tranfverfely; all the quills of the tail are terminated
with light rufous; the throat and all the under
fide of the body are barred tranfverfely with
blackifh brown, on a rufous ground; there is an
oblong fpot of light rufous below the eyes; the
eyes are yellowifh rufous, the bill horn colour, and the legs brown gray.

In the female the upper fide of the head and
neck is lefs dotted, and the under fide of the
body is of a lighter rufous.

This Cuckoo is much larger than the ordinary kind, and almoft equal to a Roman pigeon.

Total length fixteen or feventeen inches; the
bill feventeen lines; the tarfus the fame; the
alar extent twenty-three inches; the tail eight
inches and a half, confifting of ten tapered quills,
and exceeding the wings four inches and one
third.

The fubject defcribed by Sonnerat * had not
the rufous fpot under the eyes, and what is ftill
more different, the quills of the tail were equal,
as in the fpotted Chinefe Cuckoo; fo that it

* *Cuculus Panayus*, Linn. and Gmel.
The *Panayan Spotted Cuckoa*, Lath.

Specific character: " It is black brown, with fpots of rufous
yellow; below rufous, with black tranfverfe ftripes; its throat
black; its tail equal."

ought

ought perhaps to be viewed as a variety of the above species [A].

XII.

The CHINESE SPOTTED CUCKOO.

Cuculus Maculatus, Gmel.

WE know nothing more of this bird than its external shape and its plumage: it is one of the small number of Cuckoos in which the tail is not tapered; the upper side of the head and neck is of an uniform blackish, except a few whitish spots that occur on the front and above the eyes; all the upper side of the body, including the quills of the wings and their coverts, is of a greenish deep gray, variegated with white, and enriched by brown gold reflections; the quills of the tail are barred with the same colours; the throat and breast variegated pretty regularly with brown and white; the rest of the under surface of the body and the thighs are variegated with the same colours, and also the feathers which fall from the lower part of the thigh on the tarsus and as far as the origin of the

[A] Specific character of the *Cuculus Punctatus:* "Its tail is wedge-shaped; its body blackish, dotted with rufous; below rufous, with black streaks; its tail quills barred with rufous."

toes; the bill is blackifh above, yellow below, and the legs yellowifh.

Total length about fourteen inches; the bill feventeen lines; the tarfus one inch; the tail fix inches and a half, confifting of ten nearly equal quills, and exceeding the wings by four inches and a half [A].

XIII.

The BROWN and YELLOW CUCKOO,
With a radiated Belly.

Cuculus Radiatus, Gmel.
The *Panayan Cuckow*, Lath.

THE throat and the fides of the head refemble the colour of the lees of wine; the upper part of the head is blackifh gray; the back and the wings are of a dull dark brown; the under fide of the quills of the wings next the body is marked with white fpots; the tail is black, radiated, and terminated with white; the breaft is of a dull orpiment colour; the belly is light yellow; the belly and breaft radiated with black; the iris pale orange; the bill black, and the legs reddifh.

[A] Specific charaćter of the *Cuculus Maculatus:* " Its tail is wedge-fhaped; its body above greenifh gray, with white fpots; below variegated with brown and white."

This

This Cuckoo is found in Panay, one of the Philippines; it is almoſt as large as the common Cuckoo, and its tail is compoſed of ten equal quills [A].

XIV.

The CRESTED JACOBINE
of COROMANDEL.

Cuculus Melanoleucus, Gmel.
The *Coromandel Creſted Cuckoo*, Lath.

THIS bird is termed *Jacobine*, becauſe it is black above and white below; its creſt, conſiſting of ſeveral long narrow feathers, lies on the crown of the head, and projects a little backwards. It is probable that the bird erects its creſt when excited by paſſion.

With regard to the colours of its plumage, we might ſay that there is a ſort of black hood thrown over a white tunic; the white on the lower part is pure and unmixed, but the black on the upper part is interrupted, on the edge of the wing, by a white ſpot immediately below the ſuperior coverts, and by ſpots of the ſame

[A] Specific character of the *Cuculus Radiatus:* " It is brown black; below yellow, lineated with black; its throat and cheeks wine-colour; its top blackiſh gray."

Y 4   colour,

colour, which terminate the quills of the tail;
the bill and legs are black.

This bird is found on the coaft of Coromandel; it is eleven inches in total length, its tail
is compofed of ten tapered quills, and exceeding
the wings by one half its length [A].

There is in the King's Cabinet a Cuckoo
brought from the Cape of Good Hope, and
pretty much like this; the only differences are,
that is an inch longer, that it is entirely black
both above and below, except the white fpot on
the wing, which occupies its exact place, and
that, of the ten middle quills of the tail, eight
are not tapered, the remaining outer pair being
eighteen lines fhorter than the reft. It is probably a variety produced by climate.

XV.

The LITTLE CUCKOO,
with a Gray Head and Yellow Belly.

Cuculus Flavus, Gmel.
The *Yellow-bellied Cuckoo,* Lath.

THIS fpecies is found in the ifland of Panay;
and we owe the account of it to Sonnerat:

[A] Specific character of the *Cuculus Melanoleucus:* " It is
black, below white; its tail wedge-fhaped, tipt with white; a
white fpot on its wings; its head fomewhat crefted."

the

the upper fide of the head and throat is light
gray; the upper fide of the neck, back, and
wings, umber coloured; the belly, the thighs,
and the inferior coverts of the tail, are pale yellow, tinged with rufous; the tail is black, ftriped with white; the legs are pale yellow, and fo
is the bill, only blackifh at the point.

This bird is as large as a blackbird, not fo
bulky, but much longer, being eight inches and
fome lines, and the tail, which is tapered, occupies more than one half of this length [A].

XVI.

The COUKEELS.

I FIND in fyftems of ornithology, three birds
of different fizes, which are made into as
many diftinct fpecies; but they appear fo analogous in their plumage, that I am difpofed to
regard them as the fame, only varying in bulk,
efpecially as they are all natives of the eaftern
parts of Afia. Edwards infers from the refemblance between their names, that the call of the
Bengal Coukeel muft refemble that of the European Cuckoo.

[A] Specific character of the *Cuculus Flavus:* " It is brick-co-
loured, below yellowifh; its top and throat dilute gray; its tail
wedge-fhaped, and black, lineated with white."

The

The firſt and largeſt * of the three is very
near equal in ſize to the pigeon ; its plumage is
entirely of a ſhining black, changing into green
and alſo into violet, but only under the quills of
the tail ; the under and inner ſide of the quills
of the wings black ; the bill and legs brown
gray ; and the nails blackiſh.

The ſecond comes from Mindanao, and is
hardly larger than our Cuckoo ; it is intermediate in regard to ſize between the preceding and
the following one ; all its plumage is blackiſh,
verging on blue ; the bill is blackiſh at the baſe
and yellowiſh at the point ; the firſt of the
quills of the wings is almoſt one half ſhorter
than the third, which is one of the longeſt ; the
tail is generally ſpread.

The third Coukeel, and the ſmalleſt †, is
ſcarcely equal to the blackbird ; it is entirely
black, like the two foregoing, without mixture
of any other permanent colour ; but, according
to the various angles of the incidental rays, the
plumage glows with all the fugitive tints of the

* *Cuculus Orientalis*, Linn. and Gmel.
Cuculus Indicus Niger, Briſſ.
The *Eaſtern Black Cuckow*, Lath.
Specific character : " Its tail is rounded ; its body gloſſy green-
iſh black ; its bill brown."

† *Cuculus Niger*, Linn. Gmel. and Klein.
Cuculus Bengalenſis Niger,· Briſſ. and Gerini.
The *Black Indian Cuckow*, Lath.
Specific character : " It is wedge-ſhaped ; its body ſhining
black ; its bill bright yellow."

rainbow.

rainbow. It is thus that Edwards, the original obferver, defcribes it, and I know not why Briffon mentions only green and violet. As in the firft Coukeel, the inner and under fide of the quills of the wing are black; the bill is bright orange, rather fhorter and thicker than in the European Cuckoo; the tarfus is thick and fhort, and of a reddifh brown, which is alfo the colour of the toes.

We muft obferve that the name *Coukeel*, beftowed in Bengal, is peculiarly applicable to this bird, and therefore the inferences drawn between the fimilarity of names and the refemblance of calls, will be more conclufive with regard to it than with regard to the other two. The edges of the upper mandible are not ftraight, but undulated.

In all the three the tail confifts of ten tapered quills. Their dimenfions are as follow:

FIRST COUKEEL.	Inches.	Lines.	SECOND. In.	Lin.	THIRD. In.	Lin.
Total length - - -	16	0	14	0	9	0
Bill - - - - - -	0	16	0	15	0	10
Tarfus - - - - -	0	17			0	7
Alar extent - - - -	23	0	0	16	wings pretty long.	
Tail - - - - - -	8	0	7	0	4	3
Length beyond the wings	4	0	3	6	2	9

XVII.

The GOLD-GREEN and WHITE CUCKOO.

Cuculus Auratus, Gmel.
The *Gilded Cuckoo,* Lath.

ALL that we know of this bird is that it is
found at the Cape of Good Hope, and carries its tail fpread like a fan. It is a new fpecies.

All the upper furface from the bafe of the bill
to the end of the tail is gold-green, waving and
rich, whofe uniformity is interrupted on the
head by five white bars, one on the middle of
the front, two others above the eyes, like eye-
brows, and ftretching backwards, and two
others, which are narrower and fhorter, below
the eyes; moft of the fuperior coverts alfo, and
the middle quills of the wings, all the quills of
the tail, and its two largeft fuperior coverts, are
terminated with white; the two outermoft of
the quills of the tail, and the outermoft of thofe
of the wings are ftreaked with white on the
outfide; the throat is white, and alfo the whole
under furface of the body, except fome green
ftripes on the flanks, and the ruffles which fall
from the lower part of the thigh upon the tarfus; the bill is brown green, and the legs gray.

This Cuckoo is nearly of the fize of a thrufh;
its

its total length about feven inches; the bill feven or eight lines; the tarfus the fame, clothed
with white feathers as far as the middle; the
tail three inches and a few lines, and confifting of ten tapered quills, which in their natural pofition diverge; it exceeds by only fifteen lines the wings, which are very long in
proportion.

XVIII.

The LONG-SHAFTED CUCKOO*.

Cuculus Paradifeus, Linn. and Gmel.
Cuculus Siamenfis Criftatus Viridis, Briff.
The *Paradife Cuckoo*, Lath.

THE plumage of this bird is entirely of a dark
green, which appears on the head, the
body, the wings, and the tail. Yet it has not
been neglected by nature, but, on the contrary,
it is decorated by unufual feathers. Befide the
tuft which diftinguifhes the head, the tail has a
remarkable form; the outermoft pair of quills
is near fix inches longer than all the reft; and
they have no webs except for the fpace of three
inches at their extremity. It was on account
of thefe two fhafts that Linnæus gave the bird
the appellation of *Paradife Cuckoo*: for the fame
reafon the two following might be termed *Wi-*

* This fpecies is new, and introduced by M. Poivre.

dow

dow Cuckoos. The iris is of a fine blue; the bill blackish; and the legs gray: it is found at Siam, where Poivre saw it alive; it is nearly as large as the jay.

Total length seventeen inches; the bill fourteen lines; the tarsus ten; the tail ten inches and nine lines, and rather forked, and exceeding the wings about nine inches [A].

───────

XIX.

The COLLARED CRESTED-CUCKOO*.

Cuculus Coromandus, Linn. and Gmel.
Cuculus Coromandelanensis Cristatus, Briss.
The *Collared Cuckow,* Lath.

THIS bird is also ornamented with a crest, and is remarkable for the length of two quills of its tail; but these are the middle ones, which project beyond the lateral ones, as also obtains in some species of widow-birds.

All the upper side is blackish, from the head inclusively to the end of the tail, except a white collar about the neck, and two round spots of

[A] Specific character of the *Cuculus Paradiseus:* "The two outermost quills of its tail are very long, and dilated at the tip; its head crested, and its body green."

* This species is new, and was observed and figured in its native country by M. Poivre.

a gray

a gray colour behind the eyes, one on each fide, which in fome meafure refemble ear-pendants: we muft alfo except the wings, of which the quills and their middle coverts are variegated with rufous and blackifh; and fo are the fhoulders, though their great quills and coverts are entirely blackifh; the throat and the thighs are blackifh; all the reft of the under fide of the body is white; the iris yellowifh; the bill deep cinereous; the legs alfo cinereous, but lighter coloured. This Cuckoo is found on the coaft of Coromandel, and is nearly of the fize of the red-wing.

Total length twelve inches and a quarter; the bill eleven lines; the tarfus ten; the wings fhort; the tail fix inches and three quarters, confifting of ten quills, the two middle ones much longer than the lateral ones, which are tapered; and it exceeds the wings by five inches and a half [A].

[A] Specific charaĉter of the *Cuculus Coromandus:* "Its tail is wedge-fhaped; its body black, below white; a bright white collar."

XX.

The S A N-H I A of C H I N A*.

Cuculus Sinensis, Linn. and Gmel.
Cuculus Cæruleus, Briss.
The *Chinese Cuckow*, Lath.

THIS Cuckoo resembles the preceding species, and consequently the widow birds,
by the length of the two middle quills of the
tail; its plumage is remarkable, though it consists of only two principal colours, blue on the
upper side, and snowy white on the under side;
but it would seem that nature, happy always
in her careless mood, has suffered a few drops
that are snowy white to fall from her pallet
upon the crown of the head, where it forms a
spot through which the blue appeared in a thou-
sand points; and also upon the cheeks a little
behind, where two sorts of ear-pendants are represented, as in the preceding species; and upon
the quills and coverts of the tail, each of which
is marked by a round white speck near the tip:
besides, it appears melted into the azure of the
rump and of the base of the great quills of the
wing, and has considerably diluted the tinge;
the whole is set off to advantage by the dark
blackish colour of the throat and the sides of the
head; and lastly, the beautiful colour of the iris,

* This species is new, and seen and figured by M. Poivre in its
native country.

                                          the

THE BLUE CUCKOO.

the bill, and the legs, adds to the richnefs of the garb.

Total length thirteen inches; the bill eleven lines, and there are fome hairs about its upper bafe; the tarfus ten lines and a half; the tail feven lines and a half, confifting of ten very un-equal quills, of which the two middle ones ex-ceed the two adjacent lateral ones by three inches and a quarter; they exceed the outer-moft by five inches and three lines, and the wings by almoft their whole length [A].

XXI.

The T A I T - S O U.

Cuculus Cæruleus, Linn. and Gmel.
Cuculus Madagafcarienfis Cæruleus, Briff.
The *Blue Cuckoo*, Lath.

As ufual, I retain the favage name, which is generally the beft and the moft charac-teriftic.

The Tait-Sou, fo called in its native country, is wholly of a fine blue, and the general uni-formity is relieved by very dazzling tints of vio-let and green, reflected by the quills of the wings; and by tints of pure violet, without the

[A] Specific character of the *Cuculus Sinenfis*: " Its tail is wedge-fhaped and long; its body blue, below white; a white fpot on the tips of the tail-quills."

leaſt caſt of green, which are reflected by the quills of the tail; laſtly, the black colour of the legs and bill gives ſhade to the picture.

Total length ſeventeen inches; the bill ſixteen lines; the tarſus two inches; the alar extent near twenty inches; the tail nine inches, and compoſed of ten quills, of which the two middle ones are a little longer than the lateral ones; it exceeds the wings ſix inches [A].

XXII.

The POINTER CUCKOO.

Cuculus Indicator, Gmel. and Bor.
The *Honey-Guide*, Sparrman.
The *Honey Cuckoo*, Lath.

IT is in the interior part of Africa, at ſome diſtance from the Cape of Good Hope, that this bird is found, which is known by its ſingular inſtinct of pointing out the neſts of wild bees *. In the morning and evening it utters its cry, *cherr, cherr, cherr,* which is very ſhrill, and

[A] Specific character of the *Cuculus Cæruleus:* " Its tail is rounded; its body blue."

* According to ſome travellers, the cry of this bird is *wieki, wieki*; and this word *wieki* ſignifies honey in the language of the Hottentots. It ſometimes happens that the hunter in following the call of this Cuckoo, is devoured by wild beaſts; which has given occaſion to ſay that the bird concerts with them to conduct their prey.

feems to invite the hunters and others, who fearch for honey in the wildernefs; they an- fwer it in a more hollow tone, and continue al- ways advancing: as foon as it perceives them it flies onward, and hovers over the hollow tree where the fwarm is lodged; if the hunters are too tardy in following it, it redoubles its cries, returns back to them, ftops and flutters about, to roufe their attention. It omits nothing to in- duce them to profit by the little treafure which it difcovers, but which it probably could not enjoy without the affiftance of man; either becaufe the entrance into the neft is too narrow, or on ac- count of fome other circumftances. While the party are employed in plundering the honey, the bird fits in a neighbouring bufh, watching ea- gerly and expecting its fhare of the booty, which is commonly left for it, though never in fufficient quantity to fatiate its appetite, or extinguifh or blunt its ardour in this kind of purfuit.

This is not the idle tale of a common tra- veller; it is the obfervation of an enlightened man, who affifted at the deftruction of feveral republics of bees, betrayed by this little fpy, and who communicates an account of what he faw to the Royal Society of London. I fhall add the defcription of the female, which he made from the only two fubjects that he could pro- cure, and which he fhot to the great fcandal of

the

the Hottentots; the exiftence of a ufeful creature is every where precious.

The upper fide of the head is gray; the
throat, the fore part of the neck, and the breaft,
are whitifh with a green tinge, which grows
more dilute as it fpreads, and is fcarcely perceptible on the breaft; the belly is white; the
thighs the fame, marked with an oblong black
fpot; the back and rump are rufty gray; the
fuperior coverts of the wings are brown gray,
thofe next the body marked with a yellow fpot,
which, on account of its fituation, is often concealed under the fcapular feathers; the quills of
the wings are brown; the two middle quills of
the tail are longer and narrower than the reft,
and of a brown verging on ruft colour; the two
following pairs are blackifh, their infide dirty
white; thofe which fucceed are white, terminated with brown, and marked with a
white fpot near their bafe, except the laft pair,
where this fpot almoft vanifhes; the iris is
rufty gray; the eye-lids black; the bill brown
at its bafe, yellow at the end, and the legs are
black.

Total length fix inches and a half; the bill
about fix lines, and there are fome briftles about
the bafe of the lower mandible; the noftrils are
oblong, with a projecting margin, placed near
the bafe of the upper mandible, and feparated only by its ridge; the tarfus is fhort; the

8   nails

nails flender ; the tail tapered, and compofed of twelve quills ; it exceeds the wings by three-fourths of its length [A].

XXIII.

The VOUROU-DRIOU*.

Cuculus Afer, Gmel.
Cuculus Madagafcarienfis Major, Briff.
The *African Cuckow*, Lath.

THIS fpecies and the preceding differ from all the reft, in the number of quills in the tail; thefe amounting to twelve, though commonly they are only ten. The differences peculiar to the Vourou-driou confift in the fhape of its bill, which is longer, ftraighter, and not fo convex above; in the pofition of the noftrils, which are oblong, and placed obliquely near the middle of the bill; and in a character which belongs alfo to the birds of prey, viz. that the female is larger than the male, and of a very

[A] Specific character of the *Cuculus Indicator*: " It is ferruginous gray, below white; its eye-lids naked and black; a bright yellow fpot on its fhoulders; its tail wedge-fhaped and ferruginous."

* The natives of Madagafcar call it *Vouroug-driou*. We are indebted to M. Briffon for the account of this fpecies, which is not the largeft in that ifland, witnefs the Greenifh Cuckoo already noticed.

different

different plumage. This bird is found in the
ifland of Madagafcar, and no doubt on the cor-
refponding part of Africa.

In the male the crown of the head is blackifh, with reflections of green and rofe copper;
there is a ftreak of black placed obliquely between
the bill and the eye; the reft of the head, the
throat, and the neck, are cinereous; the breaft,
and all the reft of the under fide of the body,
are of a handfome white gray; the upper fide
of the body, as far as the end of the tail, is of
a green colour, changing into rofe-copper; the
middle quills of the wing are nearly of the fame
colour; the large ones blackifh, verging on
green; the bill is deep brown; and the legs
reddifh.

The female is fo different from the male, that
the inhabitants of Madagafcar have called it by
a different name, *Cromb.* The head, the throat,
and the upper fide of the neck, are ftriped tranfverfely with brown and rufous; the back, the
rump, and the fuperior coverts of the tail, are
of an uniform brown; the fmall fuperior coverts of the wings are brown, edged with rufous; the great ones dull green, edged and terminated with rufous; the quills of the wing
are the fame as in the male, only the middle
ones are edged with rufous; the fore part of the
neck, and all the reft of the under fide of the
body, are variegated with blackifh; the quills
of the tail are of a gloffy brown, terminated
with

with rufous; the bill and legs are nearly as in the male [A].

Their relative dimenſions are as follow:

	MALE.			FEMALE.	
	Inch.	Lin.		Inch.	Lin.
Total length - - - -	15	0	- - -	17	6
Bill - - - . - .	2	0	- - -	2	4
Tarſus - - - - -	1	3	- - -	1	3
Alar extent - - -	25	8	- - -	29	4
Tail - - - - - -	7	0	- - -	7	9
Exceſs above the wings	2	4	- - -	2	7

[A] Specific charaƈter of the *Cuculus Afer:* " It is copper-green, below ſhining gray; its head and neck cinereous; its top copper blackiſh; its tail equal and gold green, below black."

AMERICAN BIRDS,

WHICH ARE RELATED TO THE CUCKOO.

I.

The OLD-MAN, or RAIN-BIRD.

Cuculus Pluvialis, Gmel.
Cuculus Jama censis, Briff.
Picus Major Leucophœus, Ray.
Cuculus Jamaicensis Major, Sloane, Brown, and Klein.
The *Rain Cuckoo*, Lath.

THE name of *Old-Man* has been given to this bird, becaufe, under its throat, there is a fort of white down or beard, the attribute of age: it is alfo called the *Rain-Bird*, becaufe it never calls, except before rain. It continues the whole year in Jamaica, and haunts not only the woods, but the ftraggling bufhes: it fuffers the hunters to approach very near before it takes flight. Its ordinary food confifts of feeds and worms.

The upper fide of the head is covered with downy or filky feathers of a deep brown; the reft of the upper fide of the body, including the wings and the two middle quills of the tail, is olive cinereous; the throat is white, and the fore part of the neck the fame; the breaft, and the reft of the under furface of the body, rufous; all the

lateral

lateral quills of the tail are black tipt with white, and the outermoſt is edged with white; the upper mandible is black; the lower one is almoſt white; the legs bluiſh black: it is ſomewhat larger than the blackbird.

The ſtomach of the one diſſected by Sloane was very large in proportion to the ſize of the bird, in which reſpect it reſembles the European ſpecies; it was lined by an exceedingly thick membrane; the inteſtines were twiſted like a ſhip's cable, and covered with a quantity of yellow fat.

Total length ſix inches and three quarters; the bill one inch; the tarſus thirteen lines; the alar extent equal to the total length; the tail from ſeven and a half to eight inches, compoſed of ten tapering quills, and projecting almoſt entirely beyond the wings [A].

VARIETIES of the RAIN CUCKOO.

I. The Rufous-winged Old Man *. The ſame colours as in the preceding appear on the

[A] Specific character of the *Cuculus Pluvialis:* " It is cinereous olive, below rufous ; its throat white."

* *Cuculus Americanus,* Linn. and Gmel.
Cuculus Carolinenſis, Briſſ. and Klein.
The *Carolina Cuckow,* Cateſby, Penn. and Lath.

Specific character: " It is wedge-ſhaped ; its body cinereous above, and white below ; its lower mandible yellow."

upper

upper furface and on the tail, and almoft the
fame on the bill; but the white of the under
furface of the body, which in the Rain-bird was
confined to the throat and breaft, extends in this
bird over all the lower part: the wings have a
rufty caft, and are longer in proportion; laftly,
the tail is fhorter and of a different fhape.

This Cuckoo is folitary; it refides in the
darkeft forefts, and on the approach of winter
it leaves Carolina to find a milder air.

Total length thirteen inches; the bill fourteen lines and a half; the tail fix inches, confifting of ten quills, of which the three middle
ones are longer than the reft, but equal to each
other, and the two lateral pairs are fhorter, and
the more fo in proportion to their diftance from
the centre; the longeft project four inches beyond the wings.

II. The Little Old Man*, known at
Cayenne by the name of Mangrove Cuckoo
(Coucou des Paletuviers). This bird, the female efpecially, refembles the Jamaica Rain-bird
fo much, both in its colours and in its general
conformation, that the defcription of the one
may ferve for the other; the only difference
confifts in the fize, the Cayenne bird being

* Cuculus Minor, Gmel.
  Cuculus Seniculus, Lath. Ind.
  The Mangrove Cuckow, Lath. Syn.

Specific character: " It is cinereous olive, below tawny, its chin
white."

much

much fmaller, its tail is alfo rather longer in
proportion; but we may ftill fuppofe that it is
a variety refulting from climate. It feeds on infects, and particularly on the larger caterpillars*
that gnaw the leaves of the mangroves; and
hence it is fond of lodging among thefe trees,
where it renders an ufeful fervice.

Total length one foot; the bill thirteen lines;
the tarfus twelve; the tail five inches and a
half, confifting of ten tapering feathers, and
exceeds the wings three inches and one third.

II.

The TACCO†.

Cuculus-Vetula, Linn. Gmel. and Bor.
Cuculus Jamaicenfis, Briff. Klein, and Ger.
Picus, feu Pluviæ Avis canefcens, Ray and Sloane.
The *Long-bellied Rain Cuckoo*, Lath.

SLOANE pofitively afferts that, except the bill,
which in the Tacco is longer, more flender,
and whiter, it refembles the Rain-bird precifely; he afcribes to it the fame habits, and ap-

* Thefe large caterpillars are four inches and a half long, and
feven or eight lines broad. In the years 1775 and 1776, they multiplied fo exceffively, that they devoured almoft entirely moft of the
mangroves and many other plants. It was then that the iflanders
regretted their not having multiplied this fpecies of Cuckoo.

† In the Antilles it is named *Tacco* from its cry; the negroes
call it *Cracra* and *Tacra-Bayo*. In St. Domingo it is termed *Coli-
vicou*.

plies

plies the fame names. But Briffon, refting probably on this remarkable difference in the length
and conformation of the bill, has made this bird
a diftinct fpecies. This feparation is the more
proper, as it appears from clofer infpection that
the plumage is not the fame, and that even the
white beard is wanting, which gave name to
the preceding fpecies: befides, the Chevalier
Lefebre Defhayes, who has obferved the Tacco with attention, finds that its habits are different from thofe afcribed by Sloane to the Rain-
bird.

Tacco is the ufual cry of this Cuckoo, but is
feldom heard. It pronounces the firft fyllable
hard, and defcends a whole octave on the fecond; it never utters this till after it has given
a jerk with its tail, which it commonly does
when it fhifts its place, or perceives any one
approach. It has alfo another cry *qua, qua,
qua, qua,* but which is never heard unlefs it be
alarmed by the fight of a cat, or fome other
dangerous enemy.

Sloane fays that this Cuckoo, like the one
which he terms *Rain-bird,* forebodes rain by its
loud calls; but the Chevalier Defhayes difco-
covered no fuch habit *.

Though the Tacco lives generally in cultivated grounds, it alfo frequents the woods, becaufe it there finds its proper food, which con-

* To the Chevalier Defhayes I owe my information with regard
to the habits and economy of the Tacco.

fifts

fifts of caterpillars, beetles, worms and vermin, *ravets* *, wood-lice, and other infects, which unfortunately are too common in the Antilles, both in the cleared lands and in the forefts. It alfo preys upon fmall lizards, called *anolis* †, fmall fnakes, frogs, young rats, and fometimes, it is faid, upon fmall birds. It furprifes the lizards when they are eagerly watching on the branches for flies, and therefore off their guard. With regard to fnakes, it feizes them by the head, and in proportion as the part fwallowed digefts, it fucks up the reft of the body, which hangs out from the bill. It is thus ufeful, fince it deftroys the pernicious animals: it would prove of ftill greater utility, could it be domefticated; and this might be poffible, for it is not fhy, but even fuffers the young negroes to catch it in the hand, though it has a ftrong bill, and could make a ftout defence.

Its flight is never lofty; it begins flapping with its wings, and, then fpreading its tail, it fhoots along, or rather fkims than flies. It flutters from bufh to bufh, and hops from bough to bough: it even fprings upon the trunks of trees, to which it clings like the wood-peckers; and fometimes it alights on the ground, and hops about like the magpye, always in purfuit of

---

* A fort of cock-chaffers, very offenfive and pernicious, frequent in the Weft Indies. T.

† Written alfo *anoulys*. They have a fine, fleek fkin, and are fometimes eaten by the people of the French Weft India iflands. T.

infects

infects or reptiles. It is said to exhale conti‑
nually a rank smell, and that its flesh is unpa‑
latable; which is very probable, considering the
kind of substances upon which it feeds.

These birds retire in the breeding season into
the depth of the forests, and remain so well con‑
cealed, that no person has ever seen their nest.
One might almost be induced to suppose that
they have none, and that, like the European
Cuckoo, they lay their eggs in other birds'
nests; but if this were the case, they would
differ from all the other American Cuckoos,
which themselves build and hatch.

The Tacco has no brilliant colours in its
plumage, but it has always a neat and becom‑
ing air: the upper side of its head and body, in‑
cluding the coverts of the wings, is gray, which
is pretty deep, with greenish reflections on the
great coverts only; the fore side of the neck
and breast is ash gray, and over all these shades
of gray there is spread a faint reddish tint; the
throat is light fulvous; the rest of the under side
of the body, including the thighs, and the in‑
ferior coverts of the wings, are more or less of
a lively fulvous; the ten first quills of the wing
are of a bright rufous, terminated with green‑
ish brown, which, in the following quills, ap‑
proaches constantly to a rust colour; the two
middle quills of the tail are of the colour of the
back, with greenish reflections; the eight others
are the same about their middle, dark brown,

       with

with blue reflections, near their base, and terminated with white ; the iris is brown yellow;
the eyebrows red ; the bill blackish above, and
of a lighter colour below; the legs are bluish.
This Cuckoo is not so large as the European
one; it is found in Jamaica, in St. Domingo,
&c.

Total length fifteen inches and a half (seventeen and one-third, according to Sloane); the
bill is eighteen lines according to Sloane; twenty-one according to the Chevalier Deshayes, and
twenty-five according to Brisson; the tongue is
cartilaginous, terminated by filaments; the tarsus about fifteen lines ; the alar extent equal to
the total length of the bird; the tail eight
inches, according to Deshayes, and eight inches
and three quarters, according to Brisson, and
consisting of ten tapering quills; the intermediate ones overlap the lateral ones; it projects about five inches and a half beyond the
wings [A].

[A] Specific character of the *Cuculus Vetula:* " Its tail is wedge-
shaped ; its body duskish, below brick coloured ; the eye-lids
red."

III.

The GUIRA-CANTARA.

Cuculus-Guira, Gmel.
Cuculus Brafilienfis Criftatus, Briff.
Guira Acangatara, Ray and Will.
The *Brazilian Crefted Cuckow,* Lath.

THIS Cuckoo is very noify; it lives in the forefts of Brazil, and makes them echo to its cry, which is louder than pleafant. It has a kind of tuft, whofe feathers are brown, edged with yellowifh; thofe of the neck and wings are, on the contrary, yellowifh, edged with brown; the upper and under fides of the body are of a pale yellow; the quills of the wings are brown; thofe of the tail brown alfo, but terminated with white; the iris is brown; the bill dun-yellow; the legs fea-green. It is as large as the European magpye.

Total length fourteen or fifteen inches; the bill about an inch, a little crooked at the end; the tarfus one inch and a half, and clothed with feathers; the tail confifting of eight quills, according to Marcgrave; but were not fome of them wanting? they appear equal in the figure.

IV.

The QUAPACTOL, or the LAUGHER.

Cuculus Ridibundus, Gmel.
Avis Ridibunda Quapachtototl, Will. and Ray.
Cuculus Mexicanus, Briff.
The *Laughing Cuckow*, Lath.

THIS Cuckoo is called the *Laughing-bird*, on account of its call; and for the fame reafon, fays Fernandez, it was reckoned unlucky by the Mexicans before the true religion was introduced among them. With regard to the Mexican name *Quapachtototl*, which I have contracted and foftened, it alludes to the fulvous colour which is fpread over all the upper furface of its body, and even on the quills of the wings; thofe of the tail are alfo fulvous, but of a darker caft; the throat is cinereous, and alfo the forepart of the neck and breaft; the reft of the under fide of the body is black; the iris is white, and the bill bluifh black.

The fize of this Cuckoo is nearly equal to that of the European kind; it is fixteen inches in total length, and the tail alone occupies the one half of this.

[A] Specific character of the *Cuculus Ridibundus*: " It is fulvous; its throat and breaft cinereous; its belly, its thighs, and the lower coverts of its tail, black."

V.

The HORNED CUCKOO,
Or the ATINGACU of BRAZIL.

Cuculus Cornutus, Linn. and Gmel.
Cuculus Brafilienfis Cornutus, Briff.
Atinga guacu mucu, Ray and Will.

THE fingular property of this Brazilian Cuckoo
is, that there are long feathers on the head,
which it can erect at pleafure, and form a double tuft; and hence the epithet of *horned*, which
has been beftowed by Briffon. The head is
large, and the neck fhort, as ufual in this genus; all the upper furface of the head and body
is footy; the wings are the fame, and even the
tail, though this has a darker caft, and the feathers at its extremity are marked with a rufty
white fpot, fhaded with black, which melts into a pure white; the throat is cinereous, and fo
is all the under fide of the body; the iris is blood
coloured; the bill yellowifh green, and the legs
cinereous.

This bird is diftinguifhed too by the length
of its tail; for though not larger than a field-
fare or large thrufh, and its body only three
inches long, its tail is nine; it confifts of ten
tapering quills, the intermediate ones overlap-
ping the lateral ones; the bill is a little hooked

at

at the end; the tarfufes are rather fhort, and feathered before [A].

VI.

The BROWN CUCKOO,
variegated with Rufous.

Cuculus Nævius, Linn. and Gmel.
Cuculus Cayanenfis Nævius, Briff.
The *Spotted Cuckow,* Lath.

THE upper fide of the body is variegated with
brown and with different fhades of rufous;
the throat is light rufous variegated with brown;
the reft of the under fide of the body is rufty
white, which affumes a diftinct light rufous on
the inferior coverts of the tail; its quills and
thofe of the wings are brown, edged with light
rufous, having a greenifh caft, particularly on
the lateral quills of the tail; the bill is black
above, rufous on the fides, rufty below, and the
legs cinereous. It is obferved as a fingular property, that fome of the fuperior coverts of the
tail extend almoft to two-thirds of its length.
With regard to fize, this Cuckoo is compared
to the red-wing.

Total length ten inches and two thirds;
the bill nine lines; the tarfus fourteen lines;

[A] Specific character of the *Cuculus Cornutus:* " Its tail is
wedge-fhaped; its creft cleft; its body footy."

the

the alar extent above an inch ; the tail about fix inches, confifting of ten tapering quills, and exceeding the wings by four inches.

The Cuckoo known at Cayenne by the name of the *Barrier-bird* is nearly as large as the preceding, and very fimilar in regard to plumage : in general, it has rather lefs rufous, gray occupying its place, and the lateral quills of the tail are tipt with white ; the throat is light gray, and the under fide of the body white ; the tail too is longer. But notwithftanding thefe trifling differences, we muft confider it as a variety of the preceding, perhaps only fexual.

The name *Barrier-bird* alludes to its habit of perching upon the palifades round plantations ; in that fituation it continually fhakes its tail.

Thefe birds, though not very wild, never gather in flocks ; yet many live in the fame diftrict at once ; they feldom haunt the forefts : they are more common, we are affured, than the *Piaye* Cuckoos, both in Cayenne and Guiana.

[A] Specific charafter of the *Cuculus Nævius:* " Its tail is wedge-fhaped; its body brown and ferruginous; its throat marked with brown furrows; its tail-quills tipt with tawny."

VII.

The ST. DOMINGO CUCKOO.

Le Cendrillard*, Buff.
Cuculus Dominicus, Linn. Gmel. and Briff,

THE prevailing colour of its plumage is afh-gray, which is more intenfe above, as far as the two middle quills of the tail inclufively, more dilute below, and intermixed with more or lefs rufous on the quills of the wings ; the three pairs of lateral quills in the tail are black-ifh, terminated with white, and the outermoft pair is edged with the fame white colour; the bill and legs are dun gray. This bird is found in Louifiana and St. Domingo, in different fea-fons, no doubt : it is faid to be nearly of the fize of the red wing.

I have feen in M. Mauduit's Cabinet a variety named the *Little Gray Cuckoo,* which differed not from the preceding, except that all the un-der furface was white, that it was rather larger, and that its bill was not fo long.

Total length from ten to twelve inches; the bill fourteen or fifteen lines, the two mandibles bent downwards ; the tarfus one inch ; the alar extent five inches and a half; the tail five inches and one third, confifting of ten tapered feathers;

* So termed by M. Montbelliard, on account of its cinereous plumage.

it

it exceeds the wings from two inches and a half
to three inches [A].

VIII.

The PIAYE CUCKOO.

Cuculus Cayanus, Linn. and Gmel.
Cuculus Cayanensis, Briff.
The *Cayenne Cuckoo*, Lath.

I ADOPT the epithet *Piaye*, applied to this
Cuckoo in the ifland of Cayenne; but I
adopt not the fuperftition which gave it birth:
Piaye fignifies *devil* in the language of the natives, and alfo *prieft*, that is among an idolatrous
people, *minifter* or *interpreter of the devil*. This
obvioufly fhews that it is looked upon as an unlucky bird: for this reafon the Indians and negroes are faid to have an abhorrence of its flefh;
but may not its perpetual leannefs and unpalatable quality account for their averfion?

The Piaye is not fhy; it allows a perfon to
get very near it, and does not fly away till the
moment he is about to feize it: its flight is
compared to that of the king-fifher; it commonly frequents the banks of rivers, and lodges

[A] Specific character of the *Cuculus Dominicus:* " Its tail is
wedge-fhaped; its body gray-brown, below partly white; its three
lateral tail-quills tipt with white."

beneath

beneath the low branches of trees, where it probably watches the infects which conftitute its food; when perched it wags its tail, and perpetually fhifts its place. Perfons who have lived at Cayenne and feen this Cuckoo feveral times in the fields, have never yet heard its call. It is nearly as large as a blackbird; the upper fide of its head and body is purple chefnut, including the wings of the tail, which are black near the end, terminated with white, and the quills of the wings, which are terminated with brown; the throat and the fore part of the neck are alfo purple-chefnut, but of a lighter tinge, and which varies in different individuals; the breaft and all the under fide of the body are cinereous; the bill and legs are brown gray.

Total length fifteen inches and nine lines; the bill fourteen lines; the tarfus fourteen lines and a half; the alar extent fifteen inches and one third; the tail ten lines, confifting of ten tapered and very unequal quills; it exceeds the wings about eight inches. N. B. The fpecimen in Mauduit's Cabinet is rather larger.

I have feen two varieties of this fpecie. The one nearly of the fame fize, but of different colours; the bill was red; the head cinereous; the throat and breaft rufous; and the reft of the under fide of the body blackifh afh-colour.

The other variety has nearly the fame colours, only the cinereous of the under fide of the body is fhaded with brown; it has alfo the fame

A a 4　　　　natural

natural habits, the only difference confifting in the fize, which is almoft equal to that of the red-wing.

Total length ten inches and a quarter; the bill eleven lines; the tarfus eleven lines and a half; the alar extent eleven inches and a half; the tail near fix inches, confifting of ten equal quills, and exceeding the wings about four inches [A].

IX.
The BLACK CUCKOO of CAYENNE.

Cuculus Tranquillus, Gmel.

ALMOST the whole plumage is black, except the bill and iris, which are red, and the upper coverts of the wings, which are edged with white; but the black itfelf is not uniform, for it is lighter below than above.

Total length about eleven inches; the bill fe-venteen lines; the tarfus eight lines; the tail compofed of ten quills, a little tapered, and ex-ceeding the wings about three inches.

M. de Sonini affures me that this bird has a tubercle on the fore part of its wing. It lives folitary and tranquil, generally perched upon

[A] Specific character of the *Cuculus Cayanus:* " Its tail is wedge-fhaped; its body purplifh-chefnut, below cinereous; al its tail-quills tipt with white."

the

the trees which grow on the fides of creeks, and it is by no means fo active as moft of the Cuckoos: in fhort, it may be regarded as the intermediate fhade between thefe and the barbets.

X.

The LITTLE BLACK CUCKOO of CAYENNE *.

Cuculus Tenebrofus, Gmel.
The *White-rumped Black Cuckow*, Lath.

THIS Cuckoo refembles the preceding, both in the colour of its plumage, and in its habits and economy. It does not frequent the woods, yet it is no lefs wild; it remains whole days perched upon a detached branch in a cleared fpot, without making any exertion beyond what is neceffary to catch the infects on which it feeds; it neftles in hollow trees, and fometimes in the ground, when it finds holes ready formed.

This Cuckoo is entirely black, except on the hind part of the body, which is white, and this white, which extends to the legs, is feparated from the back of the fore part by a fort of orange cincture. In the fpecimen which I

---

* We are indebted to M. de Sonini for the account of this bird.

faw

faw at Mauduits', the white did not ftretch fo far.

Total length eight inches and a quarter; the bill nine lines; the tarfus very fhort; the tail is not three inches, it is a little tapered, and projects not much beyond the wings [A].

[A] Specific character of the *Cuculus Tenebrofus*: " It is black; its belly and thighs ferruginous; its rump and creft white; its tail equal."

The A N I S.

ANI is the name which the natives of Brazil give to this bird*, and which we retain, though the French travellers† and our modern nomenclators call it *Tobacco-end* ‡, a ridiculous appellation beſtowed on account of the reſemblance of its plumage to the colour of a tobacco roll. Father Dutertre aſſerts, indeed, as the reaſon of that denomination, that it ſeems to articulate the words *petit bout de petun*, which is falſe and improbable; eſpecially as the Creoles of Cayenne have an appropriated deſignation for its ordinary warble, *Canary boiler*, becauſe it reſembles the noiſe of a kettle boiling. It has alſo the name *Devil*, and one of the ſpecies is called the *Savanna devil*, and the other the *Mangrove devil*; the former living conſtantly in the ſavannas, and the latter frequenting the ſea ſhores and the margins of ſalt marſhes, where the mangroves grow.

Their generic characters are theſe:—Two toes before and two behind, the bill ſhort, hooked, thicker than broad; the lower mandible ſtraight, the upper one raiſed into a ſemicircle at its origin, and this remarkable convexity extends over all the upper part of the bill till

---

* Marcgrave. † Dutertre.

‡ *Bout de Petun,* or *Bout de Tabac.*

within

within a little diftance of its extremity, where
it is hooked; this convexity is compreffed on
the fides, and forms a fort of fharp ridge quite
along the upper mandible; below and round
there rife fmall ragged feathers as ftiff as hogs'
briftles, about half an inch long, and all pointed forwards: this fingular conformation of the
bill is fufficient to difcriminate thefe birds, and
feems to conftitute a feparate genus, though it
includes only two fpecies.

The SAVANNA ANI.

FIRST SPECIES.

Crotophaga-Ani. Linn. Gmel. and Bor.
Crotophagus, Briff. and Gerini.
Pfittaco congener Ani, Ray and Will.
Monedula tota nigra major, Sloane and Brown,
Cornix garrula major, Klein.
The *Razor-billed Blackbird*, Catefby.
The *Leffer Ani*, Lath.

THIS Ani is as large as a blackbird, but its
large tail gives it a longer form; for this
is feven inches, which is more than half the total length of the bird: the bill is thirteen lines
long, and rifes nine lines and a half; it is black,
and fo are the legs, which are feventeen lines
in height. The defcription of its colours fhall
be very fhort: all the body is black, faintly
fhaded with fome violet reflections, except a
                                    fmall

fmall edge of deep fhining green, which borders
the feathers on the upper part of the back and
the coverts of the wings, and which cannot be
perceived at a certain diftance, for then the bird
appears entirely black. The female differs not
from the male; they conftantly keep in troops,
and are of fo focial a difpofition that they lodge and
lay their eggs together in the fame neft. They
conftruct it with dry fticks, but ufe no lining;
it is exceedingly wide, often a foot in diameter,
and its capacity is faid to be proportioned to the
number of fellow-lodgers which they intend to
admit. The females hatch in company, and
five or fix are often feen in the fame neft. This
inftinct, which would prove ufeful in the cold
countries, feems to be at leaft fuperfluous in the
fouthern regions, where the neft will eafily preferve its heat. It originates entirely from the
impulfe of focial temper; for they are conftantly
together, both when they fly and when they
repofe and fettle on the branches of trees as near
as poffible to each other. In this fituation they
all warble in concert, and almoft through the
whole day; and their fmalleft troops confift of
eight or ten, and they fometimes amount to
twenty-five or thirty. They fly low, and to
fhort diftances; and hence they oftener alight
among bufhes and thickets than upon trees.
They are neither timorous nor fhy, and never
make any remote retreat. They are hardly
fcared by the report of fire arms, and it is eafy

to

to kill many, one after another. But they are
in no requeſt, for their fleſh cannot be eaten,
and the birds have an offenſive ſmell : they feed
on ſeeds and ſmall ſerpents, lizards, and other
reptiles ; they alſo alight upon oxen and cows
to feed on the ticks, maggots, and inſects,
which neſtle in their ſkin [A].

The MANGROVE ANI.

L'Ani des Palituviers, Buff.

SECOND SPECIES.

Crotophaga Major, Gmel.
Crotophagus Major, Briſſ.
The *Great Ani,* Lath.

THIS bird is larger than the preceding, and
almoſt equal to the jay ; it is eighteen
inches long, including the tail, which occupies
the half of that extent : its plumage is nearly of
the ſame browniſh black colour as that of the
former, only it is ſomewhat more variegated
with brilliant green, which terminates the feathers of the back and the coverts of the wings ;
inſomuch that if we reſted our opinion ſolely on
the difference of ſize and colours, we might regard theſe two birds as only varieties of the ſame

[A] Specific character of the *Crotophaga-Ani:* " It is ſmaller ;
its feet ſcanſory."

ſpecies.

THE ANI.

fpecies. But what proves that they are really
two diftinct fpecies is, that they never inter-
mingle; the one kind conftantly inhabits the
open favannas, the other lodges among the mangroves only : yet the latter have the fame natural habits with the former; they likewife
keep in flocks; they haunt the brinks of falt
marfhes; they lay and hatch, many of them
together in the fame neft, and feem to be only
a different race accuftomed to live in more wet
fituations, where the abundance of infects and
reptiles affords an eafier fubfiftence.

Since writing the above, I have received a
letter from the Chevalier Lefebre Defhayes concerning thefe St. Domingo birds, and I fhall
here extract what he fays with regard to the
Mangrove Ani.

" This bird," fays he, " is one of the moft
common in the ifland of St. Domingo . . . .
The negroes give it different appellations, *Tobacco-end*, *Amangoua*, *Black Parrot*, &c. . . . .
If we attentively confider the ftructure of the
wings of this bird, the fhortnefs of its flight,
and the weight of its body compared to its bulk,
we fhall not hefitate to conclude that it is a native of the new world : how, with its feeble
narrow wings, could it traverfe the vaft ocean
that divides the two continents? . . . The kind
is peculiar to fouth America. When it flies it
fpreads its wings; but its motion is not fo quick
nor fo continued as the parrot's . . . It cannot
withftand

withftand the violence of the wind, and the hurricanes deftroy numbers of thefe birds.

" They inhabit the cultivated grounds, or fuch as have once been in the ftate of cultivation, and they are never found in the lofty forefts. They feed on various forts of feeds and fruits, fuch as fmall millet, maize, rice, &c. and when reduced to want, they eat caterpillars and fome other infects. We cannot fay that they have a fong or warble; it is rather a whiftling or chirping: fometimes, however, this becomes more varied, but it is always harfh and difagreeable; it receives different inflexions according to the paffions which incite it. If the bird perceives a cat, or other dangerous animal, it informs its companions by a very diftinct fcream, which it prolongs or repeats until its apprehenfions are quieted: its fears are moft remarkable when it has young, for then it flutters and beats about its neft. Thefe birds live in fociety, though they do not form into fuch large flocks as the ftares; they feldom part from one another .... and even previous to their hatching, we fee feveral males and females working together at the conftruction of the neft, and afterwards the females hatch befide each other, each fitting on her eggs and rearing her young. This harmony is the more admirable, fince love commonly diffolves all other ties but what it forms ... Their amours commence early: in February the males ardently court
the

the females, and in the following month the
happy couple are bufy in collecting materials
for the neft . . . . Thefe birds are more lafci-
vious than even fparrows; and, during the
whole feafon of their ardour, they are much
more lively and cheerful than at any other time
. . . They breed in fhrubs, coffee-trees, bufhes,
and hedges; and they place their nefts in the
cleft where the ftem divides into feveral branches
. . . When feveral females affociate together,
the one readieft to lay does not wait till the neft
be completed, but fits on her eggs while the
reft are employed in enlarging the fabric. They
employ a precaution which is unufual with
other birds, viz. to cover their eggs with leaves
and grafs-ftalks, as faft as they lay them . . . .
And during incubation, they cover the eggs in
the fame manner, if they are obliged to leave
them in queft of food . . . . The females which
thus hatch befide each other are not quarrel-
fome, like hens that breed in the fame crib;
they take their ftations in order: fome, however, before they lay, make a partition in the
neft with ftalks of herbs, to contain their own
eggs; but if the eggs happen to be jumbled together, one female hatches them indifcrimi-
nately; fhe collects them, heaps them, and covers the whole with leaves, fo as to diffufe the
heat equally, and prevent its diffipation . . . Yet
each female lays feveral eggs . . . Thefe birds
build their neft very folid, though rude, with

the fmall ftems of filamentous plants, the branches of the citron trees, and other fhrubs; the infide only is covered with tender leaves that foon wither; and upon this bed the eggs are depofited: thefe nefts are wide, and much raifed at the margin; fometimes the diameter is more than eighteen inches, but its fize depends on the number of females which it is deftined to receive. It would be difficult to decide with accuracy whether all the females contained in the fame neft have each their male; perhaps thefe birds are polygamous, in which cafe it would, in fome meafure, be neceffary to enlarge the nefts, and thus, even without any friendly focial principles, they might be conftrained to unite in performing the work . . . The eggs are as large as thofe of a pigeon; they are of an uniform beryl, and have none of thofe little fpots on the ends, which are ufual on moft of the eggs of wild birds . . . It is probable that the females hatch twice or thrice a year, according to circumftances; if the firft fucceeds, they do not make another till autumn; if on the contrary, the eggs are robbed, or eaten by fnakes or cats, they make a fecond, and towards the end of July, or during the courfe of Auguft, they hatch a third time: certain it is, that their nefts are found in the months of March, May, and Auguft . . . They are gentle, and eafily tamed; and it is faid, that if they are taken young, they may be educated and taught to fpeak,

fpeak, though their tongue is flat, and terminates in a point, while that of the parrot is
flefhy, thick, and round . . .

" The fame friendfhip and concord which
appears during incubation, continues after the
broods are hatched; when the mothers have
covered together, they feed fucceffively all the
little family . . . The males affift in bring-
ing fupplies; but when the females hatch feparately, they rear their young apart, yet
without fhewing any jealoufy or ill temper;
they carry the food by rotation, and the young
ones receive it from all the mothers. The
nature of the food depends upon the feafon, fometimes confifting of caterpillars, maggots, and infects, and fometimes of fruits and
feeds, fuch as millet, maize, rice, and wild
oats, &c. . . . . In a few weeks the young
ones are able to try their wings, but they do
not venture far; foon afterwards they perch
befide their parents among the bufhes, and
then are expofed to the ravages of the birds of
prey . . .

" The Ani is an innocent bird; it does not
plunder the rice plantations, like the blackbird;
it does not feed upon the nuts of the cocoa-tree,
like the woodpecker; nor does it confume the
patches of millet, like the parrots or parrakeets."

[A] Specific character of the *Crotophaga Major*: " It is larger;
its feet fcanfory."

The HOUTOU or MOMOT, *Buff.*

Ramphaſtos-Momota, Linn. Gmel. and Bor.
Momotus Braſilienſis, Lath. Ind.
Motmot & Yayaubquitotl, Fernandez.
Motmot & Avis Caudata, Nieremberg.
Guiraguainumbi Braſilienſibus, Johnſt.
Iſpidæ, ſeu Meropis affinis, Ray and Will.
The *Braſilian ſaw-billed Roller*, Edw.
The *Braſilian Motmot*, Lath. Syn.

WE retain the name *Houtou*, which has
been given by the natives of Guiana,
ſince it is expreſſive of the cry. Whenever the
bird makes a ſpring, it briſkly and diſtinctly articulates *Houtou*; the tone is deep, and reſembles a man's voice: that character alone ſuffi-
ciently diſcriminates the living bird, whether it
be in the ſtate of freedom or of domeſtication.

Fernandez, who firſt noticed the Houtou,
has inadvertently mentioned it by two different
names, and this miſtake has been copied by all
the nomenclators: Marcgrave is the only naturaliſt who has not been miſled. It would
ſeem that Fernandez was deceived by the ſight
of a mutilated ſpecimen, which induced him to
admit two ſpecies; for the ſingle naked quill
which he obſerved could not be natural, ſince
in all birds the feathers grow conſtantly by pairs,
juſt as other animals have two legs or two
arms.

4   The

THE BRASILIAN MOTMOT.

The Houtou is about the fize of the magpie; it meafures feventeen inches and three lines from the point of the bill to the end of the great quills of the tail; its toes are placed as in the king-fifhers, the manakins, &c. But it is diftin-guifhed from thefe, and even from all other birds, by the form of its bill, which, though proportioned to the body, is conical and incur-vated, and the edges of the two mandibles in-dented. This character would difcriminate the Houtou; but it has another more fingular one peculiar to itfelf; to wit, near the ends of the two long quills of the middle of the tail there is a fpace of about an inch, abfolutely bare or fhaved, fo that the fhaft is naked in that part. This appearance, however, belongs to the adult; for when the bird is young thefe quills are, like the other feathers, webbed their whole length. It has been fuppofed that this naked fpace is not a natural production, and that it is perhaps ow-ing merely to the caprice of the bird, which plucks the feathery fibres. But it is obferved that in young fubjects the webs are continuous and entire, and as they grow up thefe become fhorter by degrees, fo as at laft to difappear. We fhall not ftop to defcribe more particularly the plumage of this bird, for the colours are fo much intermingled that it would be impoffible by words to convey a diftinct idea of them; they are alfo affected by age or fex.

They

They are difficult to rear, though Pifo afferts the contrary; and as they feed upon infects, it is not eafy to choofe what will fuit
their tafte. Thofe caught old cannot be bred;
they are extremely fhy, and refufe all fufte-
nance. The Houtou is a wild folitary bird, never found but in the gloomy receffes of forefts;
it affociates not in flocks, or even in pairs; it
is almoft continually on the ground, or among
the low branches, for it never properly flies,
but leaps nimbly, pronouncing fmartly *houtou*. It is early in motion, and its cry is heard
before the warble of the other birds. Pifo was ill informed when he faid that it
builds on lofty trees; for it never conftructs
a neft, nor does it rife to any confiderable
height: it is contented with fome hole of the
armadillos, of the *cavies*, or of other fmall quadrupeds, which it finds on the furface of the
ground; it lines this with dry herb ftalks, and
there lays its eggs, which are generally two in
number. The Houtous are common in the interior parts of Guiana; but they feldom frequent
the neighbourhood of plantations. Their flefh
is hard and unpalatable food. Pifo is miftaken
too, in faying that they live upon fruits. As
this is the third time he has been mifled, it is
probable that he has applied the attributes of
another bird to the prefent, which he defcribes
only from Marcgrave, and with which he was
perhaps

perhaps unacquainted; for it is certain that the *Houtou* is the fame bird with the *Guira-guai-numbi* of Marcgrave, which is difficult to tame, which is unfit for eating, and which neither perches nor neftles upon trees, nor feeds on fruits.

[A] Specific character of the *Ramphaftos-Momota:* " Its feet are greffory (*i. e.* the toes difpofed three before and one be-hind)."

The HOOPOES, the PROMEROPS, and the BEE-EATERS.

COMPARISON is the great source of knowledge. When objects have many common properties, their contrast throws mutual light; it points out the real differences which obtain, and destroys those false analogies which are apt to be formed when they are viewed separately. For this reason, I have ranged in a single article the general facts with regard to the three contiguous genera of the Hoopoes, the Promerops, and the Bee-eaters.

Our Hoopoe is well known by its beautiful double tuft, which is almost *unique* in its kind, since it resembles no other, except that of the cockatoo: its bill is long, slender, and incurvated, and its legs are short. The black and white Hoopoe of the Cape differs from ours in several particulars, and especially because its bill is shorter and more pointed, as will be found in the descriptions. But it ought to be referred to that genus, being more related to it than to any other.

The Promerops resembles the Hoopoes so much that, were we for a moment to adopt the principles of the system-makers, we should say that they are Hoopoes without the crest *. But

* *Huppes sans Huppe.*

the

the fact is that they are rather taller, and their tail is much longer.

The Bee-eaters resemble, in the shortness of their legs, the Hoopoe and king-fisher, more especially the latter, by the singular disposition of their toes, of which the middle one adheres to the outer as far as the third phalanx, and to the inner one as far as the first phalanx only. The bill of the Bee-eaters, which is pretty broad and strong at its base, holds a middle rank between the slender bills of the Hoopoes and Promerops on the one hand, and the long, straight, thick, and pointed bills of the king-fishers on the other; but, on the whole, it rather inclines to the former, since the Bee-eaters live upon insects like the Hoopoes and the Promerops, and not upon small fish like the king fishers; and it is well known how much the force and conformation of the bill serve to regulate the choice of the food.

There are also some traces of analogy between the genus of the Bee-eaters and that of the king-fishers. In the first place, the beautiful beryl, which is by no means common in the European birds, decorates alike the plumage of our king-fisher and of our Bee-eater. In the second place, the greatest number of the species of Bee-eaters have their two middle quills of the tail projecting far beyond the lateral ones; and the genus of the king-fisher contains also some species in which these two middle quills project also.

alfo. And in the third place, there are fome fpecies of king-fifhers in which the bill is a little incurvated, which, in this refpect, refembles that of the Bee-eaters.

On the other hand, how clofe foever the Bee-eaters and Promerops be related, nature, ever rich and unexhaufted, has ftill feparated them ; or rather fhe has melted them into one another by imperceptible fhades. Thefe intermediate birds incline fometimes more to the one genus, and fometimes more to the other; I fhall denominate them *Merops*.

All thefe different birds, which refemble each other in fo many refpects, are fimilar alfo in point of fize. The largeft fpecies exceed not the thrufhes, and the leaft are fcarcely fmaller than the fparrows and the warblers. The exceptions are few, and obtain equally in the different genera.

With regard to climate, a difcrimination takes place. The Promerops inhabit Afia, Africa, and America ; and never occur in Europe : if they are natives of the old continent, they muft have migrated into the new by the north of Afia. The Hoopoe is peculiar to the old world, and I may affert the fame thing of the Bee-eaters, though there is a bird termed the *Cayenne Bee-eater:* for ornithologifts who have frequently vifited that ifland have never feen this bird. And with regard to the two Bee-eaters depicted by Seba, the one from Brazil and the other from
Mexico,

THE COMMON HOOPOE.

Mexico, the authority of that compiler is too
fufpicious to have much weight; particularly
as thefe would be the only two fpecies of Bee-
eaters that are natives of the new continent.

The HOOPOE*.

La Huppe, Buff.
Upupa-Epops, Linn. and Gmel.
Upupa, Frif. Briff. Scop. Kram. Klein, Mul. Sibb. &c.
The *Dung-bird, Hooper, or Hoopoop*, Charleton.

A RESPECTABLE ornithologift, Belon, fays,
that this bird has derived its name from its
large beautiful tuft *(huppe)*; but a little attention would have convinced him that it is really
formed from the Latin *Upupa.*

This

---

* In Arabic, *Al Hudud Garefol:* In Egyptian, *Cucufa:* In
Hebrew, *Kaath, Cos, Hakocoz, Ataleph, Racha, Anapha, Chafi-
da, Dukiphat:* In Greek, Εποψ: In Latin, *Upupa*; which name,
according to Plautus and St. Jerome, was given alfo to girls of
pleafure: In Italian, *Buba, Upega, Gallo di Paradifo, Galletto di
Maggio, Puppula, Criftella, Putta:* In Spanifh, *Abubilla:* In Portuguefe, *Popa:* In German, *Wyd-Hopff, Wede-Hoppe, Kathaan:* In
Flemifh, *Hupetup:* In Brabantifh, *Hueron:* In Norwegian, *Ær-
fugl:* In Danifh, *Her-fugl:* In Swedifh, *Hær-fogel:* In Scanian,
Popp.
   Varro, *Lingua Lat.* lib. IV. fays, that the Latin name *Upupa* is
formed from the cry of the bird, *poo, poo*; and a fable explains the
origin of this cry. Tereus, king of Thrace, having ravifhed Philomela, the fifter of his wife Progne, the latter, in revenge, killed
her fon by him, and ferved up the flefh at her hufband's table.
Upon

This tuft, in its ordinary pofition, reclines
backwards, both when the bird flies or feeds;
in fhort, whenever it is free from the agitation
of paffion *. I had occafion to fee a Hoopoe,
which was caught in a net, and which was old,
or at leaft grown up, and confequently had acquired its natural habits. Its attachment for its
miftrefs was already ftrict and ardent; it feemed
uneafy unlefs it alone enjoyed her company ; if
ftrangers happened to break in upon its domeftic fociety, it erected its tuft, through furprife
or difquietude, and fled to the top of a bed which
was in the fame room ; fometimes it had the refolution to defcend from its afylum, but then it
flew directly to its miftrefs, who enjoyed exclufively all its regard and affection. It had two very
different kinds of cries ; the one foft and tender,
flowing from fentiment, and directed to its miftrefs ; the other harfh and fhrill, and expreffing
anger or fear. It was never confined in its
cage, either by day or night, but ran about the
houfe ; and, though the windows were often
open, it never fhewed any defire of effecting its
efcape. At laft, happening to be fcared, it difappeared fuddenly; it flew but a fhort diftance,
and not being able to find its way back again, it

Upon the difcovery of this horrid repaft, Progne was changed into
a fwallow, Philomela into a nightingale, and Tereus into a *Hoopoe*;
who, ftill bemoaning his lofs, fcreams πυ, πυ, or *where*, *where*;
where, *my fon*.

* It is faid alfo to feek to get near the fire, and to be fond of
fleeping before the chimney.

threw itfelf into a nun's cell, where the window had been left open; fo neceffary was human fociety become to its exiftence and comfort! It died in this retreat, where it could only be fed, and where its proper mode of treatment was unknown. Yet it lived three or four months in its firft condition, its fole fubfiftence being a little bread and cheefe. Another Hoopoe was fed for eighteen months upon raw flefh*; it was exceffively fond of this, and haftened to eat it out of the hand; it rejected, on the contrary, what had been cooked. This predilection for raw flefh feems to indicate an analogy to the rapacious birds and thofe which live upon infects.

The ordinary food of the Hoopoe is infects in general, and efpecially fuch as grovel on the furface†, either their whole life, or during a part of it; beetles, ants ‡, worms, wild bees, and many kinds of caterpillars §, &c. Hence this

---

* Gefner fed one with hard eggs: Olina with worms, or with the hearts of oxen and fheep, cut into little longifh fhreds, nearly like worms; but, above all, he advifes not to fhut it up in a cage.

† The Hoopoe feldom perches upon trees; but, when it does, it prefers oziers, willows, and probably all fuch as grow in wet grounds.

‡ Frifch fays that it digs with its long bill into the ant-hills, to extract the eggs: and, in fact, the one fed by Gefner was very fond of the eggs or nymphs of ants, but rejected the ants themfelves.

§ Salerne adds that it clears the houfe of mice; but this is undoubtedly by driving them away, for with a bill fo flender, with

claws

this bird haunts wet grounds *, where its long
and flender bill can eafily penetrate; and hence,
in Egypt, it follows the retreat of the Nile :
for, in proportion as the waters fubfide †, the
plains are left covered by a coat of flime, which,
being heated by a powerful fun, quickly fwarms
with immenfe numbers of all kinds of infects ‡.
Accordingly, the migratory Hoopoes are very
fat and delicious. I fay the migratory Hoopoes,
for ,there are others in the fame country often
feen on the date trees, in the neighbourhood of
Rofetta, which are never eaten : the fame is
the cafe with thofe which are very frequent in
Grand Cairo §, where they breed with full fe-

claws fo weak, and with a throat fo narrow, it could neither feize
nor devour them, ftill lefs fwallow them entire. It alfo eats vegetable fubftances, and among others, myrtle-berries and grapes.
See Olina and the ancients. I found in the gizzard of thofe which
I diffected, befides infects and worms, fometimes grafs, fmall feeds,
and buds, fometimes round grains of an earthy matter, fometimes
fmall ftones, and fometimes nothing at all.

  * It is becaufe it runs thus in the mud that its feet are almoft
always bedaubed.

  † Hence the appearance of the Hoopoe in Egypt announced the
retreat of the waters of the Nile, and confequently the feed time :
This bird is accordingly reprefented often in the Egyptian hiero-
glyphics.

  ‡ Among others a kind of infect peculiar to Egypt, and which
refembles a wood-loufe. The Nile leaves alfo, in its retreat, the
young and fpawn of frogs, which, in cafe of want, may fupply the
place of infects.

  § They are eaten in Bologna, Genoa, and in fome other parts of
Italy and of France. Some prefer them to quails. It is true that
all our Hoopoes are birds of paffage.

                                                curity

curity on the houfe-tops *. It is eafy, indeed, to conceive that Hoopoes which live remote from man, in forfaken plains, are better food than fuch as haunt the ftreets or the environs of a large city: the former fubfift upon the infects that lodge among the clay or mud; the latter prowl among all forts of filth, which abounds wherever vaft numbers of men are collected; a circumftance which cannot fail to beget an averfion to the city Hoopoes, and even communicate an offenfive odour to their flefh †. There is a third intermediate clafs, which fettling in our gardens, live upon caterpillars and earth worms ‡. It is univerfally agreed that the flefh of the Hoopoe, which feeds fo naftily, has no fault but that of tafting ftrongly of mufk, which is perhaps the reafon that cats, which are generally fo fond of birds, will not touch it §.

In Egypt the Hoopoes gather, it is faid, in

* Thefe two laft notes were communicated to me by M. de Sonini, in two letters, dated from Cairo and Rofetta, the 4th and 5th of September, 1777.

† It is to thefe ftationary, city Hoopoes that we muft refer what Belon afferts, perhaps with too great latitude, " that their flefh is good for nothing, and that no perfon in any country will tafte it." They were alfo held to be unclean by the Jews.

‡ Olina, *Uccelleria*. Albin fpeaks of a Hoopoe that lived in a garden in the middle of Epping Foreft.

§ Several expedients are mentioned for removing this favour of mufk; the moft general advice is to cut the head from the bird the moment it is killed: yet the hind parts tafte more of mufk than the fore parts.

fmall

fmall flocks, and if one happens to ftray, it calls
on its companions with a very fhrill cry of two
notes, zi, zi*. In moft other countries they
appear either fingle, or at moft in pairs: Sometimes in the feafon of their paffage numbers are
found in the fame diftrict; but thefe are folitary
individuals, unconnected by any focial tie, fo
that when they are hunted, one rifes after another. Yet as they have all the fame organization, they muft be actuated by the fame views;
hence they direct their flight towards the fame
country, and follow nearly the fame rout. They
are fcattered through almoft the whole of the
ancient continent, from Sweden, where they inhabit the great forefts, and even from the Ork-
neys and Lapland †, as far as the Canaries and
the Cape of Good Hope on the one hand, and
the iflands of Ceylon and Java ‡ on the other.
They are migratory in every part of Europe,
and even in the delicious climates of Greece and
of Italy §: they are fometimes found at fea‖;
and excellent obfervers ⱶ clafs them with thofe
birds which pafs the ifle of Malta twice a year.
It muft, however, be confeffed, that they do not
conftantly hold the fame courfe; for it often
happens, that though they appear numerous in a

* Note communicated by M. de Sonini.
† Schœffer.   ‡ Edwards.   § Belon and Pliny.
‖ " On the 18th of March, while we were paffing through the
Canaries, a Hoopoe alighted on our veffel, and flew towards the
weft." *Voyage à l'Ifle de France & de Bourbon, par un Officier du
Roi.* Merlin, 1773, t. I.
ⱶ Among others, Comamnder *Defmazys.*

place

place one year, very few or none of them can be found there in the following year. In some countries too, such as England, they are very rare, and never nestle; in others, as in Bugey, they never occur at all. And since Bugey is mountainous, it follows that they are not attached to mountains, at least not to that degree which Aristotle supposed *. But this is not the only fact which contradicts the assertion of that philosopher; for the Hoopoes settle in the midst of our plains, and are frequently seen on the straggling trees which grow on sandy islands, such as those of Camargue in Provence †. Frisch says that they can creep on the bark of trees like the woodpeckers; which is perfectly consistent with analogy, since, like these birds, they nestle in the hollow trunks. In these they usually lay their eggs, and also in the holes of walls upon the mould or dust which is usually collected at the bottoms of such cavities, but do not line it with straw, as Aristotle says. Yet there are some exceptions, at least what are apparently such: of six hatches that were brought to me, four of them had no litter, but the two others had a very soft bedding composed of leaves, moss, wool, feathers ‡, &c. These seeming

* Hist. Anim. *Lib.* I. 1.
† Note communicated by the Marquis de Piolenc.
‡ In the bottom of one of these nests was more than two litrons of moss (a litron is a measure, the 16th part of a bushel) fragments

feeming difparities may be reconciled ; for it is very probable that the Hoopoe fometimes lays her eggs in nefts that were, in the preceding year, occupied by woodpeckers, wrynecks, tit-mice, and other birds, which had lined them according to their different inftincts.

It has been long faid and often repeated, that the Hoopoe befmears her neft with the excre-ments of the wolf, of the fox, of the horfe, of the cow, and of all forts of animals, not ex-cepting man *; and that fhe does this with the view to defend her young by the loathfome ftench †. But the fact is not more true than the intention ; for the Hoopoe never plafters the mouth of its neft like the nuthatch. At the fame time, the neft is indeed very dirty and of-fenfive,

of May-flies, and fome worms that had no doubt dropt from the bill of the mother or of her young. The fix trees in which thefe nefts were found were three black cherries, two oaks, and a pear-tree ; the loweft of thefe nefts was three or four feet above the ground, the higheft ten.

* See Salerne, Gerini, &c. It is pretty fingular that the an-cients, who regarded the Hoopoe as an inhabitant of the mountains, of the forefts, and of the deferts, fhould impute to it the employing human excrements for its neft. This is another particular fact in-judicioufly generalized : the mother, in collecting the infects for her young among filth, might dirty herfelf, and fo pollute her neft ; and fuperficial obfervers would thence conclude that this was a habit common to the whole fpecies.

† It has alfo been faid that her object was to difpel the charms that might be caft upon her brood ; for the Hoopoe was reckoned very fkilful in this way. She knew all the plants that defeat faf-cinations, thofe which give fight to the blind, thofe which open

barred

fenfive, the neceffary confequence of its great
depth, which is often twelve, fifteen, or even eighteen inches : the young ones cannot throw out
their excrements, and therefore grovel a long time
among filth *. Hence undoubtedly the proverb,
" Nafty as a Hoopoe." But it is only in rearing its young that this bird can be accufed of
naftinefs; at other times it is very cleanly. The
one which I before fpoke of never foiled its
miftrefs, nor the chairs, nor even the middle of
the room, but always retired to the top of the
bed, which was the remoteft and moft concealed place.

The female lays from two to feven † eggs,

barred gates; which laft is propped by a fable equally abfurd.
Ælian gravely relates that a man having three times in fucceffion
clofed the neft of a Hoopoe, and having remarked the herb with
which the bird opened it, he employed the fame herb with fuccefs
to charm the locks of the ftrongeft coffers. Death even does but
heighten its virtues, and give them new energy; its heart, its liver,
its brain, &c. eaten, with certain magical incantations applied, fufpended to different parts of the body, occafioned pleafant or frightful dreams, &c. In England, it was formerly held an unlucky
bird; and even at prefent, the people of Sweden regard its appearance as a prefage of war. The ancients had better reafon, me-
thinks, to believe that when it was heard to fing before the time
when they ufually began to drefs the vine, it promifed a good vintage : in fact, its early fong would imply a mild fpring and a forward feafon, which is ever favourable to the maturity of the vine,
and to the quality of its fruit.

* When Schwenckfeld was a child, he had his fingers dirted in
taking a brood of Hoopoes out of a hollow oak.

† Linnæus and the authors of the Britifh Zoology mention only
two eggs. But this cafe is as rare, at leaft in our climates, as that
of feven eggs. In the more northern countries, fuch as that of
Sweden, the Hoopoe may be lefs prolific.

but

388　　　H O O P O E.

but more commonly four or five; these eggs are
greyiſh, ſomewhat larger than thoſe of the partridge. They do not all hatch at the ſame time;
for three young Hoopoes, taken out of the ſame
neſt, differed very much in ſize; in the largeſt
one, the quills of the tail had ſprouted ſeventeen
lines, and in the ſmalleſt only ſeven lines. The
mother has often been ſeen carrying food to the
neſtlings, but I never heard that the father paid
them that attention. As theſe birds hardly ever
appear in knots, it is moſt likely that the family
diſperſes as ſoon as the brood are fledged; and
this is the more probable, if, as the authors of
the *Italian Ornithology* aſſert, each pair makes
two or three hatches in the courſe of the year,
thoſe of the firſt hatch might fly as early as the
end of June.—Theſe are the few facts and conjectures that I am able to offer in regard to the
incubation of the Hoopoe and the education of
its young.

The cry of the male is *bou, bou, bou*; it is
moſt frequent in the ſpring, and may be heard
at a great diſtance *. Thoſe who have liſtened
attentively to theſe birds, pretend to have noticed different inflections and accents, correſponding to their different circumſtances: ſome-

* Ariſtophanes thus expreſſes the cry of theſe birds: *epopoe, popopo,
popoe, popoe, io, io, ito, ito, ito, ito.* I ſuſpect he inclines to make
them ſpeak Greek. Of all the names that have been given to
them, that which imitates their ſong the beſt is *bou, bou*; by which
they are known in Lorrain, and in ſome other provinces of France.
Εποπίζειν in Greek, from εποψ, ſignifies *to ſing like a Hoopoe.*

7   time

times a hollow moaning, which foreboded rain; fometimes a fhriller cry, indicating a fox in fight, &c. This character bears fome analogy to the two voices of the tame Hoopoe mentioned above. That bird feemed fond of mufic; whenever its miftrefs played on the harpfichord or the mandoline, it kept as near the inftruments as poffible during the whole time.

It is faid that this bird never drinks at fprings or brooks; and that, for this reafon, it is feldom caught in fnares. It is true that the Hoopoe killed in Epping Foreft in England fhunned the numerous decoys laid for the purpofe of taking it alive; but the one which I have frequently mentioned had been caught in a net, and drank, from time to time, by plunging its bill with a brifk motion, and without repeatedly lifting up its head, like many birds: it had probably a power of raifing the water into its gullet by a kind of fuction. The Hoopoes retain that brifk motion of the bill even at other times, when they neither eat nor drink; this habit muft arife from their mode of living in the favage ftate; catching infects, cropping buds, boring into the mud for worms, or perhaps for earthy liquor alone, and fearching ants' nefts for the eggs. If they be difficult to enfnare, they are eafy to fhoot; for they fuffer a perfon to come very near them *, and, though

* Thofe who have judged of the Hoopoe from mythology, have reprefented it as very fhy, and as feeking the heart of forefts and

the

though they fly with fudden jerks and in a tor-
tuous courfe, their motion is flow. They flap
their wings in launching off, like the lapwing *; and when they alight on the ground
they walk with an even pace, like common
hens.

They leave our northern climate about the
end of Auguft, or the beginning of autumn,
and never ftay till the cold fets in. But though
they are birds of paffage in Europe, it may happen in certain cafes that fome remain through
the winter; fuch, for inftance, as are wounded,
or fick, or too young, or in fhort too feeble to
undertake the diftant voyage. Thefe Hoopoes,
which are thus left behind, will continue to lodge
in the fame holes where they neftled; they will
pafs the winter in a half-torpid ftate, requiring
little food, and being hardly able to repair the
lofs of feathers occafioned by moulting. Some
hunters, difcovering them in that condition,
have afferted that all the Hoopoes winter in hollow trees, benumbed, and divefted of plumage †, as has been faid of the cuckoo, with as
little foundation.

the fummits of mountains, to avoid man. Sportfmen affure me
that this bird will not fuffer them to get quite fo near it in autumn;
it having then, no doubt, acquired a little more experience.

　* Its refemblance, in its flight and in its creft and its fize, to the
lapwing, is certainly the caufe why the fame name Hoop has been
applied to both birds.

　† Albertus, and Schwenckfeld. It is for this reafon, fays Agri-
cola, that they are feen in the fpring almoft featherlefs.

According

According to fome, the Hoopoe was among the Egyptians efteemed an emblem of filial piety; they took care, it was faid, of their aged parents, cherifhed them under their wings, and in cafe of a tedious moulting, lent them af-fiftance in plucking the old feathers; they blew into their fore eyes, and applied healing herbs, and in a word repaid all the endearments they had received in their tender infancy. Some-thing of this kind has been alleged of the ftork. Would to God that we could give the fame amiable character of all other fpecies of ani-mals.

The Hoopoe lives only three years, accord-ing to Olina; but this muft be in the domeftic ftate, where the term of life is abridged by im-proper food. It would be difficult to determine the extreme age of the free wild Hoopoe, par-ticularly as it is a bird of paffage.

As it has a great abundance of feathers, it appears thicker than in reality. It is about as large as a thrufh, and it weighs from two ounces and a half to three or four, more or lefs, ac-cording to its plumpnefs *.

Its creft is longitudinal, confifting of two rows of equal and parallel feathers; thofe in the middle of each row are longer than the reft, fo that when they are erect they form a kind of

---

* " With all its feathers," fays Belon, " it looks like a very large pigeon, but when plucked it appears fcarce bigger than a ftare."

femi-

femi-circle of two inches and a half in height *.
All thefe feathers are rufous, terminated with
black; the middle ones, and thofe next them,
have a fhade of white between thefe two colours. There are alfo fix or eight feathers behind, which belong to the creft, and which are
entirely rufous, and are fhorter than the others.

The reft of the creft, and all the fore part of
the bird, are gray, verging fometimes on wine
colour, and fometimes on rufous; the fore part
of the back is gray, and the hind part is ftriped
tranfverfely with dirty white on a dark ground;
there is a white fpot on the rump; the fuperior
coverts of the tail are blackifh; the belly and
the reft of the under fide of the body are tawny
white; the wings and tail are black ftriped with
white; the ground of the feathers is flate coloured.

All thefe different colours, thus fpread over
the plumage, form together a fort of regular
picture, which has a good effect when the bird
erects its creft, expands its wings, and raifes
and difplays its tail; the part of the wings next
the body then fhews on each fide a black and
white crofs ftripe, perpendicular to the axis of
the body; the higheft of thefe ftripes has a
rufty caft, and joins a horfe-fhoe of the fame
colour traced on the back, the convex part of
which approaches the white fpot on the rump;

* Pliny, *Lib. X.* 29.

the

the loweft, which hems one half of the circumference of the wing, runs into another broader
bar, which croffes the fame wing two inches
from its tip, and parallel to the axis of the
body; this laft white ftripe correfponds alfo to
a crefcent * of the fame colour that interfects
the tail at an equal diftance from the end, and
forms the frame of the picture: laftly, if we
conceive the whole crowned by a raifed tuft of
gold colour edged with black, we fhall have a
much better idea of the plumage than could be
got by defcribing each feather feparately.

All the white bars which appear on the upper face of the wing appear alfo on the lower
face, fo that the bird has the fame afpect when
feen flying over head, except that the white is
lefs tarnifhed or mixed with rufty.

I have feen a female, difcovered to be fuch
by diffection, which had all the fame colours,
and thofe equally diftinct; perhaps it was of an
advanced age. It was rather larger than the
male, though the authors of the Italian Ornithology affert the contrary.

Total length about eleven inches; the bill
two inches and a quarter (more or lefs according to the age of the bird) flightly arched; the
tip of the upper mandible projects a little beyond that of the lower mandible, and they are
both pretty foft; the noftrils are oblong, and

+ When the tail is entirely fpread, this crefcent changes into a
ftraight bar.

hardly

hardly fhaded; the tongue is very fhort, almoft
loft in the gizzard, and forming a fort of equi-
lateral triangle, whofe fides are not three lines
in length; the ears are placed five lines from
the opening of the bill, and in the fame continuation; the tarfus is ten lines; the middle toe
is joined to the outer toe by the firft phalanx;
the hind toe is longer and ftraighter, efpecially
in old fubjects; the alar extent above feventeen
inches, the tail near four inches, confifting of
ten equal quills (and not twelve, as Belon afferts), and projecting twenty lines beyond the
wings, which have nineteen quills, the firft being the fhorteft, and the nineteenth the longeft.

The inteftinal tube, from the gizzard to the
anus, is twelve or eighteen inches; the gizzard
is mufcular, lined with a loofe membrane which
projects like a fcabbard into the *duodenum*; the
great diameter of the gizzard is from nine to
fourteen lines; the fmaller diameter from feven
to twelve lines, and thefe parts are larger in the
young birds than in the old ones. They have
all a gall bladder, though but flight veftiges of
a *cæcum*; at the angle of the bifurcation of the
trachea arteria, there are two holes covered by
a very fine membrane; the two branches of
the *trachea arteria* are formed behind by a fimilar
membrane, and before by cartilaginous femicircular rings; the elevator mufcle of the creft
is implanted between the crown of the head and
the bafe of the bill; when it is drawn back,
the

the tuft rifes, and when drawn towards the bill it collapfes.

In the female which I opened on the 5th of June, there were eggs of different fizes, the largeft of which was a line in diameter [A].

VARIETIES of the HOOPOE.

THE ancients faid that this bird was liable to change its colour in different feafons, which might be occafioned by moulting. But people who have reared Hoopoes have not perceived this alteration.

Belon mentions his knowing two fpecies, though he does not affign their difcriminating qualities; unlefs, perhaps, *the handfome collar, partly black and partly white*, and *the reverted neck*, which do not belong to our fpecies, were intended to mark the diftinction.

Commerfon and Sonnerat have brought a Hoopoe from the Cape of Good Hope very like ours, and which the traveller Kolben had found long before in the neighbourhood of the Cape. It has, upon the whole, the fame plumage, the fame fhape, the fame cry, the fame gait, and eats nearly the fame food; but on a clofer infpection it will be perceived that it is rather fmaller, its legs

[A] Specific character of the *Upupa-Epops:* " It is variegated and crefted."

longer,

longer, its bill fhorter in proportion, its tuft
lower, and that there is no trace of white on
the feathers that form the tuft; and in general,
there is lefs variety in its plumage.

In another fubject brought from the fame
country, the top of the back was of a pretty
deep brown, and the belly variegated with white
and brown; it was certainly a young one, for
it was fmaller than the reft, and its bill five lines
fhorter.

Laftly, the Marquis Gerini faw at Florence,
and again on the Alps, near the town of Ronta,
a very beautiful variety, whofe tuft was edged
with fky-blue.

FOREIGN BIRD,

WHICH IS RELATED TO THE HOOPOE.

The BLACK and WHITE HOOPOE
of the CAPE OF GOOD HOPE *.

Upupa Capenfis, Gmel.
The *Madagafcar Hoopoe*, Lath.

THIS bird is diftinguifhed from our Hoopoe, and its varieties, by its fize; by its fhort and pointed bill; by its creft, of which the fea- thers are lower in proportion, and alfo loofe, as in the tufted cuckoo of Madagafcar; by the number of quills in its tail, of which there are twelve; by the fhape of its tongue, which is pretty broad, and the extremity divided into many threads; and laftly, by the colours of its plumage. The creft, the throat, and all the under fide of the body, are white, without any fpots; the upper fide of the body, from the creft exclufively to the end of the tail, is brown, whofe fhades vary, and are much lefs intenfe on the fore parts; there is a white fpot on the

* The bird of Madagafcar, which Flacourt names *Tivouch*, feems to have fome affinity to this: its head is ornamented with a beauti- ful creft, and its plumage confifts of two colours, black and gray; we may fuppofe that this is light gray.

wing;

wing; the iris is of a bluiſh brown; the bill, the legs, and even the nails, are yellowiſh.

This bird inhabits the great foreſts of Mada-gaſcar, of the iſle of Bourbon, and of the Cape of Good Hope. In its ſtomach are found the ſeeds and berries of the *pſeudo-buxus*; its weight is four ounces, but varies much, and muſt be more conſiderable in the months of June and July, at which time the bird is very fat.

Total length ſixteen inches; the bill twenty lines, very pointed, the upper mandible having its edges ſcolloped near the tip, and its ridge very obtuſe; it is longer than the lower mandi-ble, which is as broad; on the palate, which in other reſpects is very ſmooth, there are ſmall tuberoſities, varying in number; the noſtrils are like thoſe of the ordinary Hoopoe; and ſo are the feet, except that the hind nail, which is the largeſt of all, is very hooked; the alar extent is eighteen inches; the tail four inches ſix lines, conſiſting of nearly equal quills, but the two middle ones are rather ſhorter; it pro-jects about two inches and a half beyond the wings, which have eighteen quills.

[A] Specific character of the *Upupa Capenſis:* " It is creſted and duſky, below white; a white ſpot on its wings."

The PROMERUPE.

Upupa Paradifea, Linn. Gmel. and Bor.
Upupa Manucodiata, Klein.
Promerops Indicus Criftatus, Briff.
Avis Paradifiaca, crifiata, orientalis, rarifima, Seba.
The *Crefted Promerops,* Lath.

THIS fpecies naturally affumes a place between the Hoopoes and the Promerops, fince it bears on its head a tuft of long feathers reclined, but which feem capable of being erected like thofe of the Hoopoe; while, on the other hand, the exceffive length of its tail marks an affinity with the Promerops.

Seba fays that it comes from the eaftern part of our continent, and that it is very rare; its throat, its neck, its head, and its beautiful large creft, are of a fine black; its wings and its tail are of a light bay colour; its belly light cinereous; its bill and legs lead colour: and the bird is nearly as large as a ftare.

Total length nineteen inches; the bill thirteen lines, a little arched, and very fharp; the tarfus about nine lines; the wings fhort; the tail fourteen inches and a quarter, confifting of very unequal quills, the two middle ones exceeding the lateral ones by eleven inches and the wings by thirteen.

[A] Specific character of the *Upupa Paradifea:* " It is crefted and chefnut; its two tail-quills very long."

5

The BLUE-WINGED PROMEROPS.

Upupa Mexicana, Gmel.
Promerops Mexicanus, Briff.
Avis Ani Mexicana, cauda longiffima, Seba.
The *Mexican Promerops*, Lath.

THIS Promerops is attached to lofty mountains : it feeds on caterpillars, flies, beetles, and other infects. The prevailing colour
of the upper part of its body is dull gray, changing into fea-green and purplifh-red ; the tail is
of the fame colour, but of a deeper fhade, and
having fine gold reflections; the quills of the
wings are of a light brilliant blue; the belly
light yellow ; the bill blackifh, edged with yellow.   The bird is of the fize of a thrufh.

Total length eighteen inches and three quarters; the bill twenty lines, fomewhat arched;
the tarfus eight lines and a half; the wings fhort ;
the tail twelve inches and a half, confifting of
very unequal quills, the four middle ones being
longer than the lateral ones; it exceeds the
wings eleven inches,

The BROWN PROMEROPS
with a Spotted Belly.

Upupa Promerops, Linn. and Gmel.
Merops Cafer, Linn. and Gmel.
Promerops, Briff.
The *Cape Promerops*, Lath.

THE belly is fpotted with brown upon a whitifh ground, and the breaft fpotted with brown upon an orange-brown ground; the throat is dirty white, having on each fide a brown line, which rifes from the opening of the bill, paffes under the eye, and defcends upon the neck; the crown of the head is brown, variegated with rufty gray; the rump and the fuperior coverts of the tail are olive green; the reft of the upper fide of the body, including the quills of the tail and of the wings, are brown; the thighs are brown; the inferior coverts of the tail are of a fine yellow; the bill and legs black.

The one figured, No. 637, *Planches Enluminées*, appears to be the male, fince it is more fpotted, and its colours better contrafted; there is a very narrow gray ftripe on the wings, formed by a fucceffion of fmall fpots that terminate the upper coverts. The fubject defcribed by Briffon wants this ftripe, its colours are feebler, and the under fide of its body is lefs fpotted; I fuppofe it to be a female; it was an eighteenth part lefs than the male, and was fcarcely larger than a lark.

Total length of the male eighteen inches; the bill fixteen lines; the tarfus ten lines and two-thirds; the wings fhort; the alar extent thirteen inches; the tail thirteen inches, confifting of twelve quills, of which the fix middle ones are much longer than the fix lateral ones, which are tapered; it exceeds the wings eleven inches [A].

The STRIPED-BELLIED BROWN PROMEROPS.

Merops Fufca, Gmel.
The *New Guinea Brown Promerops*, Lath.

THIS bird was brought from New Guinea by Sonnerat. In the male the throat, the neck, and the head, are of a fine black, that on the head gloffed like burnifhed fteel; all the upper part of the body is brown, with a tinge of deep green on the neck, back, and wings; the tail is of a more uniform and lighter brown, except the laft of the lateral quills, which is black on the infide; the breaft and all the under fide of the body are ftriped tranfverfely with black and white; the iris and legs are black.

[A] Specific chara&er of the *Upupa Promerops:* " It has fix tail-quills, the middle ones very long."

I have

THE RING-BELLIED PROMEROPS

I have feen one which had a rufous fhade on the head. In the female, the throat, the neck, and the head, are of the fame brown with the upper fide of the body, and without any reflections; in every other refpect it refembles the male.

Total length twenty-two inches; the bill two inches and a half, ftraight, round, and very much arched ; the tail is thirteen inches, confifting of twelve tapered quills, very unequal, the fhorteft being four inches, and the longeft exceeding the wings nine lines.

The GREAT PROMEROPS, with Frizled Flounces *.

Upupa Magna, Gmel.
Upupa Superba, Lath. Ind.
The *Grand Promerops*†, Lath. Syn.

THE frizled flounces which at once decorate and characterize this fpecies‡, confift of two thick tufts of frizled foft feathers, painted

---

* *Paremens,* i. e. Protuberant decorations in general.

† *Voyage a la Nouvelle Guinée,* p. 166. The name of *four-winged,* which has been given by voyagers to an African bird of prey, would agree very well with this Promerops.

‡ The whiftler, defcribed in a former part of this work, has alfo a fort of flounces, but neither their form, nor the feathers of which they confift, are the fame ; and thofe of the fuperb paradife bird have a contrary direction.

with

with the moſt beautiful colours, which projeĉt
on either ſide of the body, and give the bird a
diſtinguiſhed figure. Theſe bunches of plumage are compoſed of the long coverts of the
wings, which are nine in number, that riſe
bending on their upper ſide, where the feathery
fibres are very ſhort, and diſplay with more advantage the long fibres of the under ſide, which
now becomes the convex ſide; the middle coverts of the wings, of which there are fifteen,
and even ſome of the ſcapular feathers, partake
of this ſingular arrangement, and riſe into a fan-
ſhape, their extremity ornamented with an edging of brilliant green, changing into blue and
violet, which forms a kind of garland on the
wings, ſpreading ſomewhat as it riſes to the back.

In all the reſt of the plumage the prevailing
colour is gloſſy black, enriched with blue and
violet refleĉtions; and all the feathers, ſays Sonnerat, have the ſoftneſs of velvet, not only to
the eye, but to the touch: he adds that the
body, though of a long ſhape, appears ſhort and
exceedingly little, compared with the great extent of its tail; the bill and legs are black. Sonnerat brought this bird from New Guinea.

Total length three feet and a half (four according to Sonnerat); the bill near three inches;
the wings ſhort; the tail twenty-ſix or twenty-
ſeven inches, conſiſting of twelve tapered quills,
which are broad and pointed, the ſhorteſt being
ſix or ſeven inches, the longeſt exceeding the
wings about twenty inches.

The ORANGE PROMEROPS.

Upupa Aurantia, Gmel.
Promerops Barbadenfis, Briff.
Avis Paradifiaca Americana elegantiffima, Seba.

THE prevailing colour is orange, which receives different tints in different parts ; a
gold tint on the throat, the neck, the head, and
the bill; a reddifh tint on the quills of the tail
and on the great quills of the wings ; and laftly, a yellow tint on all the reft of the plumage;
the bafe of the bill is furrounded with fmall red
feathers.

Such, I conceive, to be the male of this fpecies, which is nearly as large as the ftare; I
reckon the cochitototl * of Fernandez to be the
female, which is of the fame fize, inhabits the
fame country, and whofe plumage differs not more
from the Orange Promerops than in many fpecies
the plumage of the male differs from that of the
female. The throat, the neck, the head, and
the wings, are variegated, without any regu-
larity, with cinereous and black ; all the reft of
the plumage is yellow; the iris is pale yellow ;
the bill is black, flender, arched, very pointed ;
and the legs are cinereous. The bird lives upon
feeds and infects, and is found in the hotteft

* *Upupa Aurantia*, Var. Gmel.
Promerops Mexicanus Luteus, Briff.

D d 3   parts

parts of Mexico, where it is neither efteemed
for the beauty of its fong nor the delicacy of its
flefh. The orange Promerops, which I fuppofe to be the male of the fame fpecies, occurs
in the north of Guiana, in the fmall iflands
formed at the mouth of the river Berbice *.

Total length of the bird about nine inches
and a half; the bill thirteen lines; the tarfus
ten; the tail near four inches, confifting of
equal quills, and exceeds the wings about an
inch [A].

* Seba fays, *in infulis Barbicenfibus,* which I think fhould be
tranflated the iflands of Berbice, and not the iflands of Barbadoes.

[A] Specific charaƈter of the *Upupa Aurantia:* " It is fulvous;
its head and neck gold-coloured; its tail equal."

The B A K E R.

Le Fournier, Buff.
Merops Rufus, Gmel.
The *Rufous Bee-eater*, Lath.

THIS is the name which Commerſon has given to this American bird, which forms the ſhade between the Promerops and the Bee-eaters. It differs from the Promerops, as its toes are longer and its tail ſhorter: it differs from the Bee-eaters, becauſe it has not, like them, its outer toe joined and as it were ſoldered into the middle toe almoſt its whole length. This bird is found in Buenos Ayres.

Rufous is the prevailing colour of its plumage, which is deeper on the upper parts, much lighter and verging on pale yellow on the lower parts; the quills of the wing are brown, with ſome rufous tints, more or leſs intenſe, on the outer edge.

Total length eight inches and a half; the bill twelve or thirteen lines; the tarſus ſixteen lines; the hind nail the ſtrongeſt; the tail rather leſs than three inches, and exceeds the wings about an inch.

[A] Specific charaĉter of the *Merops Rufus*: " It is rufous; its wing-quills brown, rufous on their outer edge."

The POLOCHION*.

Merops Moluccensis, Gmel.
The *Molucca Bee-eater*, Lath. Syn.

POLOCHION is the name, and the inceſſant cry, of this Molucca bird; it ſits on the higheſt branches and continually repeats it, and this word, in the language of thoſe iſlands, invites to love and pleaſure. I range it between the families of the Promerops and of the Bee-eaters, becauſe it has the bill of the latter, and the feet of the former.

All its plumage is gray, but this colour is deeper on the upper parts, and lighter on the under; the cheeks black; the bill blackiſh; the eyes encircled by a naked ſkin; the back of the head variegated with white; the feathers of the tuft make a re-entrant angle on the front, and thoſe at the origin of the neck terminate in a kind of ſilk. The ſubject which Commerſon deſcribed came from the iſland of Bouro, one of the Moluccas belonging to the Dutch; it weighed five ounces, and was nearly as large as the cuckoo.

Total length fourteen inches; the bill very pointed, two inches long, five lines broad at its

* This word, in the language of the Moluccas, ſignifies *let us kiſs*; and M. Commerſon therefore propoſes to call it *Philemon*, or *Philedon*, or *Deoſculator*. I think it better to retain the original name, eſpecially as it expreſſes the cry of the bird.

MEROPS. 409

base, two lines at its middle, and seven lines
thick at its base, three and a half at the middle,
its edges scallopped near the point ; the nostrils
oval and open, invested by a membrane behind,
and placed nearer to the middle of the bill than
to its base ; the tongue equal to the bill, terminated by a pencil of hair ; the middle toe
joined at its base to the outer toe ; the hind one
strongest ; the alar extent eighteen inches ; the
tail five lines and two thirds, consisting of twelve
quills, which are equal, except that the outer
pair are rather shorter than the rest ; it projects
three inches beyond the wings, which consist
of eighteen quills ; the outer one is one half
shorter than the three following, which are the
longest of all [A].

The RED and BLUE MEROPS.

Merops Brasiliensis, Gmel.
Apiaster Brasiliensis, Briss.
Pica Brasiliensis amænissimis coloribus, Seba.
The *Brasilian Bee-eater*, Lath.

SEBA, from whom we borrow the account of
this bird, seems to have been charmed with
its plumage. Ruby colour sparkles on its head,
on its throat, and on all the under side of the

[A] Specific character of the *Merops Moluccensis*: " It is gray ;
its orbits naked ; its cheeks black ; its tail nearly equal."

body ;

body; it alſo appears on the upper coverts of the wings, but of a deeper hue; a light brilliant blue is ſpread on the quills of the wings and on thoſe of the tail; the luſtre of theſe fine colours is heightened by the contraſt of darker ſhades, and by black and white ſpaces ſcattered on the upper ſurface; the bill and legs are yellow, and the wings are lined with the ſame colour; the red feathers of the under ſide of the body are of a ſilky nature, as ſoft to the feel as they are brilliant to the eye.

This bird is a native of Brazil, if we believe Seba, who in matters of this kind can hardly ever be relied on. It is nearly as large as the Bee-eater; its legs too are as ſhort, but I can perceive nothing either in the deſcription or figure that ſhews the toes to be placed in the ſame way: its bill is more analogous to that of the Promerops, for which reaſon I make it an intermediate ſpecies.

[A] Specific character of the *Merops Braſilienſis:* " It is fire-coloured, above variegated with brown and black; its tail and wing-quills pale blue."

THE COMMON BEE-EATER.

The BEE-EATER.

Le Guepier *, Buff.
Merops Apiaster, Linn. and Gmel.
Merops, Gesner. Aldrov. Ray, &c.
Merops Galilæus, Haffelquist.
Apiaster, Briff.
Ispida cauda molli, Kram.
Gnat-Snapper, Kolben †.

THIS bird feeds not only upon common bees and wasps, but also upon humble-bees, locusts, gnats, flies, and other insects, which it catches like the swallows on the wing. Such are the prey to which it is most attached, and which serve the boys of the island of Candia as baits for lines to fish it in the air ; they pass a bent pin through the body of a living locust, and fasten to it a long thread ; the Bee-eater flies at it, and swallows it with the hook. When insects fail, it contents itself with small seeds, and even wheat ‡ ; and, in collecting
that

---

* i. e. *Wasp-eater.*

† Aristotle calls the Bee-eater Μεροψ, which Pliny writes in Roman characters *Merops :* it was also termed Αεροψ, Φλωρος, Μελισσοφας, contracted for Μελισσοφαγος (honey-eater), and in Latin *Apiaster,* from *Apis,* a bee. In Italy it has the appellations, *Dardo, Dardaro, Barbaro, Gaulo, Jevolo, Lupo dell' Api* (bee-wolf) : In Sicily, *Piccia Ferro* (iron-bill) : In Spain, *Juruco :* In Germany, *Bienen-Fresser* (bee-eater) ; *Heu-Vogel* (hay-bird) ; and, *Gelber Bienen-Wolf* (the yellow bee-wolf) : In Austria, *Meer-schwalbe* (the sea-swallow) : In Poland, *Zotna, Zotcawa.*

‡ The only one I had occasion to open with Dr. Remond had
five

that food on the ground, it feems alfo to gather
fmall pebbles like all the granivorous birds, and
with the fame view. Ray fufpects, from many
analogies, both internal and external, that the
Bee-eater, as well as the king-fifher, feeds
fometimes on flefh.

The Bee eaters are very common in the ifland
of Candia, infomuch that Belon, who was an
eye-witnefs, fays that they are feen flying in
every part of it. He adds that the Greeks on
the main land are unacquainted with it, which
he could accurately learn from his travelling in
that country; but he afferts, on too flight foun-
dations, that they are never feen in Italy; for
Aldrovandus, who was a citizen of Bologna, affures that they were common in the neighbourhood of that city, where they were ufually
caught both with nets and lime-twigs. Willughby faw them frequently at Rome, expofed
to fale in the public markets; nor is it probable
that they are ftrangers to the reft of Italy, fince
they are found in the fouth of France, where
they are not regarded even as birds of paffage *.
Thence they fometimes penetrate in fmall flocks

five large drones in its throat: Belon found, in the ftomach of
thofe which he opened, rape, parfley, and colewort feeds, wheat,
&c.

  * Belon doubts whether they remain the whole winter in the ifle
of Candia, but he had no obfervation on that head. What I have
faid of thofe of Provence was communicated by the Marquis de
Piolenc. I know not why Frifch fays that thefe birds are fond of
deferts.

of

of ten or twelve into the more northern provinces ; and we faw one of thefe flocks that had
arrived in the vale of Sainte-Reine in Burgundy,
on the 8th of May 1776 : they kept conftantly
together, and called inceffantly on each other ;
their cry was very noify but agreeable, and re-
refembled fomewhat the whiftling that one
might make with a bored nut * ; they emitted
it both when perched and when on the wing ;
they preferred the fruit trees which were then
in bloffom, and confequently frequented by the
bees and wafps ; they often dived from the
branch to catch the little winged prey ; they
appeared always very timorous, and fcarce fuffered a perfon to get near them : however one
was fhot feparate from the others, perched upon
a fir ; the reft of the flock, which were in a
neighbouring vineyard, frighted at the report,
flew away all fcreaming together, and took
fhelter among fome chefnuts that were at a little diftance ; they continued to harbour among
the vineyards, but in a few days they took their
final departure.

* Belon compares it to " the found that a man would make by
contracting his mouth into a round aperture and whiftling *grulgru-*
rurul as loud as an oriole." Others pretend that it feems to fay
crou, crou, crou. The author of the poem Philomela reprefents its
fong as refembling much that of the gold crefted wren and of the
fwallow.

  Regulus atque Merops & rubro pectore Progne
   Confimili modulo zinzibulare folent.

But it is well known that almoft always the naturalift muft in fome
meafure modify the expreffions of the poet,

Another

Another flock was feen in June 1777, in the
vicinity of Anfpach *. Lottinger informs me
that thefe birds feldom appear in Lorraine, that
there are never more than two together, that
they fit on the longeft branches of trees and
fhrubs, and feem to feel embarraffed, as if they
had ftrayed. They appear ftill feldomer in Sweden, where they haunt the fea-coaft †. But
they hardly ever vifit England, though not fo far
north as Sweden, and to which they could eafily
pafs from Calais ‡. In the eaft, they are fpread
through the temperate zone, from India § to
Bengal ‖, and undoubtedly farther, though their
courfe has not been traced.

Thefe birds neftle, like the fhore fwallow and
the king-fifher, in the bottom of holes, which
they form with their fhort and ftrong feet and
their iron bill, as the Sicilians term it, in little
hillocks where the foil is loofe, and fometimes
in the fhelving fandy brinks of large rivers +:
thefe holes are made more than fix inches deep,
and as wide. The female depofits, on a bedding
of mofs, four or five, or even fix or feven white
eggs, rather fmaller than thofe of the blackbird.
But their economy in thefe dark caverns cannot
be obferved; we know only that the young fa-

---

* La Gazette d'Agriculture, No. 55, *année* 1777.
† Fauna Suecica.   ‡ Charleton and Willughby.
§ M. Haffelquift fays that they occur in the woods and plains
between Acre and Nazareth.
‖ Edwards.   + Ariftotle, and Kramer.

mily

mily does not difperfe; indeed feveral families
muft unite to form thofe numerous flocks which
Belon faw in the ifland of Candia, fettled among
the ridges of the mountains, where the abundance of thyme affords rich pafture to the bees
and wafps.

The flight of the Bee-eater has been compared to that of the fwallow, which we have
feen to refemble it in many other refpects; it is
alfo analogous to the king fifher, particularly in
the beautiful colours of its plumage, and in the
fingular conformation of its feet; and laftly Dr.
Lottinger, who is a clofe and accurate obferver,
finds that, in fome particulars, it is akin to the
goat-fucker.

A property which, were it well afcertained,
would diftinguifh this bird from every other, is
the habit, afcribed to it, of flying backwards.
Ælian mightily admires this *; but he had better called it in queftion, for it is an error arifing
from fome overfight. Such too is the filial piety
that has been fo liberally beftowed on birds, but
moft remarkably on the Bee-eater; fince, if
we believe Ariftotle, Pliny, Ælian, and thofe
who have copied them, the young ones do
not wait till the parents need their affiftance;
as foon as they are flown they give a cheerful
attendance, and carry provifions to their holes.
It is eafy to fee that thefe are fables, but the
moral at leaft is good.

* *De Nat. Anim.* Lib. I. 49.

The

The male has fmall eyes, though of a vivid
red, and which derive additional luftre from a
black bar; the front is of a fea-green; the upper fide of the head is chefnut tinged with
green; the hind part of the head and of the
neck is chefnut, without any admixture, but
which grows continually more dilute as it approaches the back; the upper fide of the body
is of a pale fulvous, with green and chefnut reflections, which are more or lefs apparent, according to the pofition; the throat is of a fhining gold-yellow, terminated in fome fubjects by
a blackifh collar; the fore part of the neck, the
breaft, and the under fide of the body, are of a
blue beryl, which grows lighter on the hind
parts; the fame colour is fpread over the tail
with a light rufous tinge, and on the outer edge
of the wing without any admixture; it runs into
green, and receives a fhade of rufous on the
part of the wings next the back; almoft all the
quills are tipt with black, their fmall fuperior
coverts are tinged with dull green, the middle
ones with rufous, and the great ones fhaded with
green and rufous: the bill is black, and the legs
reddifh brown (black according to Aldrovandus); the fhafts of the quills of the tail are
brown above and white below. Befides, all
thefe different colours are very variable, both in
their tint and their diftribution; and hence the
difference among defcriptions.

This bird is very nearly as large as the
redwing,

redwing, its fhape longer, and its back rather
more convex. Belon fays that nature has made
it hunch-backed.

Total length ten or twelve inches; the bill
twenty-two lines, broad at its bafe, a little
arched; the tongue thin, terminated by long
threads, the noftrils fhaded by a fort of rufty
hairs; the tarfus five or fix lines, and pretty
thick in proportion to its length; the outer toe
adheres to the middle one almoft its whole
length, and to the inner one by its firft phalanx only, as in the king-fifher; the hind nail
is the fhorteft of all and the moft hooked; the
alar extent fixteen or feventeen inches; the
tail four inches and a half, confifting of fix pairs
of quills, of which the five lateral ones are
equal; the middle pair projects nine or ten lines
beyond them, and about eighteen lines beyond
the wings, which confift of twenty-four quills,
according to fome, and of twenty-two according to others: the one I obferved contained
twenty one quills.

The œfophagus three inches long, and dilates
at its bafe into a glandulous bag; the ftomach
is rather membranous than mufcular, and of
the fize of an ordinary nut; the gall bladder is
large and of an emerald colour; the liver is pale
yellow: there are two *cæca*, the one fifteen
lines, the other fixteen and a half; the intef-

tinal tube could not be meafured, being too
much injured by the fhot [A].

YELLOW and WHITE BEE-EATER.

Merops Flavicans, Gmel.
Manucodiata Secunda Aldrovandi, Ray. and Will.
Apiafter Flavicans, Briff.
The *Yellow Bee-eater,* Lath.

ALDROVANDUS faw this fpecies at Rome; it
is remarkable for the length of the two
middle quills of its tail, and the proportional
fhortnefs of its bill; its head is white, variegated with yellow and gold colour; its eyes yellow; its eye-brows red; its breaft reddifh; its
neck, its belly, and the under fide of its wings,
are whitifh; its back yellow; its rump, its tail,
and its wings, are of a bright rufous; its bill is
greenifh-yellow, fomewhat arched, two inches
long; and its tongue is long, and pointed nearly
like that of woodpeckers.

This bird was much larger than the ordinary Bee-eater, and its alar extent was twenty
inches; the two middle quills projected eight
lines beyond the lateral ones. The Signior
Cavalieri, to whom it belonged, was uncertain
what country it commonly inhabits.

[A] Specific character of the common Bee-eater, *Merops Apiafter:* " Its back is ferruginous, its belly and tail bluifh-green, two
of the tail-quills longer than the reft, its throat yellow."

The GRAY-HEADED BEE-EATER.

Merops Cinereus, Linn. Gmel. and Klein.
Apiafter Mexicanus, Briff.
Avicula de Quaubcilui, Seba.
The *Cinereous Bee-eater,* Lath.

PERHAPS this bird has nothing elfe American but the Mexican name *quaubicilui,* which Seba has been pleafed to beftow upon it. It is as large as the fparrow of Europe, and is included in the genus of the Bee-eaters on account of the length and fhape of its bill, the length of the two middle quills of the tail, and by the thicknefs and fhortnefs of its legs. It probably. refembles alfo in the difpofition of its toes.

Its head is of a fine gray ; the upper fide of its body the fame, variegated with red and yellow ; the two long middle quills of its tail are pure red ; its breaft and all the under fide of its body are orange yellow, and the bill is of an handfome green.

Total length nine or ten inches ; the bill and tail occupy the one half of it.

[A] Specific charaćter of the *Merops Cinereus*: " It is variegated with red and yellow, below reddifh yellow ; two of its tail-quills very long and red."

THE

GRAY BEE-EATER of ETHIOPIA.

Merops Cafer, Linn. and Gmel.

LINNÆUS is the only naturalist who has taken notice of this species, which he does from a drawing of Burmann. His indication, to which I can add nothing, is, that the plumage is gray; that there is a yellow spot near the anus; and that its tail is very long.

THE

CHESNUT and BLUE BEE-EATER.

Merops Badius, Gmel.
Merops Castaneus, Lath. Ind.
Apiaster ex Franciæ Insula, Briss.
The *Chesnut Bee-eater,* Lath. Syn.

CHESNUT predominates on the anterior parts of the upper side of the body, including the top of the back, and beryl on the rest of the upper side of the body, and on all the lower part, but which is much more beautiful and more conspicuous on the throat, the fore part of the neck, and the breast, than any where else; the wings are green above, fulvous below, terminated with blackish; the tail is of a pure blue; the bill black; and the legs reddish.

This

This bird is found in the Ifle of France ; it is hardly larger than the crefted lark, but much longer.

Total length near eleven inches; the bill nineteen lines ; the tarfus five and a half; the hind toe the fhorteft of all ; the alar extent fourteen inches ; the tail five inches and a half, confifting of twelve quills, of which the two middle ones project two inches and two lines beyond the lateral ones, and three inches and a half beyond the wings ; thefe confift of twentyfour quills, of which the firft is the fhorteft, and the third the longeft.

VARIETY.

THE Chefnut and Blue Bee-eater of Senegal is a variety produced by climate. No more than thefe two colours are found in the whole of its plumage, but their diftribution is different from that of the preceding. The chefnut is fpread on the coverts and the quills of the wings, except the quills next the back, and on the quills of the tail, except the projecting part of the two middle ones, which is blackifh.

This Bee-eater is found in Senegal, whence it was brought by Adanfon. Its total length is about a foot, and it has nearly the fame proportions as that from the Ifle of France.

The PATIRICH.

Merops Superciliosus, Linn. and Gmel.
Apiaster Madagascarienfis, Briff.
The *Supercilious Bee-eater,* Lath.

THE natives of Madagafcar call this bird *Patirich tirich,* which is manifeftly formed from its cry, and which I have fhortened and retained. The principal colour of its plumage is dull green, changing into brilliant chefnut on the head, not fo dark on the upper fide of the body, growing more dilute on the hind parts, ftill lighter on the lower parts, and continually melting away towards the tail: the wings are terminated with blackifh; the tail is dull green; the throat is yellowifh white at its origin, and fine chefnut at its lower part. But what beft characterizes this bird and gives it a fingular afpect, is a broad blackifh bar, edged round its whole circumference with greenifh white; this border bends about the bafe of the bill and grafps the origin of the neck, affuming a yellowifh tinge, as I have before faid; the bill is black, and the legs are brown. This bird is found in Madagafcar; it is rather larger than the chefnut and blue Bee-eater.

Total length eleven inches and one-third; the bill twenty-one lines; the tarfus five lines; the hind toe the fhorteft; the alar extent fif-

teen

teen inches and two thirds; the tail five inches
and a half, confifting of twelve quills; the two
middle ones projeƈt more than two inches beyond the lateral ones, and two inches and three
quaƈters beyond the wings, which confift of
twenty-four quills, of which the firft is very
fhort, and the twelfth is the longeft.

I have feen another Bee-eater from Madagafcar, much like this in regard to the fize, the
colours of the plumage, and their diftribution,
though lels contrafted; the bill was weaker, and
the two middle quills of the tail exceeded not
the lateral ones. It was undoubtedly a variety
occafioned by age or fex; its bar was edged with
beryl, and the rump and tail were of the fame
colour as in the fubjeƈt brought home by Sonnerat; but in the latter, the two middle quills
of the tail were very narrow and much longer
than the lateral ones [A].

[A] Specific charaƈter of the *Merops Superciliofus :* " It is green,
" a white line on its front above and below the eyes, its throat
" yellowifh, two of its tail quills elongated."

The GREEN BLUE-THROATED BEE-EATER.

Merops Viridis, Linn. Gmel. and Bor.
Apiafter Madagafcarienfis Torquatus, Briff.
The *Indian Bee-eater,* Edw. Penn. and Lath.

A LITTLE accident which happened to a bird of this fpecies, long after it was dead, affords an inftance of the miftakes which are apt to embarrafs the nomenclature. It belonged to Mr. Dandridge, and was defcribed, delineated, engraved, and coloured by two Englifh naturalifts, Edwards and Albin: a Frenchman, well fkilled in ornithology, and though he had a fpecimen befide him, has fuppofed that thefe two figures have reprefented two diftinct fpecies, and has in confequence defcribed them feparately and under different denominations.

The bird of Mr. Dandridge obferved by Edwards was one-third fmaller than the European Bee-eater, and the two middle quills of its tail were much longer and narrower; the front was blue, there was a great fpot of the fame colour on the throat, included in a fort of black frame formed below by a half-collar like a reverfed crefcent, and above by a bar which paffed over the eyes and defcended on both fides of the neck, ftretching towards the two extremities of the half collar; the upper furface of the head and neck was orange; the back, the fmall coverts, and the laft
quills

quills of the wings, were green, like the plumage
of the parrot; the fuperior coverts of the tail
were beryl blue; the breaft and belly were light
green; the thighs reddifh brown; the inferior
coverts of the tail dull green; the wings variegated with green and orange, and terminated
with black; the tail of a fine green above and
dark green below; the two middle quills exceeding the lateral ones by more than two
inches, and the projecting part deep brown and
very narrow; the fhafts of the quills of the tail
very brown, and fo were the legs; the bill,
black above, and whitifh below, at its bafe.

In the fubject defcribed by Briffon, which is
alfo delineated in the *Planches Enluminées*, there
was no blue on the front, and the green of the
under fide of the body partook of the beryl
caft; the upper fide of the head and of the
neck was of the fame gold green as the back;
in general, there was a tint of gold yellow thrown loofely on the whole of the plumage, except on the quills of the wings and
the fuperior coverts of the tail; the black
bar did not extend acrofs the eyes, but below them. Briffon has remarked befides, that
the wings were lined with fulvous, and that the
fhafts of the tail, which were brown above, as
in Edwards' bird, were whitifh beneath. Laftly,
there were feveral quills and coverts of the
wings, and many quills of the tail, edged near
the end and tipt with gold yellow. But it is

obvious that all thefe minute differences are not more than might be expected in individuals of even the fame fpecies, but only diverfified by age or fex; the flight variation of fize may be imputed to the fame caufes.

The bird called by Briffon *the little Philippine Bee-eater*, is of the fame fize and plumage with the collared Bee-eater of Madagafcar; the chief difference remarked between them is, that in the former the two middle quills of the tail, inftead of being longer than the lateral ones, are, on the contrary, rather fhorter. But Briffon himfelf fufpects that thefe middle quills were not yet fully grown, and that in thofe fubjects where they were complete they projected far beyond the lateral ones: this is the more probable, as thefe two middle quills appear, in the prefent cafe, to be different from the lateral ones, and even nearly akin to the projecting part of the middle quills in the blue-throated green Bee-eater. The other differences are thefe; that the bar was not black, but of a dull green, and that the legs were brown red: but ftill it ought to be referred to the fame fpecies. This bird is fpread from the coaft of Africa to the moft eaftern of the Afiatic iflands; it is nearly as large as our fparrow.

Total length fix inches and a half, (probably it would be about eight inches and three quarters, as in the blue-throated green Bee-eater, if the two middle quills had been fully grown) the bill

bill fifteen lines; the tarfus four lines and a half; the alar extent ten inches; the ten lateral quills of the tail two inches and a half, exceeding the wings fourteen lines [A].

The GREEN and BLUE YELLOW-THROATED BEE-EATER, *Buff*.

Merops Chryfocephalus, Gmel.
The *Yellow-throated Bee-eater*, Lath.

THIS is a new fpecies introduced by Sonnerat. It is diftinguifhed from the preceding in its plumage, its proportions, and above all, in the length of the middle quills of the tail; its throat is of a fine yellow, which extends on the neck under the eyes, and even farther, and is terminated with blue in its lower part; the front, the eve-brows, and all the under part of the body, are glaucous; the quills of the wings are green, edged with glaucous from their middle; their fmall fuperior coverts are dun green, fome fnuff-coloured, the longeft next the body are of a light yellow; the upper fide of the head and neck is fnuff-coloured; all the upper fide of the body gold green; the fuperior coverts of the tail green.

[A] Specific character of the *Merops Viridis*: " It is greenifh, a black ftripe on its breaft, its throat and tail blue, two of its tail quills elongated."

Total

Total length ten inches; the bill twenty
lines; the tarſus ſix lines; the hind nail the
ſhorteſt and moſt hooked; the tail four inches
and a quarter, conſiſting of twelve quills, the
ten lateral ones nearly equal to each other; the
two middle ones exceed the lateral ones by ſeven
or eight lines, and the wings by eighteen.

The LITTLE GREEN and BLUE TAPER-TAILED BEE-EATER.

Merops Angolenſis, Gmel.
Apiaſter Angolenſis, Briſſ.
The *Angola Bee-eater*, Lath.

ITS ſmallneſs is not the only property that
diſtinguiſhes this from the preceding; it
differs alſo in the colour of its head, in its proportions, and, above all, by the conformation of
its tail, which is tapered, and of which the
two middle quills do not project much. With
regard to its plumage, the upper ſurface is gold
green, the under beryl blue; the throat is yellow; the fore part of the neck, cheſnut; there
is, acroſs the eyes, a zone dotted with black;
the wings and tail are of the ſame green as the
back; the iris is red; the bill black, and the legs
cinereous:—Theſe are the chief colours of this
bird, which is the ſmalleſt of the Bee-eaters. It
is

is found in the kingdom of Angola in Africa; it is the only one of the genus that has a tapered tail.

Total length about five inches and a half; the bill nine lines; the tarfus four lines and a half; the hind toe the fhorteft; tail two inches and more, confifting of twelve quills; it exceeds the wings about an inch.

The AZURE-TAILED GREEN BEE-EATER, *Buff*.

Merops Philippinus, Gmel.
Apiafter Philippinenfis Major, Briff.
The *Philippine Bee-eater*, Lath.

ALL the upper furface of the head and body is of a dull green colour, changing into rofe copper; the wings are of the fame colour, terminated with blackifh, lined with light fulvous; the nineteenth and twentieth quills, marked with glaucous on the outfide, and the twenty-fecond and twenty-third, on the infide. All the quills and coverts of the tail are of a beryl blue, which is lighter on the inferior coverts; there is a blackifh bar on the eyes; the throat is yellowifh, verging on green and fulvous; this laft tint is more intenfe below; the under fide of the body and the thighs are of a yellowifh green changing into fulvous; the bill is
black,

black, and the legs brown.   This bird is found
in the Philippines, and is larger than the common Bee-eater.

Total length eight inches and ten lines; the
bill twenty-five lines; the angle of its aperture
at a confiderable diftance from the eye; the
tarfus five lines and a half; the hind toe the
fhorteft; the alar extent fourteen inches and ten
lines; the tail three inches and eight lines, confifting of twelve quills nearly equal, and it projects eleven lines beyond the wings, which have
only twenty-four quills, the firft being the
fhorteft, and the fecond the longeft of all [A].

THE

BLUE-HEADED RED BEE-EATER.

Merops Nubicus, Gmel.
The *Blue-headed Bee-eater*, Lath.

A FINE beryl glows on the head and on the
throat, where it becomes deeper, and alfo
on the rump and on all the coverts of the tail;
the neck, and all the reft of the under fide of
the body, as far as the legs, are crimfon, fhaded
with rufous; the back, the tail, and the wings,
are brick colour, which is dunner on the coverts

[A] Specific character of the *Merops Philippinus:* " It is green,
below yellowifh, its rump blue, its tail equal."

of

of the wings; the three or four quills of the
wings neareſt the back are of a brown green,
with bluiſh reflections; the great quills terminated with bluiſh gray, melted with red; the
middle ones are of a blackiſh brown; the bill
black, and the legs light cinereous. This is a
new ſpecies found in Nubia, where it was delineated by Mr. Bruce; it is not quite ſo large
as the European ſpecies.

Total length about ten inches; the bill twenty-
one lines; the tarſus ſix lines; the hind toe the
ſhorteſt; the tail about four inches, a little
forked, and it exceeds the wings about twenty-
one lines [A].

The RED and GREEN BEE-EATER of SENEGAL *.

Merops Erythropterus, Gmel.
The *Red-winged Bee-eater*, Lath.

THE upper ſurface of the head and body, including the ſuperior coverts of the wings
and thoſe of the tail, is dun-green, browner on
the head and back, lighter on the rump and the
ſuperior coverts of the tail; there is a dark ſpot

[A] Specific character of the *Merops Nubicus*: " It is blue green,
below red; its back, its wings, and its forked tail, brick colour."
* We owe this ſpecies to M. Adanſon. The deſcription and
figure are as accurate as they could be made from the ſkin of the
bird dried and prepared between two leaves of paper.

8 behind

behind the eye; the quills of the tail and of the wings are red, terminated with black; the throat is yellow; all the under furface of the body is dirty white; the bill and legs black.

Total length about fix inches; the bill one inch; the tarfus three lines and a half; the tail two inches, and it exceeds the wings about one inch [A].

The RED-HEADED BEE-EATER.

Merops Erythrocephalus, Briff.
Apiafter Indicus Erythrocephalus, Briff.

IF the name, *cardinal*, can ever be applied to any of the bee-eaters, it certainly belongs to the prefent; for it has a fort of hood that covers, not only the head, but alfo a part of the neck: it has alfo a black bar on the eyes; the upper fide of the body is of a fine green; the throat yellow; the under fide of the body light orange; the inferior coverts of the tail yellowifh, edged with light green; the tail is green above, cinereous below; the iris red, the bill black, and the legs cinereous.

This bird is found in the Eaft Indies, and is

[A] Specific character of the *Merops Erythropterus:* " It is olive, below partly whitifh, its throat bright yellow; its wings and tail red, tipt with black."

nearly

nearly as large as the blue-throated green bee-eater.

Total length fix inches; the bill fixteen lines; the tarfus five lines; the hind toe the fhorteft; the tail twenty-one lines, confifting of twelve equal quills, and exceeding the wings by ten lines [A].

The GREEN BEE-EATER with RU-FOUS WINGS and TAIL.

Merops Cayanenfis, Gmel.
The *Cayenne Bee-eater*, Lath.

THE denomination which we have beftowed on this fpecies almoft defcribes it: we need only to add, that the green is deeper on the upper part of the body and lighter below the throat than on any other part; that the quills of the wings are white at their origin; that their fhafts as well as thofe of the tail quills are blackifh; that the firft are of a yellowifh brown, and rather longer than ufual in this genus of birds, and the bill black.

This Bee-eater refembles much the yellow and white-headed one in the colour of its tail and wings; but the reft of its plumage is en-

[A] Specific charaƈter of the *Merops Erythrocephalus*: "It is green, below yellowifh, its head and neck red, its throat bright yellow; its wings and tail equal, and cinereous below."

tirely

tirely different. It is befides much fmaller, and the two middle quills of the tail do not project.

I am affured that it is not found in Cayenne; and am the more inclined to think that this is really the cafe, as the genus of the bee-eaters appears to me peculiar to the ancient continent, as I have already faid. But M. de la Borde, who is at prefent in Cayenne, will foon fend me the folution of this little problem.

The ICTEROCEPHALE, or YEL-LOW-HEADED BEE-EATER *.

Merops Congener, Linn. and Gmel.
Apiafter Icterocephalus, Briff.
The *other Bee-eater of Aldrovandus*, Will.

THE yellow colour of the head is only inter-rupted by a black bar, and extends on the throat and all the under fide of the body; the back is of a fine chefnut; the reft of the upper fide of the body is variegated with yellow and green; the fmall fuperior coverts of the wings are blue; the middle ones variegated with yellow and blue, and the great ones entirely yellow; the quills of the wings are black, ter-

* In German it is called *See-Schwalm*, or fea-fwallow; which name is, in parts of Italy, given to the king-fifher: Nor is this furprifing, when we confider the analogy between that bird and the Bee-eaters.

minated

minated with red; the tail has both colours, black at its bafe and green at its extremity; the bill is black, and the legs yellow.

This bird is rather larger than the ordinary Bee-eater, and its bill is more hooked. It is feen very feldom near Strafburg, according to Gefner [A].

[A] Specific charaĉter of the *Merops Congener :* " It is yellowifh, its rump greenifh, its wing-quills tipt with red, its tail-quills yellow at the bafe."

The EUROPEAN GOAT-SUCKER*.

L'Engoulevent, Buff.
Caprimulgus Europæus, Linn. Gmel. &c. &c.
The *Night Raven,* Sibbald.
The *Dorr-hawk, Goat-sucking Owl,* or *Night Jarr,*
　　Charleton.
In Shropshire, the *Fern-Owl,* and in Yorkshire, the
　　Churn-Owl, Ray.
The *Night Hawk,* Edwards.
The *Nocturnal Goat-sucker,* Penn.

THE Goat-sucker feeds chiefly on nocturnal
insects †. It begins to wheel only a little
before

* Aristotle calls it Αιγοθηλας, from Αιξ, a goat, and θηλαζω, to
milk : The name which Pliny bestows is a literal translation of this,
Caprimulgus. Hence, too, are derived many of its designations in
the modern languages: In Italian, *Succhia Capre*; in French, *Tette-
Chevre* ; in German, *Geifs-Melcher, Milch-Ziegen Suger, Kinder-
Melcher*; and in Norwegian, *Gede-Melcher.* As it never appears
but in the twilight, this circumstance has also procured it a class of
names. In Greek, Νυκλικοραξ; in Latin, *Fur Nocturnus*; in Eng-
lish, *Night-Raven*; in Italian, *Nottola*; in German, *Nacht-Schade,
Nacht-Raeblin, Nacht-Vogel*; in Danish, *Nat-Raun, Nat-Skade*; in
Swedish, *Nattskraefwa, Nattskiarra.* It is also called *Cova-Terra*
(ground-hatcher) in Italian; *Chasse-Crapaud* (hunter-toad) in
French ; *Nacht - Schwalbe* (night - swallow) and *Grofs -Bartige
Schwalbe* (great-bearded swallow) in German.
　　M. de Montbeillard, author of this article, remarks with great
justice, that the names [1] *Goat-sucker,* [2] *Flying-toad,* [3] *Great Blackbird,* [4] *Night Crow, and* [5] *Square-tailed Swallow,* ought to be rejected as founded on prejudice and inaccurate observation. The
first of these appellations, though ancient and generally admitted,

[1] *Tette-Chevre*; [2] *Crapaud Volant*; [3] *Grand Merle*; [4] *Corbeau de Nuit*; [5] *Hirondelle a Queue Carrée.*

is

† Such as moths, gnats, dorrs or chaffers, beetles, may-bugs,
and no doubt night-flies.

THE GOAT SUCKER.

before fun-fet*, and it never takes wing in the middle of the day, except in dark cloudy weather, or when obliged to make its eicape. Its eyes are fo delicate as to be dazzled and overpowered by the meridian effulgence, and they perform their office only in a weak light. But we muft not fuppofe that it can diftinguifh objects and fly in total darknefs; the proper time for its excurfions, and indeed for thofe of all the other nocturnal birds, is the dufk of the evening.

The Goat-fucker needs not fhut its bill to fecure the winged infects; for a fort of glue oozes from the palate, which entangles them †.

The Goat-fuckers are widely fcattered, yet in no place are they common. They are found in almoft all the countries of our continent,

is highly improbable, and contradicted by fact; for Schwenckfeld made particular enquiries in a country where numerous flocks of goats are kept in folds, but could never difcover that they were fucked by any bird whatever. The other names ought equally to be rejected; it is furely not a toad, or a blackbird, or a crow, or an owl. Nor is it even a fwallow, though much akin to it; for its external figure and its habits are different: its legs are fhort, its bill fmall, its throat wide; its food too, and its mode of preying, are not the fame. M. de Montbeillard adopts the appellation *Engoulevent* (guttler), given in fome provinces of France, which, though vulgar, conveys a diftinct idea of the bird in its ftate of activity; its wings fpread, its look haggard, its throat extended to its utmoft width, and wheeling with a hoarfe buzzing noife in purfuit of infects, which it feems to *guttle (engouler)* by drawing in its breath.

* Hence Ariftotle calls it a lazy bird.
† Note communicated by M. Hebert.

from

from Sweden, and even the more northern
tracts, to Greece and Africa, on the one hand;
and to India, and, no doubt, ftill farther, on the
other.   Sonnerat has fent a fpecimen for the
Royal Cabinet, from the coaft of Coromandel;
which is certainly either a young one or a female, fince it, in no refpect, differs from the
common kind, except that it wants thofe white
fpots on the head and wings which Linnæus regards as the peculiar character of the adult male.
The commander de Godeheu informs us, that, in
the month of April, the fouth weft wind brings
thefe birds to Malta *; and the Chevalier Def-
mazis, an excellent obferver, writes to me that
they repafs in as great plenty in autumn.   They
occur both in flat and in mountainous countries;
in Brie, in Bugey, in Sicily †, and in Holland,
and almoft always under a bufh, or in young
copfes, or about vineyards; they feem to prefer the dry ftony tracts, the heaths, &c.   In
the cold countries they arrive later, and retire
earlier ‡.   They breed on their progrefs, as the

* See Savans Etrangers, t. III. 91.

† A well-informed traveller informs me, that on the mountains
of Sicily thefe birds appear an hour before fun-fet, and fpread in
fearch of food in company with the bee-eaters, and that fometimes
five or fix fly together.

‡ In England they appear about the end of May, and retire
about the middle of Auguft, according to the Britifh Zoology.
In France M. Hebert faw them in the month of November;
and a fportfman affured me that he has met with them in winter.

fituations

fituations invite *; fometimes more foutherly,
at other times more northerly. They are at
little trouble in forming their neft; they are
content with any fmall hole which they happen to find in the earth, or among fmall ftones,
at the foot of a tree or the bottom of a rock.
The female lays two or three eggs, larger than
thofe of the blackbird, and of a darker colour †;
and though the affection of parents is in general
proportioned to the care beftowed in providing
for their accommodation, the Goat-fucker is
not wanting in tender attentions: on the contrary, I am affured that fhe hatches with the
greateft folicitude, and, when fhe perceives the
threats or keen obfervation of an enemy, fhe
changes her fite, pufhing the eggs dexteroufly,
it is faid, with her wings, and rolling them
into another hole, which, though not better
fafhioned, will, fhe imagines, afford a fafer con-
cealment.

The feafon when thefe birds appear moft frequent, is autumn; they fly nearly like the
woodcock, and they have the geftures of the
owl. Sometimes they teafe and difturb fportf-
men who are on the watch. They have an odd

* The fowlers whom I have confulted affirm that they never
breed in the canton of Burgundy which I inhabit (l'Auxois), and
that they appear there only in the time of vintage.

† They are oblong, whitifh, and fpotted with brown, fays M.
Salerne; marbled with brown and purple on a white ground, fays
the Count Ginanni, in the Italian Ornithology: the latter adds,
that the fhell is extremely thin.

F f 4   fort

fort of habit, which is peculiar to them; they
wheel an hundred times in fucceffion round
fome large naked tree, with a very irregular
and rapid motion; at intervals they dive brifkly, as if to catch their prey, and then rife as
fuddenly.   In fuch cafes they are undoubtedly engaged in purfuit of the infects that flutter about the aged trunks; but it is then difficult
to get within gun-fhot of them, for they quickly difappear, nor can their retreat be difcovered.

As the Goat-fucker flies with its bill open,
and with confiderable rapidity, the air continually ftrikes againft the fides of its throat, and
occafions a fort of buzzing, like the noife of a
fpinning-wheel : this whirring infallibly takes
place whenever the bird is on the wing, but it
varies according to the celerity of the flight.
Hence the name of *wheel-bird,* by which it is
known in fome counties of England *.   But is
this noife generally regarded unlucky, as Belon,
Klein, and others who have copied them, affert ? Or is it not rather a miftake occafioned
by confounding the Goat-fucker with the white
owl ? When it fits, it utters its true cry, which
is a plaintive tone repeated three or four times
in fucceffion; but we are not quite certain
whether this is ever heard while the bird is on
the wing.

It feldom perches, and when it does, it is faid

* Our author means Wales, where this bird is called *Aderyn y
droell,* which in fact fignifies *wheel-bird.* T.

not

not to cling acrofs the branch, like moft other
birds, but to fit lengthwife, refembling the pof-
ture of the cock in *treading (cochant* or *cho-
chant)* the hen ; and hence the name *chauche-
branche.* It is a folitary bird, and is, for the
moft part, fingle; feldom two are found together, and, even then, they are ten or twelve
paces from one another.

I have faid that the Goat-fucker flies like the
woodcock ; their plumage alfo is fimilar, for
all the upper fide of the neck, of the head, and
of the body, and even the under fide, is gaily
variegated with gray and blackifh, with more
or lefs of a rufty caft on the neck, the fcapular
feathers, the cheeks, the throat, the belly, the
coverts, and the quills of the tail and wings ;
but the deepeft fhades appear on the upper furface of the head, of the throat, of the breaft,
on the fore part of the wings, and on their tips:
there is fuch variety that the ideas would be
loft in the *minutiæ* of defcription ; I fhall therefore only add the characteriftic properties. The
lower jaw is edged with a white ftripe that extends behind the head ; there is a fpot of the
fame colour on the infide of the three firft quills
of the wing, and at the ends of the two or
three outmoft quills of the tail ; but thefe
fpots are peculiar to the male, according to
Linnæus * : the head is large ; the eyes very
protuberant ;

* Willughby obferved an individual in which thefe fpots were
of

protuberant; the hole of the ears pretty confiderable; the aperture of the throat ten times
wider than that of the bill; the bill fmall, flat,
and fomewhat hooked; the tongue fhort, pointed, not divided at the tip; the noftrils round,
and their edge projecting towards the bill; the
fkull tranfparent; the nail of the mid-toe indented, as in the heron; and laftly, the three
fore toes are connected by a membrane as far as
the firft phalanx. It is faid that the flefh of the
young Goat-fuckers is tolerable food, though it
leaves a tafte of ants.

Total length ten inches and a half; the bill
fourteen lines; the tarfus feven lines, feathered
almoft to the fole; the middle toe nine lines;
the hind toe the fhorteft of all, and it can be
turned forwards, and often has that pofition;
the alar extent twenty-one inches and a half;
the tail five inches, fquare, and compofed of
ten quills only; it exceeds the wings fifteen
lines [A].

of a pale yellow, tinged with purple, and obfcurely marked: I
perceived the fame thing in two fubjects; they are probably females, and the one, which is fmaller than the other, I judge to be
younger.

[A] Specific character of the common Goat-fucker, *Caprimulgus Europæus:* " It is black, variegated with cinereous, brown,
ferruginous, and white; its noftrils obfcurely tubulated." It is
moft frequent in the wooded and mountainous parts of this ifland.

FOREIGN BIRDS,

WHICH ARE RELATED TO THE GOAT-SUCKER.

THERE is only one species of this genus settled in the three divisions of the old continent; but ten or twelve are found in the new. We might therefore regard America as their original and chief abode, from which the European Goat-sucker has been expelled by some fortuitous event: and as the colony ought ever to be subordinate to the mother-state, the order of nature would require that the American species should precede those of Europe. This arrangement we would have followed; but a more cogent reason recommends a different plan. The order of the understanding is to proceed from what is well ascertained to what is more obscure: we therefore begin with the European birds, which are best known to us, and which will tend to illustrate those of other climates; leaving to the American philosophers to begin their natural history (and would to God that they would compose one!) with the productions of America.

The principal attributes of the Goat-suckers are these: the bill is flat at its base, the point being slightly hooked, apparently small, but having a gape wider than the head, according

to

to fome authors; large protuberant eyes, like
thofe of nocturnal birds; and long black whifkers about the bill: the effect of the whole gives
it a dull, ftupid afpect, and declares it a floth-
ful, ignoble race, allied to the martins and the
nocturnal birds, and yet fo nicely characteriftic,
that it is eafy at the firft fight to diftinguifh the
Goat-fucker from every other bird: their wings
and tail are long, the latter feldom forked, and
then in a very flight degree, and is compofed of
ten quills only: their legs are fhort, and, for
the moft part, rough; the three fore toes are
connected together by a membrane as far as the
firft joint: the hind toe is moveable, and turns
forward fometimes; the nail of the middle toe
is commonly indented on the inner edge: the
tongue is pointed, and not divided at the end:
the noftrils are tubulated, that is, the projecting brims form on the bill the beginning of a
fmall cylinder: the opening of the ear is wide,
and probably its hearing is very acute; and we
might even expect this to be the cafe in a bird
which has a weak fight and hardly any fmell,
for, the ear being thus alone capable of intimat-
ing what paffes at a diftance, the bird will naturally be led to improve that organ. The properties now enumerated are not, however, found
in all the fpecies; fome there are which have
no whifkers; others that have more than ten
quills in the tail; others in which the middle
nail is not indented; in fome it is indented, not
on

on the inner edge, but on the outer; in others
the noftrils are not tubulated; in others the
hind nail feems incapable of being turned forwards. But, what is common to all the fpecies, their organs of fight are too delicate to
fupport the light of day; and from this fingle
property are derived the chief circumftances
which difcriminate the Goat-fuckers from the
fwallows. Hence they appear not till fun-fet
in the evening, and retire in the morning a little after fun-rife; hence they live folitary, dif-
quieted by gloomy apprehenfions; hence the
difference of their cry; hence, too, in my opinion, is owing their not building a neft, for the
weaknefs of their fight does not permit them to
choofe and arrange, and interweave, the materials. In fact, I know not of any bird that
builds during the night, and the Goat-fucker
can, in our latitudes, have only three hours of
twilight, which is entirely confumed in purfuing their humble fugacious prey. Of all the
owls the eagle one is faid alone to make a neft;
and it the leaft deferves the appellation of nocturnal bird, fince it can fly to confiderable diftances in broad daylight. The little owl, which
hunts and catches fmall birds before the fetting
and after the rifing of the fun, gathers only a
few leaves, or ftalks of herbs, and upon thefe
drops the eggs in the holes of rocks or of old
walls. Laftly, the long-eared, the white, the
aluco, and the brown, owls, which of all the

S   nocturnal

nocturnal birds are the leaft capable of fupporting the light of the fun, lay alfo in fimilar crevices, or in hollow trees, but without any lining, or fometimes in the nefts of other birds which they find ready formed. And I might affert the fame thing in general of all birds whofe eye is exceffively delicate.

Another confequence refulting from the too exquifite mechanifm of the organ of fight, is that the Goat-fuckers, like the other nocturnal birds, have no brilliant colour in their plumage, and are denied even the rich varying glofs which gliftens on the fober attire of the fwallow: black, and white, and gray, arifing from the mixture of thefe, and rufous, form the whole garb of the Goat-fuckers, and thefe are fo intermingled that the general complexion is dufky and confufed. They fhun the light, and light is the fource of all the fine colours. Linnets, kept in the cage, lofe that charming red which glowed in all its beauty when in free air they imbibed the direct influence of the folar beams. It is not in the frozen tracts of Norway, or in Cimmerian fhades of Lapland, that we find the birds of paradife, the cotingas, the flamingos, the humming birds, and the peacocks; thofe dreary neglected climates never produce the ruby, the fapphire, or the topaz. And laftly, thofe flowers which are forced at great expence in the hot-houfe, acquire but a fickly hue that cannot compare with the brilliant colours which a vernal

nal fun fheds on the fpontaneous growth of the
painted meadow. The night-flies, it is true,
are fometimes decked with charming tints; but
this apparent exception feems even to corrobo-
rate my idea: for intelligent obfervers * remark
that thofe of them which flutter fometimes in the
day are more gaudily attired than fuch as appear
not until evening. I have myfelf perceived,
that in thofe infect tribes which iffue forth at
fun-fet, the colours refemble the dufky caft of
the Goat-fuckers; and, if among the vaft number there be fome with dazzling wings, we may
fuppofe that the tints were already formed in
their *larvæ*, which enjoy the enlivening influence of the fun-beams in an equal degree as
thofe of the diurnal flies. Laftly, the chryfalids of thefe, which are conftantly difclofed and
expofed to the open air, fhine for the moft part
with brilliant colours, and fome of them appear
decorated with fcales of gold and of filver, which
we fhould in vain expect to find in the chryfalids
of the nocturnal flies, enveloped, as they are,
with fhells, or buried in the earth. I conceive,
therefore, that I am warranted to infer, that if
a feries of obfervations were made upon the
plumage of birds, the wings of infects, and perhaps the hair of quadrupeds †, thofe fpecies

* Roefel. *Infecten beluftigung,* t. I. *Vorbericht zu der nacht-voe-*
gel erften claffe.

† The plumage of the king-fifher is much more brilliant between the tropics than in the temperate zone, as we learn from
Forfter, in Captain Cook's fecond Voyage.

would

would be difcovered to have the richeft and moft
brilliant colours, which, other circumftances being alike, were moft expofed to the action of
the light.

If my conjectures have fome foundation, the
intelligent reader will not be furprifed that different degrees of fenfibility in the fame organ
may produce confiderable differences in the natural habits of an animal and in its properties
both external and internal.

I.

The CAROLINA GOAT-SUCKER.

Caprimulgus Carolinenfis, Gmel. and Briff.
The *Rain Bird*, Brown.
The *Short-winged Goat-fucker*, Penn.

IF, as in all probability, Europe owes its Goat-
fuckers to America, this undoubtedly is the
fpecies which croffed the northern ftraits to
found a colony in the ancient continent. It inhabits North America, and its fize and plumage are fimilar to thofe of the European kind:
its lower jaw is edged with white, and there is
a fpot of the fame colour on the margin of the
wing.   The chief difference confifts in this,
that the under part of the body is variegated,
not with fmall crofs lines, but with fmall longitudinal

gitudinal ones, and that the bill is longer. And would not the great change of climate be fufficient to produce fuch change in the fhape and plumage of the bird?

Of the habits of this bird we learn the following particulars from Catefby: it appears in the evening, but never fo frequently as in dark cloudy weather, whence it derives the appellation of *Rain-bird*; it purfues with open jaws the infects on which it feeds, and it flies with a whirring noife ; laftly, it lays on the ground, and its eggs are like thofe of the lapwing. This account correfponds exactly with the hiftory of the European fpecies.

Total length eleven inches and a quarter; the bill nineteen lines, befet with black briftles; the tarfus eight lines ; the middle nail indented on the infide; the three fore toes connected by a membrane which does not extend beyond the firft joint; the tail is four inches, and exceeds the wings fixteen lines.

[A] Specific character of the *Caprimulgus Carolinenfis* : " It is variegated above by tranfverfe angled lines, alternately black and gray ; below rufous gray, with blackifh longitudinal lines ; its tail gray, latticed with black." In Carolina it is ufually called *Chuck, chuck Will's Widow*. It feems to have the fame habits with the *Whip-poor Will* of Virginia. Its egg is olive, with blackifh fpots.

II.
The WHIP-POOR WILL.

Caprimulgus Virginianus, Briff. and Gmel.
The *Long-winged Goat-fucker*, Penn.
The *Virginian Goat-fucker*, Lath.

THESE birds arrive in Virginia about the middle of April, particularly in the back parts
of the country. There they cry the whole
night in a voice fo fhrill and fo loud, and repeated and encreafed to fuch a degree by the
echoes of the mountains, that one can hardly
fleep in their neighbourhood. They begin a
few minutes after fun-fet, and continue till
dawn. They feldom appear near the coaft, and
ftill feldomer during the day. They lay two
eggs of a dirty green, variegated with fmall
fpots, and fmall blackifh ftreaks; the female
drops them carelefsly in the middle of a path,
without forming any neft, without gathering
mofs or ftraw, and even without fcraping the
ground; and when fhe hatches, one may approach very near before fhe takes to flight.

Many believe the Whip-poor Will to be of ill
omen. The favages are perfuaded that the fouls
of fuch of their anceftors as were maffacred by
the Englifh have paffed into the bodies of thefe
birds, and allege as a proof, their being never
feen prior to the fettlement of the colony. But
this

this fact fhews only that the ftrangers introduced new fpecies of cultivation, which invited
new tribes of birds.

The upper fide of the head and of all the
body, as far as the fuperior coverts and quills
of the tail inclufively, and even the middle
quills of the wings, are of a deep brown, radiated tranfverfely with a lighter brown, and
fprinkled with fmall fpots of the fame colour,
with a very irregular mixture of cinereous; the
fuperior coverts of the wings are the fame, only
fprinkled with a few fpots of light brown; the
great quills of the wings are black, the five
firft marked with a white fpot near the middle
of their length, and the two outer pairs of the
tail are marked fimilarly near the end; the circle of the eye is light brown, verging on cinereous; there is a feries of orange fpots, which
begins at the bafe of the bill, paffes above the
eyes, and defcends upon the fides of the neck;
the throat is covered with a broad reverfed crefcent, white at the top and tinged with orange
at the bottom, and whofe horns point on both
fides to the ears; all the reft of the lower part
is white, tinged with orange, and ftriped acrofs
with blackifh; the bill is black, and the legs
flefh coloured. This Goat-fucker is a third
fmaller than the European, and its wings are
longer in proportion.

Total length eight inches; the bill nine lines
and a half, its bafe befet with black briftles;

the

the tarfus five lines; the nail of the mid-toe is indented on its inner edge; the tail three inches and a quarter, and does not project at all beyond the wings [A].

III.

The GUIRA-QUEREA.

Caprimulgus Jamaicenfis, Gmel.
The *Wood Owl*, Sloane.
The *Mountain Owl*, Brown.
The *Jamaica Goatfucker*, Lath.

THOUGH Briffon makes no diftinction be-
tween the *guira* defcribed by Sloane and the one defcribed by Marcgrave, I conceive that they ought to be difcriminated and regarded as at leaft varieties of climate: I fhall ftate my reafons when I treat of Marcgrave's Guira. In that of Sloane the head and neck are variegated with the colour of Spanifh tobacco, and with

[A] Specific character of the *Caprimulgus Virginianus :* " It is brown, variegated tranfverfely with gray-brown, and here and there with cinereous ; below it is ftriped tranfverfely with reddifh white ; there is a triangular white fpot on its chin ; the fpace about its eyes and its neck are variegated with orange fpots." It re-ceived the name of *Whip-poor Will* on account of its note : but it really founds *Wiperi-wip*, laying the ftrefs on the laft fyllable, and fliding lightly over the fecond. It fits on the bufhes, the fence-rails, or the fteps of houfes, where the infects are moft abundant ; it makes a fpring at them as they pafs, and fettles again to renew its fong. In the ftate of New York it appears in May, and re-tires in Auguft.

black ;

black; the belly and the fuperior coverts of the
tail and of the wings, variegated with whitifh;
the quills of the tail and of the wings variegated
with deep brown and white; the lower jaw almoft featherlefs; the head, on the contrary, is
over-charged with them; the eye-balls protrude
from the focket about three lines; the pupil is
whitifh, and the iris orange.

This bird is found in Brazil; it inhabits the
woods, lives upon infects, and flies only in the
night.

Total length fixteen inches; the bill two
inches, and of a triangular fhape; its bafe is
three inches, fomewhat hooked, and edged
with long whifkers; the noftrils are placed in
a pretty large groove; the throat is wide; the
tarfus three lines; the alar extent thirty inches;
the tail eight inches; the tongue fmall and triangular; the ftomach whitifh, flightly mufcular, containing half-digefted beetles; the liver red, divided into two lobes, the one on the
right, and the other on the left; the inteftines
are rolled into many circumvolutions.

The Guira of Marcgrave has two very obvious characters which are not found in the defcription of Sloane, but which could not have
efcaped fo accurate an obferver. Thefe are
the gold collar and the two middle quills of
the tail, which are much longer than the lateral ones: befides, it is fmaller, for Marc-

grave

grave reckons it not to exceed the lark; and
it is difficult to fuppofe that fuch a bird
would meafure thirty inches acrofs the wings,
as Sloane ftates it.   There are alfo fome differences in the plumage, which confpire to
fhow that it is a variety from climate.   Its
head is broad, flat, and large; its eyes large;
its bill is fmall, with a wide aperture; its body
is round; its plumage is afh-brown, variegated
with yellow and whitifh; it has a gold collar tinged with brown; the edges of the bill,
near its bafe, are befet with long black whifkers; the fore toes are connected by a fhort
membrane; the nail of the mid-toe is indented; the wings have fix quills; the tail eight,
including the two middle ones, which project
beyond the reft.

[A] Specific character of the *Caprimulgus Jamaicenfis*: "It is
variegated with longitudinal ferruginous and black ftreaks; the
fpace about its eyes clothed with a difk of plumules; its wings
brown and fpotted; its tail cinereous, variegated with black fpots
and dark brown ftripes."

IV.

The IBIJAU.

Caprimulgus Brafilianus, Gmel.
Caprimulgus Brafilienfis Nævius, Briff.
Caprimulgus Americanus Minor, Ray.
The *Brafilian Goat-fucker,* Lath.

THIS Brafilian bird has all the characters of the Goat-fuckers: its head is broad and flat, its eyes large, its bill fmall, its throad wide; its legs fhort, the mid-toe indented on its inner edge, &c.; but what is peculiar to it, is the habit of expanding its tail from time to time. Its head and all the upper fide of its body are blackifh, fprinkled with fmall fpots, moftly white, fome of them tinged with yellow; the under fide of its body white, variegated with black, as in the fparrow-hawk, and its legs are white.

It is nearly as large as the fwallow; its tongue very fmall; its noftrils open; the tarfus fix lines; the tail two inches, and exceeds not the wings.

VARIETIES of the IBIJAU.

I. THE LITTLE SPOTTED GOAT-SUCKER OF CAYENNE *. It bears a ftrong refemblance

* *Caprimulgus Cayanenfis.* Gmel.
The *White-necked Goat-fucker,* Lath.

to

to the Ibijau in its fize, in the length of its
wings, and in the proportions of its other dimenfions, and in the blackifh caft of its plumage
fpotted with a lighter colour; thefe fpots are
rufous or gray, except on the neck, whofe fore
part has a fort of white collar, not mentioned by
Marcgrave in his defcription of the Ibijau, and
which chiefly diftinguifhes this variety; the
under fide of the body is alfo darker.

Total length eight inches; the bill fifteen
lines, black, befet with fmall briftles; the tail
two inches and a half.

II. The Great Ibijau *. The difference
of bulk is very confiderable, it being as large as
an owl, and its bill fo wide as to admit the
hand; in other refpects the colours and proportions are the fame as in the little Ibijau. Marc-
graves does not inform us whether it alfo fpreads
its tail; nor does he mention that there is a
horn on the fore part of the head and behind it
a fmall tuft, as his figure feems to reprefent.
But it is well known that Marcgrave's figures
art inaccurate, and that more reliance ought to
be had on the text.

With this fpecies we fhould alfo range the
great Goat-fucker of Cayenne, both on account
of its bulk, and of its plumage, which is fpotted
with black, with fulvous and with white,

* *Caprimulgus Grandis*, Gmel.
  Caprimulgus Brafilienfis Major Nævius, Briff.
  The *Grand Goat-fucker,* Lath.

principally

principally on the back, the wings, and the tail; the upper fide of the head and of the neck, and the under fide of the body, are ftriped tranf-verfely with different fhades of the fame co-lours; but the general caft of the breaft is browner, and forms a fort of cincture. M. de Sonini faw one whofe plumage was darker, and which had been found in the hollow of an ex-ceeding large tree; this is its ordinary abode, but it prefers thofe trees which grow near water. It is at once the largeft of the Goat-fuckers known in Cayenne, and the moft folitary.

Total length twenty-one inches; the bill three inches long, and as broad, the upper man-dible has a deep fcalloping on both fides near the point, the lower mandible fits into thefe fcallops, and its edges are reflected outwards; the noftrils are flat and fhaded by the feathers of the bafe of the bill, which grow forwards; the tarfus is eleven lines, feathered almoft to the toes; the nails are hooked, hollowed below by a furrow, which is parted into two by a longitudinal ridge; the mid-toe is not indented, but is very large, and appears even more fo on account of a mem-branous ledge on each fide; the tail nine inches, a little tapered; the wings project fome lines beyond it.

The SPECTACLE GOAT-SUCKER, or the HALEUR.

Caprimulgus Americanus, Linn. Gmel. and Bor.
Caprimulgus Jamaicensis, Briff. and Ray.
Hirundo Jamaicensis, Klein.
The *Screech-Owl*, Brown.
The *Small Wood-Owl*, Sloane.
The *American Goat-fucker*, Lath.

THE protuberant noftrils of this bird have fome refemblance to a pair of fpectacles, and hence its name of *Spectacle Goat fucker (En-joulevent à Lunettes)*: that of *Haleur* evidently alludes to its cry.

This Goat-fucker lives upon infects, like all the others; and, in its internal conformation, it refembles the guira of Sloane, with which it conforts: it inhabits both Jamaica and Guiana; its plumage is variegated with gray, with black, and with the colour of withered leaves; its bill is black; its legs brown; and there is abundance of feathers in the head and under the throat.

The length, according to Sloane, is feven inches; the bill is fmall but wide; the upper mandible fomewhat hooked, three lines long (reckoning, no doubt, from the root of the feathers on the front) edged with black whifk-

ers;

ers; the tarfus, together with the foot, eighteen lines; the alar extent ten inches [A].

VI.

The VARIEGATED GOAT-SUCKER of CAYENNE.

Caprimulgus Cayannenfis, Gmel.
The *White-necked Goat-fucker*, Lath.

ALL the birds of this genus are variegated, but
this is more fo than the reft; it is the moft
common in Cayenne; it frequents the plantations, the roads, and other cleared parts. When
on the ground it utters a feeble cry, attended
conftantly with a fhivering of the wings, and
refembling the croaking of the toad: It has alfo
another cry like the barking of a dog. It is not
fhy, and when fcared, it never flies to any
great diftance.

The head is delicately ftriped with black on
a gray ground, with fome fhades of rufous; the
upper fide of the neck is ftriped with the fame
colours, but not fo nicely: on each fide of the
head are five parallel bars, ftriped with black
on a rufous ground; the throat is white, and

[A] Specific charaĉter of the *Caprimulgus Americanus:* " Its
noftrils are tubulated and projeĉting."

alfo

alfo the fore part of the neck ; the back is ftriped acrofs with blackifh on a rufous ground ; the breaft and belly are ftriped alfo, but lefs regularly, and fprinkled with a few white fpots; the lower belly and the thighs are whitifh, fpotted with black ; the fmall and middle coverts of the wings are variegated with rufous and black, fo that rufous predominates on the fmall ones, and black on the middle ones ; the great ones are terminated with white, which forms a crofs bar of that colour; the quills of the wings are black ; the five firft marked with white two thirds or three fourths of their length; the fuperior coverts and the two middle quills of the tail are ftriped acrofs with blackifh on a gray ground, clouded with black; the lateral quills edged with white; and this edging is broader as the quill is more exterior; the iris is yellow, the bill black, and the legs yellowifh-brown.

Total length about feven inches and a half; the bill ten lines, befet with briftles ; the tarfus five lines ; the tail three inches and a half, and projecting about an inch beyond the wings.

VII.

SHARP-TAILED GOAT-SUCKER.

L'Engoulevent Acutipenne de la Guyana, Buff.
Caprimulgus Acutus, Gmel.

THIS bird differs from the preceding not only in its dimenfions, but in the fhape of its tail feathers, which are pointed. It is diftinguifhed alfo by the colours of its plumage. The upper furface of the head and neck is ftriped tranfverfely, but not delicately, with tawny brown and black; the fides of the head are variegated with the fame colours, only rufous predominates; the back is ftriped with black on a gray ground, and the under furface of the body on a rufous ground; the wings are nearly as in the preceding fpecies; the quills of the tail are ftriped acrofs with brown on a pale cloudy rufous, terminated with black, but a little white precedes this black tip; the bill and legs are black.

It is faid that thefe birds fometimes affociate with the bats; which is not very extraordinary, fince they leave their retreats at the fame hours, and purfue the fame prey. Probably thefe are the fame with the fmall fpecies mentioned by M. de la Borde, which neftle like the wood

pigeons,

pigeons, the turtles, &c. in October and November, that is, two or three months before the rainy feafon, which begins about the fifteenth of December, and during which moft of the birds breed.

Total length about feven inches and a half; the tail three inches, confifting of ten equal quills, and projecting a few lines beyond the wings.

VIII.

The GRAY GOAT-SUCKER.

Caprimulgus Grifeus, Gmel.

I saw in Manduit's cabinet a Goat-fucker from Cayenne much larger than the preceding; it had more gray in its plumage, and its proportions were fomewhat different, and the quills of the tail were not pointed. The quills of the wings were not fo black as in the preceding fpecies, and were ftriped acrofs with gray; thofe of the tail were ftriped with brown on a gray ground variegated with brown, without any white fpots; the bill was brown above, and yellowifh below.

Total length thirteen inches; the bill twenty lines; the tail five lines and a quarter; and projecting a little beyond the wings.

IX.

The MONTVOYAU of GUIANA.

Caprimulgus Guianenfis, Gmel.
The *Guiana Goat-fucker*, Lath.

MONTVOYAU is the cry of this bird, which pronounces diftinctly the three fyllables, and repeats them very often in the evening among the bufhes. Like the European Goat-fucker, it has a white fpot on each of the five firft quills of the wing, of which the ground is black, and another white fpot or bar which rifes from the corner of the bill, and ftretches forwards, but extends alfo under the neck, in which circumftance it differs; and befides it has in general more of the fulvous and rufty colours in its plumage, which is almoft wholly variegated with thefe two colours; yet thefe affume different fhades and modifications in different parts; crofs ftripes on the lower region of the body, and the middle quills of the wings; longitudinal ftripes on the upper fide of the head and neck; oblique ftripes on the top of the back; and laftly, there are irregular fpots on the reft of the upper fide of the body, where the fulvous affumes a gray caft.

Total length nine inches; the bill nine lines and a half, befet with briftles; the tarfus naked; middle nail indented on its outfide, the tail three inches, exceeding the wings one inch.

X.

The RUFOUS GOAT-SUCKER of CAYENNE.

Caprimulgus Rufus, Gmel.

RUFOUS clouded with blackifh forms almoft all the ground of the plumage; and black varioufly intenfe conftitutes its whole ornament : it is difpofed in longitudinal, oblique, irregular bars, on the head and the upper fide of the body; it makes a fine irregular tranfverfe ftriping on the throat, a little broader on the fore part of the neck, the under fide of the body, and of the legs; then a little broader on the fuperior coverts and on the inner edge of the wing near its extremity; laftly, broadeft of all on the quills of the tail. Some fpots are fcattered here and there on the body, both above and below. In general, blackifh predominates on the top of the belly; rufous on the lower belly, and ftill more on the inferior coverts of the tail; the middle part of the great quills of the wings prefents fmall fquares alternately rufous and black, checkered almoft as regularly as fpots on a chefs board; the iris is yellow, the bill light brown, and the legs flefh coloured.

Total length ten inches and a half; the bill twenty-one lines; the tail four inches and two thirds, exceeding the wings fix lines.

I have

I have feen at M. Mauduits' a Goatfucker, from
Louifiana, of the fame fize with this, and very
fimilar, only the crofs ftripes had more intervening fpaces, and the rufous was lighter, which
formed a kind of collar; the reft of the under
fide of the body was ftriped as in the preceding;
the bill black at the point, and yellowifh at the
bafe.

Total length eleven inches; the bill two
inches, edged with eight or ten ftiff briftles,
bending forward; the tail five inches, and projecting a very little beyond the wings.

The S W A L L O W S*.

WE have seen that the goat-fuckers may be reckoned night Swallows, and that the only essential difference between them and the real Swallows consists in the excessive delicacy of their eyes, and its influence on their structure and habits. In both tribes of birds the bill is small, and the throat wide ; the legs short, and the wings long ; the head flat, and the neck scarce visible ; and both live upon insects which they catch in the air. But, 1. The Swallows have no bristles about the bill ; the nail of the mid-toe is not indented ; their tail contains two more quills, and, in most of the species, it is forked ; and they are in general smaller than the goat-fuckers.

2. Though the colours are nearly the same

---

* In Hebrew, *Agur, Sus, Chauraf, Thartaf, Chatas, Chataf: in* Greek, the Swallow is denominated Χελιδων, derived perhaps from χειλος, *the check,* and δινεω, *to whirl*; alluding to their rapid flutter, and the continual motion of their bill. It had the epithets κωλιλη, *chatterer*; ολολυγων, *moaner*; αυκυπτερη, *swift-winged.* The Latin, *Hirundo,* was first written *helundo,* and evidently borrowed from χελιδων. In Italian it is termed *Rondina, Rondinella, Gesila:* In Spanish, *Golondrina, Andorinha:* In German, *Schwalbe:* In Swiss, *Schwalm:* In Flemish, *Swalwe:* In Swedish, *Swala:* In Polish, *Jaskotka.* The English word *Swallow* perhaps comes from the verb, but more probably from the German *Schwalbe,* which is softened in the parent Saxon into *Swale.* The French *Hirondelle* is evidently formed from the Latin *Hirundo.*

4

in both, confifting of black, of brown, of gray,
of white, and of rufous, they are difpofed in
large fpots on the Swallows, and better contrafted; and the plumage has a bright varying
glofs.

3. The goat-fuckers entangle the night-flies
with the vifcous faliva that trickles within
their mouth; but the Swallows, and alfo the
martins, fnap the winged infects, and the fudden clofing of their bill occafions a fort of cracking noife.

4. The Swallows are more focial than the
goat-fuckers; they often gather in numerous
flocks; and in certain circumftances they lend
mutual affiftance, as in building their nefts.

5. In this conftruction they generally difplay much attention and art; and if a few fpecies lay in the holes of walls, or in fuch as they
form in the ground, they choofe excavations of
a fufficient depth to afford protection for their
young, and they provide whatever will contribute to convenience, warmth, and eafe.

6. The manner in which the Swallows fly
differs in two principal points from that of the
goat-fuckers. It is not attended with that
whirring noife which I have before mentioned,
becaufe the bill is not kept open: and though
their wings feem not better calculated for motion, they wheel with much greater boldnefs,
celerity, and continuance; becaufe the diftinct-
nefs of their vifion permits them to exert all

H h 2   their

their force. They live habitually in the air,
and perform their various functions in that ele-
ment. The flight of the swallow is perhaps
less rapid than that of the falcon, but it is easier
and more unrestrained; the one darts forward
with vigour, the other glides smoothly through
the air: she shoots in every direction to survey,
as it were, her aerial domain; and her shrill
slender notes express the cheerfulness of her
condition: sometimes she pursues the fluttering
insects, and nimbly follows their devious wind-
ing tracks, or leaves one to hunt another, and
snaps a third as it passes: sometimes she escapes
the impetuosity of the bird of prey by the quick
flexures of her course. She can always command her swiftest motion, and in an instant
change its direction; and she describes lines so
mutable, so varied, so interwoven, and so con-
fused, that they can hardly be pictured by
words.

7. The Swallows seem not to be peculiar to
either continent, and as many species nearly
are diffused through the old as through the new.
They are found in Norway and in Japan*, on
the coasts of Egypt and those of Guinea, and at
the Cape of Good Hope †. What country is
inaccessible to their easy swift course? But seldom they remain the whole year in the same
climate; those of Europe continue only during

* Kæmpfer.   † Villaut and Kolben.

the

the fummer months, appearing at the vernal
equinox, and retiring at the autumnal. Ariftotle, who wrote in Greece, and Pliny, who
copied him in Italy, affert that the Swallows
pafs into the milder climates to winter, when
thefe are not very diftant; but that, in other
cafes, they feek a lodgment in the warm fheltered dales. Ariftotle adds that many of them
have been found thus concealed with not
a fingle feather on their body *. This opinion, countenanced by the authority of great
names, and fupported by facts, became popular, infomuch that even poets drew their com-
parifons from it †. Several modern obfervations feemed to confirm it ‡; and, with fome
modifications, it might have been brought to
the truth. But a bifhop of Upfal, *Olaus Magnus*, and a Jefuit, named *Kircher*, amplifying
the affertion of Ariftotle, already too general,
have afferted that, in the northern countries,
the fifhermen often find in their nets heaps of
Swallows grouped together and clofe entangled with each other, bill to bill, feet to feet,

---

* Arift. *Hiſt. Anim.* Lib. VIII. 12 and 16. Plin. *Hiſt. Nat.*
Lib. X. 24.

   † *Vel qualis gelidis, pluma labente, pruinis*
    Arboris immoritur trunco brumalis hirundo.
CLAUD.

‡ Albertus, Auguftinus Nyphus, Gafpar Heldelin, and fome
others, aver that they frequently found during winter, in Germany,
Swallows torpid in hollow trees, and even in their nefts, which is
not abfolutely impoffible.

H h 3   and

and wings to wings; that when thefe birds are carried to ftoves they quickly recover from their torpor, but die foon after; and that none fur-vive the renovation of their vital powers, ex-cept fuch as gently feel the growing warmth of the feafon, and, rifing flowly from the bot-tom of the lakes, are, with all the fucceffive gradations, reftored by nature to their true ele-ment. This affertion has been repeated, em-bellifhed, and loaded with more extraordinary circumftances; and, as if it were not fufficient-ly marvellous, fome have added that, about the beginning of autumn, thefe birds plunge in crowds into the wells and cifterns *. I muft confefs, that many authors and other perfons, refpectable by their character or rank, have be-lieved in this phænomenon. Linnæus himfelf has given a fort of fanction to it by his autho-rity; only he reftricts it to the chimney Swal-low and the common martin, but does not impute it to the fand martin, which was more natural. On the other hand, the number of naturalifts who reject the opinion is fully as great +; and their proofs feem to be much more cogent. I know that it is fometimes imprudent to judge of a particular fact by what are called the ge-neral laws of nature; becaufe thefe, being found-

* P. Ant. Tolentinus.

+ Marfigli, Ray, Willughby, Catefby, Collinfon, Wager, Ed-wards, Reaumur, Adanfon, Frifch, Tefdorf, Lottinger, Vallifnieri, the authors of the Italian Ornithology, &c.

ed

ed on obfervation, are true only fo far as they comprehend all the facts; but the fubmerfion of Swallows appears by no means afcertained; and I fhall here ftate my reafons.

Moft of thofe who atteft this marvellous tale*, particularly Hevelius and Schœffer, who were appointed by the Royal Society of London to examine and weigh the proofs, adduce nothing but vague reports †, and a fufpicious tradition, to which the work of Olaus Magnus might have given origin. Even thofe who affert their having feen the phænomenon, as Etmuller, Wallerius, and fome others ‡, only repeat the words of the bifhop of Upfal, without joining any circumftantial remarks which give probability to a relation.

If it were true that all the Swallows which

---

* Schœffer, Hevelius, Aldrovandus, Neander and Bartius, Gerard, Schwenckfeld, Rzaczynfki, Derham, Klein, Regnard, Ellis, Linnæus, &c. We might enlarge the lift, but the number of partifans in reality weakens the opinion which they maintain; fince among fo many obfervers not one can produce a fingle circumftantial and authentic fact.

† *See* Philofophical Tranfactions, No. 10, and judge if the Royal Society ever verified the fact, as afferted by the journalifts of Trevoux, the Abbe Pluche, and fome others.

‡ Chambers cites Dr. Colas, who fays that he faw fixteen fwallows taken out of the lake Sameroth, thirty taken out of the Royal Pool at Rofineilen, and two others at Schledeiten, the moment they came out of the water: he adds that they were very wet and feeble, and that he had obferved that thefe birds are ufually very weak on their firft appearance. But this is contrary to daily obfervation; befides Dr. Colas mentions neither the fpecies, nor the date, nor the circumftances, &c.

inhabit

SWALLOW.

inhabit a country plunge into the water or mud
annually in October, and rife from their fub-
aqueous bed in the following April, there muft
have been frequent opportunities of obferving
them, either in the inftant of their immerfion,
or, what is much more curious, in the moment
of their emerfion, or during their long repofe
at the bottom of the pool. Thefe would have
been notorious facts, confirmed by the united
teftimony of perfons of all conditions, by fifhermen, hunters, farmers, travellers, fhepherds,
mariners, &c. No one doubts that the marmots, the dormice, and the hedge-hogs, fleep
benumbed during the winter in their holes; no
one doubts that the bats pafs that cold feafon in
the fame torpid ftate, clinging to the roofs of
fubterraneous caves, and muffled in their wings.
But it is hard to believe that Swallows can live
fix months without breathing, and all that time
under water. Their emerfion has never been
obferved *, though, if it were true, it muft
happen frequently in the feafon when the pools
are fifhed. The account is fufpected even on
the fhores of the Baltic: Dr. Halmann, a Ruffian, and M. Brown, a Norwegian, who were
at Florence, affured the authors of the *Italian
Ornithology* that, in their northern climates, the
Swallows appeared and retired at the fame times

* I know that Heerkens, in his poem entitled *Hirundo,* has defcribed in Latin verfes this emerfion; but at prefent we have nothing to do with poetical defcriptions.

as

as in Italy, and that their pretended fubmerfion under water was a fable current only among the vulgar.

M. Tefdorf of Lubec, a man who joins much philofophy to extenfive and various information, has written to the Count de Buffon, that, notwithftanding forty years attention to the fubject, he could never fee a fingle Swallow drawn out of the water.

M. Klein, who has been at fuch pains in fupporting the opinion of immerfion and emerfion of Swallows, confeffes that he was never fortunate enough to catch them in the fact *.

M. Hermann, a learned profeffor of natural hiftory at Strafburgh, and who feems even to lean to Klein's idea, owns to me in his letters, that he was never gratified with a fight of the fuppofed phænomenon.

Two other obfervers of the moft undoubted authority, M. Hebert and the Vifcount Querhoent, affure me that they knew the fubmerfion of Swallows only from hearfay, and could never verify it by their own obfervations.

Dr. Lottinger, who has much ftudied the economy of birds, and who does not always coincide with me in opinion, regards this fubmerfion as an incredible paradox.

In Germany, a reward of an equal weight of

* In Nivernois, Morvand, Lorraine, and many other provinces where pools abound, the people have no idea of the immerfion of the Swallows.

filver was offered publicly to whoever fhould produce Swallows found under water; yet no perfon ever claimed the prize *.

Many perfons of learning or rank †, who believed in this ftrange phænomenon, and wifhed to perfuade others, offered to exhibit clufters of Swallows fifhed up in winter, but never fulfilled their promife.

Klein produces certificates; but almoft all of them are figned by a fingle perfon, and refer only to one occurrence, which happened long prior, and either founded upon mere report, or feen when the obferver was a child. They feem to be fervilely copied from the text of Olaus, and want thofe little minute incidents which mark an original relation; and this uncertainty alone is fufficient to overturn the affertion ‡.

But it is not enough that we invalidate the proofs on which this paradox refts, we muft fhew that they are inconfiftent with the known laws of animal economy. When any quadruped or bird has once breathed, and the *foramen ovale*, which in the *fœtus* formed the communication between the two ventricles of the heart, is fhut, refpiration becomes ever after neceffary

* Frifch.

† A Grand Marfhal of Poland and an Ambaffador of Sardinia had promifed them to M. de Reaumur; the Governor of R. and many others had promifed them to M. de Buffon.

‡ The periodical publications have alfo recorded obfervations favourable to the hypothefis of Klein; but the leaft examination of them will convince us that they are incomplete and indecifive.

to the continuance of life. Swallows kept under water, with all the due precautions, die in
a few minutes*, and even when fhut up in an
ice-houfe †, do not furvive many days; how
then could they live fix months at the bottom of a lake? I know that in fome animals
this may be poffible; but fhall we, as Klein has
done, compare the Swallows to infects ‡, to
frogs, or to fifhes, which have their internal
ftructure fo different? Shall we infer that, if
marmots, dormice, hedge-hogs, and bats, continue, as we have juft faid, torpid in winter, the
Swallows will alfo in a fimilar ftate outlive the
rigours of the feafons? But not to mention,
that thefe quadrupeds can be fupported by re-
abforption of the fuperabundant fat with which
they are provided in the autumn, and which is
wanting to the Swallows; not to mention the
low temperature of their bodies, as obferved by
the Count de Buffon, in which refpect they
differ from the Swallow §; not to mention, that
they

---

* *See* the Italian Ornithology, *t. III. p.* 6. The authors affert
pofitively that all the Swallows which they plunged into water,
even at the time of their difappearance, expired in a few minutes;
and though thefe recent drowned Swallows might have been recovered, yet if they had lain fome days, and ftill more, feveral weeks
or months, they would have been totally paft recovery.

† This experiment was made by the Count de Buffon.

‡ Caterpillars die in water after a certain time, as M. de Reaumur proved; and the fame is probably the cafe with other infects
that have *tracheæ*.

§ Dr. Martin found the heat of birds, and particularly that
of

they often perifh in their holes when the rigours
of the feafon are of uncommon duration, and
that the hedge-hogs are alfo torpid in Senegal,
where the winter is hotter than our fineft fum-
mers, but where the Swallows are perpetually
active * : I fhall only obferve, that thefe quadrupeds are in air, and not under water; that
they can ftill breathe, though numbed; and that
the circulation of the blood and of other flu-
ids, though more fluggifh than ufual, goes on
in the fame manner. Nay, according to the
obfervations of Vallifnieri, thefe functions are
performed in frogs, which fleep through the
winter in the bottom of marfhes. But circulation is effected by a different mechanifm in amphibious animals from that in quadrupeds or in
birds †. In thefe, refpiration is effential to life.
　　　　　　　　　　　　　　　　　　　There

of Swallows, to exceed two or three degrees that of the warmeft
quadrupeds. *See his Effay on Thermometers.*

　* Confult *Adanfon's Voyage to Senegal.*

　† The circulation of the blood in quadrupeds and in birds is nothing but the perpetual motion of that fluid, determined by the
fyftole (or contraction) of the heart, to pafs from its right ventricle, through the pulmonary artery, into the left ventricle; to pafs
from this ventricle, which has alfo its fyftole, through the trunk of
the *aerta* and its branches, into all the reft of the body; to return
by the branches of their veins into their common trunk, which is the
vena cava; and finally into the right auricle of the heart, where it
again begins to repeat its round. From this mechanifm it follows that, in quadrupeds and in birds, refpiration is neceffary to
open for the blood the paffage through the breaft, and confequently
is neceffary to circulation; whereas in the amphibious animals, as
the heart has only a fingle ventricle, or feveral ventricles which,
　　　　　　　　　　　　　　　　　　　communicating

There is a well-known experiment of Dr. Hook's: having ftrangled a dog, and having made incifions in the ribs, in the diaphragm, in the pericardium, and in the top of the windpipe, he renewed or ftopped, as often as he pleafed, the vital action, by blowing into the lungs or clofing the paffage. It is impoffible, therefore, that Swallows or ftorks, for they alfo have been ranked among the diver-birds *, could live fix months under water without any communication with the external air; the more fo, as this feems to be neceffary even for fifhes and frogs, which is evinced by feveral experiments that I have lately made.

Of ten frogs, which were found beneath the ice on the fecond of February, I put three of the livelieft into three glafs veffels full of water, where they could move freely, but not rife to the furface, though a part of this even was

communicating with each other, perform the function of one, the lungs afford not a paffage to the whole mafs of blood, but only receive a quantity fufficient for their nourifhment; and by confequence their motion, which is that of refpiration, is much lefs neceffary to that of circulation. This inference is confirmed by experiment: a tortoife, which had the trunk of its pulmonary artery tied, lived, and its blood continued to circulate for the fpace of four days, though its lungs were open and cut in feveral places. *See* Animaux de Perrault, *part II. p.* 196.

  * *See* Schwenckfeld, *Aviarium Silefiæ, p.* 181. Klein, Ordo *Avium, pp.* 217, 226, 288, & 229. St. Cyprian, *contra Bodinum, p.* 1459. Luther, *Comment. ad Genef. cap. I.* But Haffelquift, when in the neighbourhood of Smyrna, faw about the beginning of March the ftorks pafs in their way to the north.

immediately

immediately in contact with the external air.
Three others were thrown, at the same time,
each into an earthen pot half-filled with water,
and permitted to breathe at the surface ; and
the four remaining ones were placed together at
the bottom of a large open empty veſſel.

I had previouſly noticed their reſpiration,
both in air and in water, and found it to be very
irregular *. When ſuffered to ſwim about at
will, they often roſe to the ſurface, and even
protruded their noſtrils : I could then perceive
a vibratory motion in the throat correſponding
nearly to the alternate dilatation and contraction
of the noſtrils. As ſoon as they plunged again
into the water, both motions ſuddenly ceaſed.
If haſtily forced to deſcend, they ſhewed a manifeſt uneaſineſs, and allowed a number of air-
bubbles to eſcape. The veſſel was filled with
water to the brim, and covered by a weight of
twelve ounces; yet the frog, to get air, puſhed
off the cover. The three frogs which were
kept under water conſtantly ſtruggled hard to
gain the ſurface, and they all died, the one in
twenty-four hours, and the others in the courſe
of two days †. But of the ſeven others, five
                                        eſcaped,

---

* " Frogs, tortoiſes, and ſalamanders, ſometimes ſwell themſelves ſuddenly, and remain in that ſtate . . . a full quarter of an
hour: ſometimes they ſuddenly make an entire expiration, and remain very long in that ſtate." *Animaux de Perrault, part II.*
p. 272.

† It is proper to remark that frogs are very vivacious, that they
                                        can

efcaped, and the remaining two, which had
been allowed both air and water, are male and
female, and, at prefent (22d April, 1779) more
lively than ever; fince the fixth the female has
layed about one thoufand three hundred eggs.

The fame experiments were made with equal
attention on nine fmall fifhes of feven different
fpecies, viz. the gudgeon *, the bleak †, the
barbel ‡, the minow §, the bull-head ‖, the
dog-fifh ╋, and another known by the name of
bouziere in Burgundy. Eight of the firft fix
fpecies having been held under water died in lefs
than twenty-four hours **; but thofe which
were kept in fimilar bottles, and permitted to
rife to the furface, lived, and retained their
ufual vivacity. The *bouziere* indeed lived longer
under water than the reft, and I found that the

can endure a month's abftinence, and that they fhow motion and
life feveral hours after their heart and bowels are detached from
their body. *See* Collection Academique. *Hift. Nat. t. I. p.* 320.

   * *Cyprinus-Gobio*, Linn.   † *Cyprinus Alburnus.*
  ‡ *Cyprinus Barbus.*   § *Cyprinus-Phoxinus.*
  ‖ *Cottus Gobio.*   ╋ *Squalus-Canicula.*

   ** The bleak died in three hours, the two little barbels in fix
hours and a half, one of the gudgeons in feven hours, the other
in twelve hours, the minow in feven hours and a half, the bull-
head in fifteen hours, the dog-fifh in twenty-three hours, and the
bouziere in near four days. The fame fifhes, kept in air, die in
this order: the bleaks in thirty-five or forty-four minutes, the
bouziere in forty-four, the dog-fifh in fifty or fifty-two, the barbels in fifty or fixty, one of the minows in two hours and forty-
eight minutes, the other in three hours, one of the gudgeons in
an hour and forty-nine minutes, the other in fix hours and twenty-two minutes. The biggeft of thefe fifh did not meafure twenty
lines from the eye to the tail.

one which was not confined appeared feldom at
the furface; and it is probable that thefe fifhes
refide more conftantly than the others at the
bottom of brooks, which implies fome difference of ftructure *. However, it often tried to
reach the furface, and, on the fecond day, it
feemed uneafy and opprefled, its refpiration
grew laborious, and its fcales pale and whitifh †.

But it will appear more extraordinary, that
of two carps ‡ equal in fize, the one, which
was kept conftantly under water, lived a third
fhorter time than the other, which was not put
into water, though in its flouncing it had fallen
from a chimney-piece four feet to the ground §.
And in two other experiments compared together and made on larger barbels than employed
before, thofe kept in the air lived longer, and

* This fifh was fmaller than a little bleak; it had feven fins,
the fcales on the upper fide of the body yellowifh, edged with
brown, and thofe of the under fide refembling mother of pearl.

† Such is the general appearance of fifhes dying under water;
but it is greatly inferior to thofe fingular changes of colour exhi-
bited at the death of a fifh, known formerly to the Romans by the
name of *Mullus* (mullet), whofe hues afforded entertainment to the
gluttons of thofe days *(proceres gulæ)*. See Pliny, *Hift. Nat. Lib. IX.*
17, and Seneca, *Quæft. Nat. Lib. III.* 18. [Nothing can be more
beautiful than the fucceffive changing tints that appear on the furface of the expiring dolphin, and the gradual progrefs of the final
livid hue, from the extremities to the head; a fpectacle which I
have frequently witneffed. T.]

‡ *Cyprinus-Carpio,* Linn.

§ The firft lived eighteen hours under water, the fecond twenty-
feven hours in the air.

fome

fome twice as long as thofe confined under water *.

It may be objected, that, as frogs are found
beneath ice, they may fubfift a confiderable
time without air. But it is well known that,
when water freezes, the air ufually contained
in it is difengaged, and gathers below the fuperficial cruft ; fo that the frogs may ftill inhale
the vital breath.

If, therefore, the foregoing experiments evince
that frogs and fifh cannot exift without air, and
if the experience of all ages and nations proves
that, at certain intervals, at leaft, every amphi-

* Of the two barbels that were left to die out of the water in a
room without a fire, the thermometer being feven degrees above
nought (about 48° of Farenheit), the one was a foot long, weighed
thirty-three ounces, and lived eight hours ; the other meafured
a little more than nine inches and a half, weighed feventeen ounces,
and lived four hours and feventeen minutes : whereas the two fifhes
of the fame fpecies lived under water, the one only three hours and
forty fix minutes, and the other but three hours and a quarter.
But fuch was not the cafe with the dog-fifh, for the largeft, which
was five inches and nine lines long, lived only three hours in the
air; and the other, which was four inches nine lines, lived three
hours and three quarters under water. During the courfe of thefe
obfervations, I thought that I could perceive the agony of the fifh
marked by the ceffation of the regular motion of the gills and by
a periodical convulfion in thefe organs, which returned twice or
thrice in a quarter of an hour : the large barbel had thirteen of
thefe in feventy-feven minutes, and the laft feemed to denote the inftant of its death. In one of the fmall ones, the final moment was
marked by a convulfion of the ventral fins ; but, in moft of them,
that, of all the external and regular motions, which lafted longeft,
was the motion of the lower jaw.

bious

bious animal * whatever requires refpiration;
how could Swallows, thofe daughters of the
air, which feem deftined to circle in that fub-
tile fluid, live fix months without breathing?

An animal which has been fuffocated by
drowning, may frequently indeed be recovered
by ftimulating the lungs, and applying gentle
warmth †; but the experiment never fucceeds
unlefs the immerfion is recent. And fuch inftances are not at all analogous to the fuppofed
refufcitation of Swallows from the bottoms of
lakes. Their appearance or difappearance has
no relation to the quality of the feafon; they
leave us in autumn, when the weather is generally warmer than in fpring, the period of
their return. In the memorable year 1740 ‡,
the Swallows made their appearance during the
fevere frofts, and many perifhed for want of

---

* Beavers, tortoifes, falamanders, lizards, crocodiles, hippopota-
mufes, whales, as well as frogs, rife often out of the water in order
to refpire. Even fhell-fifh, which are, of all, the moft aquatic, feem
to require air, and mount from time to time to the furface, as in the
pool-mufcle. *See* Mery in the Memoirs of the Academy of Sciences for 1710.

† I have thought it proper to infert this fentence, and omit the
long detail which M. de Montbeillard gives of his recovering, by
the fimple application of heat, a Swallow that had fallen into a ba-
fon of water and was taken out ftiff and apparently dead. The
methods ufed in this country for the recovery of drowned perfons
are well known: warmth, gentle motion, and friction; the application of ftimulants to the noftrils, the inflation of the lungs,
&c. T.

‡ Coll. Acad. *part. etran. t. XI. Acad. of Stock. p.* 51.

food;

food; and in the mild, and even warm, fpring of 1774, they arrived no earlier than ufual.

The opinion that Swallows pafs the winter under water feems to have originated in this way: among the number which flock together at night among the rufhes and aquatic plants, on their arrival and previous to their retreat, fome may have been drowned by accident * ; and the fifhermen, finding them in their nets, would carry them to a ftove, and thus reftore them to life. And a paffage in Ariftotle induced the learned to afcribe this fubmerfion to thofe of the northern countries only †, as if the diftance of four or five hundred leagues would prove any bar to birds which can fly through the fpace of two hundred leagues in a day, and which, by advancing farther fouth, may always find a milder temperature, and a more abundant provifion of their infect food. That philofopher indeed believed that the Swallows and fome other birds lay hid during the winter; but his affertion was too general. There are inftances, however, of chimney-fwallows, fandmartins, &c. being feen in mild winters : two fand-martins were obferved to circle about the caftle of Mayac, in Perigord, the whole of the 27th of December, 1775, when there was a foutherly wind, attended with light rain : I

---

* In fummer they are fometimes found drowned in the meers.

† Hift. Anim. *Lib. VIII.* 12 & 16.

have

have a certificate signed by many respectable names to attest this fact. These had, no doubt, been detained by late hatching, or were young birds unable to perform the migration, but fortunate enough to obtain a convenient retreat, a warm season *, and the proper food. Some such occurrences, which are probably more frequent in Greece than in the north of Europe, might dispose Aristotle to think that all the species of Swallows remained concealed and dormant, during the winter months. Klein asserts, in fact, that the sand-martins lie torpid in their holes †; and these are often seen in the winter at Malta, and even in France. M. de Buffon conjectured that the sand-martins are less affected by cold than the other Swallows, since they haunt the brooks and rivers; and that, as they are probably of a colder temperament, and construct their holes like those animals which sleep during the winter, they also undergo the

---

* In this year, 1775, the autumn was fine and not cold in that part of Burgundy where I live, which is two degrees more northerly than Perigueux. Of ninety-five days till the 27th of December, there were only twenty-seven in which the sun did not shine: the thermometer never sunk more than five or six degrees below nought (20°¾ or 18°½ Farenheit), and was often five or six above that point (43°¼ or 45°½ F.), even at the end of December: on the 27th, at sun rise, it was three degrees above. (38°¾ F.)

† To these are added, the swifts, the rails, the nightingales, the warblers; and M. Klein would wish to join many others. Were his system realized, the earth could not furnish caverns enow, nor the rocks holes. And the more general this hiding be supposed, the more would it be notorious. *See* Ordo Avium, *passim*.

same

fame ftate of inaction. Befides, they may find infects in the ground at all feafons, and can therefore fubfift when other Swallows muft inevitably perifh. Inftances of this kind may happen; but we muft not infer that in winter they generally lodge thus concealed. Collinfon directed, in England, a bank which was quite bored by thefe birds to be carefully dug, in the month of October 1757, and yet not one could be found.

If, therefore, Swallows (I might fay the fame of all the birds of paffage) can never obtain under water an afylum congruous to their nature, we muft return to the moft ancient opinion, and the moft confonant to obfervation and experience. When the proper infects begin to fail, thefe birds remove into milder climates, which ftill afford that prey, fo neceffary to their fubfiftence *. This is the general and directing caufe of migrations: thofe which live upon winged infects are the firft to retire, becaufe their provifions are fooneft deficient: thofe which feed upon the *larvæ* of ants and other crawling infects, find a more lafting fupply, and are later in difappearing. Thofe birds, again, which eat berries, fmall feeds, and fruits that ripen in autumn, and hang on the trees the whole winter, do not arrive until autumn, and fettle among us the greateft part of the winter.

* Swammerdam.

Thofe

Thofe which confume the fame provifions with
man, and live upon his fuperfluities, refide conftantly in our vicinity. Laftly, when a new
fpecies of culture is introduced into a country,
it in the end occafions new migrations. Thus
after barley, rice, and wheat, were begun to be
cultivated in Carolina, the colonifts were furprifed to fee, regularly every year, numerous
flocks of birds arrive, with which they were
totally unacquainted, and hence denominated
them *rice-birds*, *wheat-birds* *, &c. It is not
unufual in the American feas to behold immenfe
troops of birds collected to prey on thofe prodigious fwarms of winged infects which fometimes darken the air †. In all cafes, it would
appear, that neither the climate nor the feafon,
but the neceffity of procuring fubfiftence ‡, directs the birds to migrate from one country to
another, to traverfe the ocean, or to fix their
permanent refidence.

There is another caufe alfo, which influences
the migrations of birds, or at leaft prompts them
to return to their natal abode. Like all other

---

* Philofophical Tranfactions, *No.* 483, *art.* 35.

† Second Voyage of Columbus.

‡ It is probable that the migrations of fifhes, and even thofe of
quadrupeds, are fubject to the fame law, or rather to a law ftill more
general, which tends to the prefervation of each fpecies and of each
individual: for inftance, I fhould fuppofe that the flying-fifh would
never have employed their gills to fly, if they had not been purfued
by the bonitos, the dorados, and other voracious fifh; and perhaps
the paffage of birds of prey, which takes place in September, has
fome influence on the departure of the Swallows.

fentient

fentient beings, they cherifh a partial tendernefs to the place that gave them birth; there
they felt their faculties firft expand; there they
tafted the frefh pleafures of the morning of life:
neceffity compelled them to leave with regret
the delicious fpot; but its image ftill dwells in
their bofom, and inceffantly awakens the ardent
craving to return and to renew the felicity of
their infant days *. But, not to enter into a
general difcuffion on the fubject, it appears that
our Swallows retire in the month of October to
the fouthern countries; fince they are obferved
about that time to leave Europe, and in a few
days are found in Africa, and have even more
than once been met with in their paffage on the
ocean. I know, fays Peter Martyr, that the
Swallows, the kites, &c. migrate from Europe
on the approach of winter, and fpend that feafon on the coafts of Egypt. Father Kircher,
that advocate for the fubmerfion of Swallows,
but who confined it to the northern climates,
affirms, from the accounts of the inhabitants of
the Morea, that great numbers of Swallows
pafs annually with the ftorks, from Egypt
and Lybia, into Europe †. Adanfon tells us
that

---

* In that part of Lybia where the Nile has its fource, the Swallows and the kites are ftationary, and remain the whole year. *He-
rodotus, Lib. II.* The fame thing is faid of fome diftricts of Ethiopia. There may be migratory and ftationary Swallows in the fame
country, as at the Cape of Good Hope.
† See the *Mundus Subterraneus* of this Jefuit. Thefe two laft facts

confirm

that the Chimney-fwallows arrive at Senegal about the ninth of October, and retire again in the fpring; and that, on the 6th of October, when he was fifty leagues off the coaft, between the ifland of Goree and Senegal, there alighted on his veffel four birds, which he found to be real European Swallows; and he adds that they were fatigued, and fuffered themfelves to be caught. In 1765, nearly in the fame feafon, the Company's fhip, *Penthievre*, was over-fpread, between the coaft of Africa and the Cape de Verd iflands, by a flight of white rumped Swallows (martins), which probably came from Europe *. Leguat, who was on the fame feas on the 12th of November, alfo obferved four Swallows which followed his veffel feven days, as far as Cape Verd. We may remark that this is precifely the time when bees fwarm profufely in Senegal, and when the gnats called *marin-gouins* are moft troublefome : in fact, this is the end of the rainy feafon, when humidity and warmth at once favour the multiplication of infects, efpecially fuch as the *maringouins*, which hover about wet places †. Chriftopher

confirm my notion, that, even in warm countries, there is a feafon for the generation of infects, of thofe at leaft which fupport the Swallows.

* Note communicated by the Vifcount de Querhoent.

† Confult *Voyage au Senegal*, par M. Adanfon, pp. 36, 82, 139, 141, 157. I fee alfo that clouds of grafshoppers fpread over thefe countries in the month of February (*ib. p.* 88). Is the generation of infects there fixed to a particular feafon?

Columbus,

Columbus, in his fecond voyage, faw one near
his veffels on the 24th of December, though
ten days before he difcovered St. Domingo *.
Other navigators have met with them between
the Canaries and the Cape of Good Hope†. In
the kingdom of Iffini, according to the miffion-
ary Loyer, multitudes of Swallows arrived from
other countries in October and the following
months ‡. Edwards affures us that the Swallows leave England in autumn, and that the
Chimney-fwallow kind are found in Bengal §.
Swallows are feen the whole year at the Cape
of Good Hope, fays Kolben, but they are more
numerous in winter; which fhews that fome
are there permanent fettlers and others migratory, for it cannot well be faid that they fleep
under water or lurk in holes during fummer.
The Swallows of Canada, Father Charlevoix
tells us, are birds of paffage as well as thofe of
Europe. Thofe of Jamaica, according to Dr.
Stubbs, leave the ifland in the winter months,

* Herrera, *Lib. II.* 1.
† Voyage aux Iles de France & de Bourbon. *Merlin,* 1773.
‡ Hift. Gen. des Voyages, *t. III. p.* 422.
§ Other obfervers, who have examined more particularly, affirm,
that the Swallows leave England about the 29th of September;
that their general rendezvous is held on the coafts of Suffolk, between Orford and Yarmouth; that they alight on the roofs of
churches, on old walls, &c.; that they remain feveral days when
the wind is not fair for croffing the fea; that if the wind changes
during the night, they all difappear at once, and not one can be
found next morning.

though

though ever fo warm *. Every body knows the fingular and happy experiment of Frifch, who faftened a dyed thread to the feet of fome of thefe birds, and faw them the following year with this thread not in the leaft difcoloured ; a fufficient proof that thefe individuals, at leaft, did not winter under water, and a ftrong prefumption that none of the fpecies ever do. We may expect that when Afia and certain parts of Africa are better known, we fhall difcover the different ftations not only of the Swallows, but of moft of the birds which the inhabitants of the iflands in the Mediterranean perceive every year advancing or retiring. They cannot undertake their diftant voyages unlefs they be affifted by a favourable breeze ; and when they are furprifed, in the middle of their courfe, by contrary winds, they become exhaufted with fatigue, and alight on the firft veffel they meet with, as feveral navigators have witneffed in the feafon of migration †. They may fometimes chance to fall into the fea and perifh in the waves ; and then, if feafonably fifhed out and properly taken care

---

* Philofophical Tranfactions, No. 36.

† Admiral Wager thus writes Mr. Collinfon : " Returning home in the fpring of the year, as I came into foundings in our channel, a great flock of Swallows came and fettled on all my rigging ; every rope was covered ; they hung on one another like a fwarm of bees ; the decks and carving were filled with them. They feemed almoft famifhed and fpent, and were only feathers and bones ; but being recruited with a night's reft, took their flight in the morning." The fame thing happened to Mr. Wright, mafter of a fhip, on his return from Philadelphia.

of,

of, they may be revived. But it is evident that
such accidents cannot happen in lakes or narrow
seas. In most countries the Swallows are held
the friends of men, and very justly, since they
destroy vast numbers of pernicious insects. The
goat-suckers are entitled to the same regard;
but themselves and their benefits are concealed
and neglected in evening shades.

My first idea was to separate the martins from
the Swallows, and to imitate nature, which has
separated them by implanting reciprocal antipa-
thies. They are never seen associated together,
though the three species of Swallows join sometimes in the same flock. The martins are distinguished too by their shape, their habits, and their
dispositions. 1. By their shape: their legs are
shorter, and entirely unfit for walking or for
rising on the wing from smooth ground; besides, their four toes are turned forward, and
each of them has only two phalanges, including the nail. 2. By their habits: they arrive
later and retire earlier, though they seem to
shun more the heat; they breed in the crevices
of old walls, and as high as they can get; they
build no nest, but line the hole well with coarse
litter, in which respect they resemble the bank-
swallows (sand-martins); when they go a-fo-
raging, they fill their craw with winged insects
of all kinds, so that they need to feed their
young only twice or thrice in the day. 3. By
their

their difpofitions : they are more fhy and timid than the Swallows; the inflections of their voice are lefs varied, and their inftinct feems more confined.

Such obvious differences, therefore, fubfift-ing between thefe birds, I fhould not hefitate to difcriminate them ; but there are many fo-reign fpecies, which it would be difficult to refer each to its proper clafs. It will be more prudent, then, not to attempt the divifion, but to arrange them as their exterior conformation moft readily fuggefts.

Nor fhall we diftinguifh the Swallows of the old and of the new world, becaufe they ex-actly refemble each other, and becaufe the ocean can prove no barrier to birds that fly fo fwiftly, and can equally endure every cli-mate.

The CHIMNEY, or DOMESTIC SWALLOW *.

Hirundo Ruſtica, Linn. Gmel Klein. &c. &c.
Hirundo Domeſtica, Ray. Will. and Briſſ. †

THE inſtinct of this bird is really domeſtic; it prefers the ſociety of man; it neſtles on our chimney-tops, and even within our houſes, eſpecially when theſe are quiet and ſtill. If the houſes be too cloſe and the vents covered above, as they are in Mantua and in mountainous countries, on account of the great falls of ſnow and raïn, it changes its lodgment, without loſing its attachment, and it finds a retreat in the roofs. But it never ſtrays far from the dwellings of men; and the weary forlorn traveller is rejoiced

* In Swediſh, *Ladu-Swala*, or Barn-ſwallow.

† Aldrovandus ſuppoſes that the Αvoπαια of Homer, *Odyſſ.* I. 320, which the commentators have been ſo much puzzled to interpret, is the common ſwallow. The lines in which the word occurs, are theſe:

'Η μεν αϱ' ὡς ειπῦσ' 'απέβη γλαυκῶπις 'Αθηνη,
Oϱνις δ'ὡς 'ανοπαια διέπ7αlο.

Euſtathius ſuppoſes that ανοπαια is a ſpecies of eagle, and Mr. Pope prudently alters the expreſſion:

" Abrupt, with *eagle-ſpeed* ſhe cut the ſky;
" Inſtant inviſible to mortal eye."

It is the Πoικιλη Χελιδων (variegated ſwallow) of Ariſtophanes; the Δαυλιδης Οϱνις (Daulian bird) of Plutarch; the *Aredula* of Cicero; the *vaga volucris* of Ovid.

to

to fee the harbinger of fafety. We fhall foon find that the fwift is more roving in its excurfions.

The common Swallow is the firft that appears in our climates, and generally a little after the equinox of fpring; but rather earlier in the fouthern countries, and later in the northern. And yet though the month of February and the beginning of March be unufually mild, or the end of March and the beginning of April uncommonly cold, they hardly ever arrive in any place before their ordinary time *, and fometimes they glide through the thick flakes of defcending fnow. In 1740, the fwallows fuffered extremely; they gathered in great numbers about a brook which fkirted a terrace then belonging to Mr. Hebert, where every minute fome fell dead †, and the water was covered with their dead bodies: nor was exceffive cold the caufe of their death; it was evidently the want of food, and thofe picked up were reduced to mere fkeletons; the walls of the terrace were their laft

---

* Pliny, *Lib. XVIII.* 26, fays that Cæfar mentions fwallows feen on the eighth of the Calends of March (22 February); but this is a fingle fact, and perhaps the birds were Sand-Martins.

† " In 1767, they were found extended and lifelefs on the brink of the pools and rivers of Lorraine." *Note communicated by M. Lottinger.* Thefe facts render very fufpicious at leaft the prefentiment of temperatures which a paftor of Nordland and fome others have thought proper to afcribe to the fwallows. *See* Collec. Acad. *Part Etran. T. XI. Acad. Stock. p.* 51.

refort,

refort, and they greedily devoured the dried
flies that hung from the old fpiders webs.

A bird, which announces the return of the
fmiling feafon, and which is innocent and even
ufeful, might be treated with gratitude; and
by the bulk of mankind, it is venerated with a
degree of affection bordering upon fuperftition *.
Yet is the fwallow often the fubject of cruel
fport; and the expert markfman is eager to difplay his fkill in fhooting it on the wing: and
what is fingular, the firing of the piece rather
attracts than fcares thefe harmlefs creatures;
this war is worfe than ridiculous, and the various infect tribes which prey in our gardens, in
our fields, and in our forefts, are thus fuffered
to extend their ravages †.

The experiment of Frifch ‡ and other fimilar
ones, prove that fwallows return to the fame
haunts. They build annually a new neft, and
fix it, if the fpot admits, above that occupied
the preceding year. I have found them in the
fhaft of a chimney thus ranged in tires; counted

* The Swallows have been faid to be under the immediate protection of the *Dii Penates:* When ill ufed, they bit the cows
udders, it was alleged, and made them lofe their milk. Thefe
were ufeful illufions.

† See Journal de Paris, *annec* 1777. It is true that they fometimes alfo deftroy ufeful infects, fuch as bees; but they can always
be prevented from building their nefts near the hives.

‡ In a caftle near Epinal in Lorraine, a few years ago, a ring
of brafs wire was faftened to the foot of one of thefe Swallows,
which it faithfully brought back on the following feafon. Heerkens
in his poem, *Hirundo,* cites another fact of this kind.

four

four one above another, and all of equal fize,
plaftered with mud mixed with ftraw and hair.
There were fome of two different fizes and
fhapes: the largeft refembled a hollow half cylinder *, open above, and a foot in height, and
attached to the fides of the chimney; the
fmalleft were ftuck in the corners of the chimney, and formed only the quarter of a cylinder,
or even an inverted cone. The firft neft, which
was the loweft, had the fame texture at the
bottom as at the fides; but the two upper tires
were feparated from the lower by their lining
only, which confifted of ftraw, dry herbs, and
feathers. Of the fmall nefts built in the corners, I could find only two in tires, and I fuppofe they belonged to young pairs; they were
not fo well compacted as the large ones.

In this fpecies, as in many others, it is the
male that fings the amorous ditty †: but the female is not entirely mute; in the love-feafon
fhe twitters more fluently, fhe warmly receives
his careffes, and fometimes, by her fportive
frifks, fhe roufes and ftimulates his paffion.
They have two hatches in the year, the firft

* Frifch fays, that the bird gives to its neft this circular or rather femicircular form, by making its foot the centre.

† The Greeks exprefs this note by thefe words, Τιθυριζειν, Τιτυ-
βιζειν, and the Latins by thefe other names, *Drinfare*, or *Trinfare*,
Zinzilulare, *Fritinnire*, *Minurifare*. M. Frifch tells us, that, of all
the fwallows, the domeftic one has a cry neareft refembling a fong,
though it confifts only of three tones, terminated by a *finale*, which
rifes to a fourth, and it is little varied.

containing

containing five eggs, the fecond three: thefe are white according to Willughby, and fpotted according to Klein and Aldrovandus: what I faw were white. While the female fits, the male fpends the night on the brim of the neft; he fleeps little, for his twittering is heard at the earlieft dawn, and he circles till almoft the clofe of the evening. After the young are hatched, both parents perpetually carry food, and are at great pains to keep the neft clean, till the brood learn to fave them that trouble. But it is pleaf- ing to fee them teaching their family to fly, encouraging them with their voice, prefenting food at a little diftance, and retiring as the young ones ftretch forward; preffing them gently from the neft, fluttering before them, and offering, in the moft expreffive tone, to re- ceive and affift them. Boerhaave tells us that a Swallow returning with provifion to its neft, and, finding the houfe on fire, rufhed through the flames to feed and protect her tender brood. How ftrong the attachment to their progeny!

It has been faid that when their young had their eyes funken or even torn out, the mo- thers cured them by the application of the herb *chelidonia* *, or fwallow-wort †, deriving its

---

* From Χελιδων, a fwallow. The common Englifh name *celandine*, feems to be only foftened from *chelidonia*. The plant is ranged by Linnæus next the poppy. T.

† *Ut quidam volunt, etiam erutis oculis.* Pliny, *Hift. Nat. Lib. XXV.* 8. Diofcorides fays nearly the fame thing, *Lib. II.* 211. Ælian reftricts it to the white Swallows, *Lib. XVII.* 20.

name

name from that imaginary quality. But the experiments of Redi and De la Hire prove, that no
fimples are needed, and that, in the infant brood,
the eyes, though burft and funken, foon fpon-
taneoufly recover *. Ariftotle knew this fact †,
Celfus repeats it ‡, and the obfervations of Redi
and De la Hire, and fome others §, inconteftibly
prove it.

Befides the different inflections of voice which
I have already noticed, the common fwallows
have their cry of invitation, their cry of pleafure,
their cry of fear, their cry of anger, that by
which the mother warns her young of the dan-
gers which threaten, and many other expreffions
compounded of thefe; a proof of their great fuf-
ceptibility of the internal fentiments.

Since the winged infects fly higher or lower
according to the greater or lefs degree of heat,
the Swallows fometimes, in the purfuit of their
prey, fkim along the furface, and gather it on
the ftems of herbs, on the grafs, and even on
the pavement of ftreets. When the fcarcity is
great, they ravifh the flies from the fpider's
web, and even devour the fpiders themfelves ‖.

* Redi made his experiments on pigeons, hens, geefe, ducks, and
turkies. *See* Coll. Acad. *Part Etran. T. IV. p.* 544. alfo *T. II.
Part. Fran. p.* 75.

† *Hift. Anim.* Lib. II. 17, and Lib. VI. 5, and *De Generatione,*
Lib. IV. 6. Ariftotle fays the fame thing of ferpents.

‡ Lib. VI. *De Re Medica.*

§ For inftance, Dr. J. Sigifmond Elfholtius. Coll. Acad. *Part.
Etran. T. III. p.* 324.

‖ Frifcn.

Their

Their ftomach is found to contain fragments of flies, grafshoppers, beetles, butterflies, and even bits of gravel *, a proof that at times they catch their prey on the ground: and in fact, though the domeftic Swallows fpend moft of their lives in the air, they often alight on the roofs of houfes, on iron bars, and even on the furface of the earth, and on trees. In our climate, they often pafs the night about the end of fummer perched on alders that grow on the banks of ri-vers; and in that feafon numbers are caught, which are eaten in fome countries †. They prefer the loweft branches under the brinks, and well fheltered from the wind ‡; and it is re-marked that the branches where they commonly fit during the night wither away.

They alfo affemble on a large tree previous to their retreat; the flocks then amount only to three or four hundred, for the fpecies is far from being fo numerous as the window Swallows (martins). In this country they commence their expedition about the beginning of October, and ufually fteal off in the night to avoid the birds of prey, which feldom fail to harafs them on their

---

* Belon and Willughby. Many abfurdities have been told of thefe fwallow-ftones and their virtues, as of eagle-ftones, cock-ftones, and other bezoars, which feem ever to have been the fa-vourite jewels of empiricifm and of credulity.

† At Valencia in Spain, at Lignitz in Silefia, &c. See Wil-lughby and Schwenckfeld.

‡ Note of Hebert. Lottinger affures me that they alfo frequent fometimes the coppices.

route.

route. Frifch faw them frequently fet out in
broad day, and Hebert, more than once, obferved, about the time of their retreat, parties
of forty or fifty gliding aloft in the air, and remarked that their flight was not only much
higher than ordinary, but more uniform and
fteady. They ftretch towards the fouth, taking advantage, as much as poffible, of favourable winds; and when no obftacles interfere,
they ufually arrive in Africa in the firft week of
October. If they be checked by a fouth-eaft
wind, they halt, like the other birds of paffage,
in the iflands that lie in their track. Adanfon
faw them arrive on the fixth of October, at half
paft fix in the evening, on the coaft of Senegal,
and found them to be real European fwallows;
he afterwards difcovered that they are never
feen in thofe countries but in autumn and winter. He tells us that they lie every night fingle,
or two by two, in the fand by the fea fhore * ;
and fometimes numbers lodge on the huts,
perching upon the rafters. Another important
obfervation he adds, that they never breed in
Senegal † ; and accordingly Frifch remarks that
young Swallows never arrive in the fpring. Hence

* This habit of lying in the fand is entirely contrary to what we
fee in Swallows in our climate; it muft depend on fome particular
circumftances that efcaped the obferver; for animals are more capable, than ufually fuppofed, of varying their mode of life according to their fituation.

† It is alfo faid that no fpecies of Swallow neftles in Malta.

we

we may infer that thefe birds are natives of more
northern climates.

Though the Swallows are in general migratory, even in Greece and in Afia, fome will remain during the winter, efpecially in the mild
climates where infects abound; for example, in
the ifles of Hières and on the coaft of Genoa,
where they fpend the night in the open country on the orange fhrubs, which they injure
greatly. On the other hand, they are faid to
appear feldom in the ifland of Malta.

Thefe birds have fometimes been employed to
convey important intelligence*: for this purpofe, the mother is taken from her eggs and
carried to the place whence the news is to be
fent, and a thread is tied to the feet, with the
number of knots and the colour previoufly con-
certed. The affectionate mother flies back to
her brood, and tranfports the billet with incredible expedition.

The chimney Swallow has its throat and front
of an orange tint, and there are two ftreaks
above the eye, of the fame colour; all the reft
of the under fide of the body is whitifh, with
an orange caft; all the reft of the upper part of
the head and body is of a brilliant bluifh black,
the only colour which appears when the feathers
are compofed, though they are cinereous at the
bafe, and white in the middle; the quills of the

* See Pliny, *Nat. Hift. Lib. X.* 24.

wings

wings are, according to their different pofitions,
fometimes of a bluifh black, which is lighter
than the upper furface of the body, and fometimes of a greenifh brown; the quills of the tail
are blackifh, with green reflections; the five
lateral pairs marked with a white fpot near the
end; the bill is black without, and yellow
within; the palate and the corners of the mouth
are alfo yellow, and the legs blackifh. In the
males the orange tint on the throat is more
vivid, and the white of the under fide of the
body has a flight caft of reddifh.

The average weight of all thofe which I have
tried is about three gros. They are apparently
larger than the window Swallows (martins),
and yet they are lighter.

Total length fix inches and a half; the bill
forms a curvilineal ifofceles triangle, whofe fides
are concave, and about feven or eight lines;
the tarfus five lines, without any down; the
nails thin, flightly curved, and much pointed,
and the hind one is the ftrongeft; the alar extent, a foot; the tail three inches and a quarter, much forked, though lefs fo in the young
birds, confifting of twelve quills, of which the
outer pair exceeds the next by an inch, and
the middle pair by fifteen or twenty lines, and
the wings by four or five lines; it is generally
longer in the male.

I have received, as varieties, fome in which
the

the colours were all fainter, and the tail little forked; thefe were probably young ones.

Among the accidental varieties I place the following. *Firſt*, The white Swallows: there is no country in Europe where thefe have not been feen, from the Archipelago to Pruffia *. Aldrovandus tells how to obtain them of that colour; according to him, we need only rub their egg with olive-oil. Ariftotle imputes this whitenefs to weaknefs of conftitution, want of food, and the action of cold. In a fubject, which I had occafion to obferve, there were fome fhades of rufous above the eyes and under the throat, and fome traces of brown on the neck and the breaft, and the tail was fhorter; perhaps its faint colours were owing to moulting, for though white Swallows are frequently feen before their paffage, it is unufual to find fuch on their return †. Some are obferved to be only partly white, as was the one mentioned by Aldrovandus, which had its rump of that colour.

In the *fecond* place, I confider as an accidental variety, the rufous Swallow, of which the orange tint of the throat and eye-lids fpreads over

---

* At Samos, according to the ancients: in Italy, in France, in Holland, in Germany, according to the moderns.

† In a hatch of five young, at the Trinitarians of la Motte, in Dauphiny, were two white Swallows which paffed the whole year in the country, but returned not the following year. *Note of the Marquis de Piolenc.*

　　　　almoft

almoſt the whole of the plumage, but grows more dilute, and verges upon pink *.

The chimney Swallows are ſcattered through the whole of the ancient continent, from Norway to the Cape of Good Hope, and in the Aſiatic regions, as far as India and Japan †. Sonnerat brought a ſpecimen from the Malabar coaſt ‡, which differs only in being rather ſmaller, owing probably to the contraction in drying. Seven other Swallows brought from the Cape of Good Hope, by the ſame gentleman, were exactly ſimilar in appearance to ours ; but on a narrow inſpection, it was found that the under part of the body was of a finer white, and the ſcalloping, which, in the ten lateral quills of the tail, divides the broad from the narrow part, was larger.

I ſhall now deſcribe ſuch as are to be regarded as varieties of climate [A].

* The Count de Riolet aſſured me that he ſaw two individuals of this colour in a flock of chimney Swallows.

† Edwards and Kæmpfer.

‡ G. I. Camel had long before inſerted the Swallow, under the name of *Layang-layang*, in the catalogue of European birds found in the Philippines. *Philoſ. Tranſ. No.* 285. *Art. III.*

[A] Specific character of the Common Swallow, *Hirundo Domeſtica :* " Its tail-quills, the two mid-ones excepted, are marked with a white ſpot." Mr. White has given a very accurate and diſtinct hiſtory of this bird. *Natural Hiſtory of Selborne, pp.* 167—172.

FIG.1.THE CHIMNEY SWALLOW FIG.2.THE MARTIN.

VARIETIES of the COMMON SWALLOW.

I. THE ANTIGUA SWALLOW, with a rufty-coloured throat*. It is rather fmaller than the common Swallow; its front bears a band of rufty yellow; under the throat there is a fpot of the fame colour, terminated below by a very narrow black collar; the forepart of the neck and the reft of the under furface of the body, white; the head, the upper fide of the neck and back, velvet black; the fmall fuperior coverts of the wings of a changeable violet; the great coverts and alfo the quills of the wing and tail, are coal black; the tail forked and projects not beyond the wings.

II. THE RUFOUS-BELLIED SWALLOW OF CAYENNE †. Its throat is rufous, and this colour extends over all the upper fide of the body, gradually fhading off; all the reft of the upper fide of the body is of a fine fhining black. It is rather fmaller than the common Swallow.

* *Hirundo Panayana*, Gmel.
The *Panayan Swallow*, Lath.
Specific character: "It is black, below white; a fpot on its front and its throat, ferruginous yellow; its collar black.

† *Hirundo Rufa*, Gmel.
The *Rufous-bellied Swallow*, Lath.
Specific character: "It is glofly black, below rufous, its front whitifh."

4   Total

Total length about five inches and a half; the bill fix lines; the tarfus four or five; the hind toe five.

Swallows of this kind alfo make their neft in houfes; they give it a cylindrical form with fmall ftalks, mofs, and feathers, and fufpend it vertically detached from the building; they lengthen the ftack in proportion as they multiply; the aperture is placed below in one of the fides, and fo nicely conftructed that it communicates with all the ftories. They lay four or five eggs.

It is not improbable that fome of our fwallows having migrated into the new continent, have there founded a colony, which ftill refembles the parent breed.

III. The Rufous-cowled Swallow *. This rufous is deepened and variegated with black; the rump is alfo rufous, terminated with white; the back and the fuperior coverts of the wings are of a fine black, verging upon blue, with the glofs of burnifhed fteel; the quills of the wings brown, edged with a lighter brown; thofe of the tail blackifh; all the lateral ones marked on the infide with a white fpot, which does not appear unlefs the tail is fpread; the throat is variegated with whitifh and brown: laftly, the under fide of the body is fprinkled

* _Hirundo Capenfis_, Gmel.
The _Cape Swallow_, Lath.

with

with fmall longitudinal blackifh fpots on a pale
yellow ground.

The Vifcount Querhoent, who had an opportunity of obferving this fwallow at the Cape of
Good Hope, informs us that it breeds in houfes
like the preceding varieties; that it fixes its neft
againft the ceilings of rooms; that it ufes earth
for the outer coat, and lines it with feathers;
that the fhape of its neft is roundifh, with a fort
of hollow cylinder fixed to it, which is the only
aperture. He adds that the female lays four or
five dotted eggs.

FOREIGN BIRDS,

WHICH ARE RELATED TO THE COMMON SWALLOW.

I.

The GREAT RUFOUS - BELLIED SWALLOW of SENEGAL.

Hirundo Senegalenſis, Linn. Gmel. and Briſſ.
The *Senegal Swallow,* Lath.

ITS tail is ſhaped like that of the common Swallow; and its plumage is marked with the ſame colours, though differently diſtributed: it is much larger, and moulded after other proportions; ſo that it may be regarded as a diſtinct ſpecies. The upper ſide of the head and neck, the back and the ſuperior coverts of the wings, are of a brilliant black, with a ſteel gloſs; the quills of the wings and of the tail are black, the rump rufous, and all the lower parts; but the throat and the inferior coverts of the wings are much diluter, and almoſt white.

Total length eight inches and ſix lines; the bill eight lines; the tarſus the ſame; the hind nail and toe the longeſt next to thoſe of the middle; the alar extent fifteen inches three lines; the tail four inches, forked, and conſiſt-

ing

ing of twenty-fix quills; it projects an inch beyond the wings [A].

II.

THE

WHITE-CINCTURED SWALLOW.

Hirundo Fafciata, Gmel.
The *White-bellied Swallow*, Lath.

IT has no rufous in its plumage, which is entirely black, except a white belt on the belly, confpicuous on that dark ground: there is
alfo a little white on the thighs; the quills of
the tail are black above and brown below.

It is a rare bird; found in Cayenne and Guiana in the interior parts of the country, on the
banks of rivers. It delights to fweep along
the furface of water, like the European Swallows; but, different from them, it alights on
the trunks that float down the ftream.

Total length fix inches; the bill black, and
meafures fix lines; the tarfus alfo fix lines;
the tail two inches and a quarter, and forked
near eighteen lines; it exceeds the wings four
lines [B].

[A] Specific character: " It is gloffy black, below rufous;
its rump rufous."

[B] Specific character: " It is black; a crofs bar on its belly,
and an external fpot on its legs, white."

III.

The AMBERGRIS SWALLOW.

Hirundo Ambrofiaca, Gmel.
Hirundo Riparia Senegalenfis, Briff.
Hirundo Marina Indigena, Seba.
Hirundo Ambram Grifeam redolens, Klein.

SEBA fays that thefe Swallows, like our fand
martins, repair to the beach when the fea is
agitated, and that they were fometimes brought
to him both dead and alive, and fmelt fo ftrongly of ambergris, that one of them was enough
to perfume a room. He thence conjectures
that they feed on infects, and other odorous
animalcules, or perhaps on ambergris itfelf.
The one defcribed by Briffon was fent from
Senegal by Adanfon; but the bird is fometimes
feen likewife in Europe.

All its plumage is of a fingle colour, which
is brown-gray, darker on the head and on the
quills of the wings than on the other parts;
the bill is black, and the legs brown; the fize
of the bird exceeds not that of the gold-crefted
wren.

I have hefitated whether I fhould not range
it with the fand martins, which it refembles in
fome refpects; but, as its economy is unknown,
and as its tail is formed like that of the domeftic
Swallow,

Swallow, I have meanwhile referred it to that fpecies.

Total length five inches and a half; the bill fix lines; the tarfus three; the hind toe the fhorteft; the alar extent above eleven inches, forked about eighteen lines, and confifting of twelve quills; it projects four lines beyond the wings.

[A] Specific character: " It is gray-brown; its bill blackifh; its legs brown."

The M A R T I N*.

L'Hirondelle au Croupion Blanc, ou L'Hirondelle de Fenetre, Buff. †
Hirundo Urbica, Linn. Gmel. Kram. Friſ. &c.
Hirundo Ruſtica, ſive Agreſtis, Ray, Will. and Briſſ.
Hirundo Sylveſtris, Geſner.
The *Martin, Martlet,* or *Martinet*, Will. Alb. Penn. and Lath.

THE epithet *rural* was by the ancients
juſtly applied to this bird, which, though
much more familiar than the ſand martin, is ſhy-
er than the domeſtic ſwallow. It delights to
build its neſt againſt the crags of precipices that
overhang lakes ‡; and it never breeds near our

---

* The Greek name, Χελιδων, we are told by Ælian, ſignified a
fig, and was transferred to the ſwallow, becauſe the appearance of
this bird announces the ſeaſon of fruits. It was alſo called
Αχαιθυλλις.

Pliny ſtyles the Martin *Hirundo Ruſtica* and *Hirundo Agreſtis*;
Lib. X. 43, &c.

In German it has a variety of names, *Kirch-Schwalbe, Mur-
Schwalbe, Berg-Schwalbe, Dach-Schwalbe, Fenſter-Schwalbe, Lau-
ben-Schwalbe, Leim-Schwalbe (i. e.* the church, wall, rock, roof,
window, leaf, lime Swallow), and *Mur-Spyren, Munſter-Spyren,
Wyſſe-Spyren (i. e.* the wall, cathedral, white Martlet). In Swediſh, *Hus Swala:* In Daniſh, *Bye-Svale, Tag-Skiægs-Svale, Hvid-
Svale, Rive-Skarſteens-Svale:* In Norwegian, *Huus-Svale.*

† *i. e.* The White-rumped or Window Swallow.

‡ This obſervation is M. Hebert's. Theſe ſwallows are well
known to neſtle on rocks. *See* Geſner, *Aves,* p. 565. M. Guys,
of Marſeilles, has aſſured me of this fact; but we muſt abate from
the exaggerated accounts of the ancients, of a very ſolid bank, a
ſtadium in length, formed entirely by theſe neſts, in the port of
Heracleum in Egypt; and of another ſimilar bank conſtructed alſo
by theſe birds in an iſland ſacred to Iſis. *See* Plin. *Lib. X.* 33.

houſes,

houfes, if it can elfewhere find a convenient
fituation.

The neft which I obferved in the month of
September, and which had been broken off from
a window, was compofed externally of earth,
particularly of the foft mould thrown up in the
morning by worms in new-delved borders; the
middle was ftrengthened by an intermixture of
ftraw chips, and the infide was bedded with a
heap of feathers *; the duft in the bottom
fwarmed with hairy worms, which writhed
and crawled nimbly in all directions, and were
moft numerous where the feathers ftuck into
the fides; there were alfo fome fleas, bigger
and browner than ordinary, and feven or eight
bugs, creeping at large, though none of thefe
could come from the houfe. The three young
ones, which were able to fly, and the parents,
I am confident, flept together at night. The
neft refembled the quarter of a hollow hemif-
pheroid of a deep fhape, its radius four inches
and a half, fticking by its two lateral furfaces
to the jamb and the window frame, and by its
upper furface to the lintel; the entrance was
near the lintel, placed vertically, very narrow
and femicircular.

The fame nefts ferve for feveral years, and
probably to the fame pair; but this is the cafe
with regard to fuch only as are built in our

* I found four or five gros of thefe feathers in a neft that weighed in all but thirteen ounces.

windows, for I am affured that thofe conftructed againft rocks are renewed annually. Sometimes five or fix days are fufficient for performing the work, and fometimes ten or twelve are
required; the birds carry the mortar both with
their little bill and with their toes, but plafter
with their bill only. It often happens that feveral Martins are feen labouring at the fame
neft*; either from their complaifance in affifting each other, or becaufe this fpecies copu-
lating only in the neft, all the males which court
the fame female are eager to haften the fabric,
and obtain the expected joy. Yet fome have
been obferved as affiduous in pulling down the
ftructure as others were forward to rear it.
Perhaps it was a difcarded lover, who gratified
his malice by retarding the fruition of his more
fortunate rivals.

The Martins arrive fooner or later, according
to the latitude: at Upfal on the 9th of May, as
Linnæus tells us; in France and England in
the beginning of April †, eight or ten days after the domeftic fwallows, which, according to
Frifch, as they fly lower, can more eafily and
earlier procure their food: they are fometimes
                                        furprifed

---

* I have counted five ftanding within the fame neft or clinging
round it, without reckoning the comers and goers: the more numerous they are, the more expeditious is the work.

† This year, 1779, the winter has been without fnow, and the
fpring very fine; yet thefe fwallows arrived not in Burgundy till
the 9th of April, and on the lake of Geneva till the 14th, It is faid
                                        that

furprifed by the fpring colds, and have been
feen fhooting through a thick fall of fnow *.
On their firft arrival they haunt the wet places;
I never faw them return to the nefts which are
in my windows before the 15th of April, and
fometimes not till the beginning of May. They
build in all afpects, but prefer fuch as look into
the fields, efpecially when the fcene is inter-
fperfed with rivers, brooks, or pools. They

that a fhoemaker in Bafil, having put a collar on a fwallow with this
infcription,

Hirondelle,
Qui es fi belle
Dis-moi, l'hiver ou vas-tu!
(Pretty fwallow, tell me whither thou goeft in winter ?)
Received, the fpring following, by the fame courier, this anfwer:
A Athenis
Chez Antoine,
Pourquoi t'en informes-tu !
(To Anthony at Athens; why doft thou inquire ?)

The moft probable part of this anecdote is, that the verfes were
made in Switzerland. Belon and Ariftotle affure us that the fwallows live only half the year in Greece, and go to pafs the winter in
Africa.

* This proves that what Hoegftroem, the paftor of Nordland,
fays of the fore-knowledge of temperatures, which he afcribes to
the fwallows, is not more applicable to that of the window than to
that of the chimney, and muft be regarded as very doubtful. " In
Lapland," fays he, " fwallows have been feen to depart, and *abandon their young* in very warm weather, and when there was no appearance of a change in the air. But this change fpeedily came,
and one might travel in a fledge by the 8th of September. In certain years, on the contrary, they have ftaid very late, though the
weather was not mild; whence it might be inferred that the cold
was diftant." In all this the reverend paftor feems to be only the
echo of popular rumour, and to have taken no pains to afcertain
the fact, which is befides contradicted by accurate obfervations.

breed,

breed, at times, within houfes; but this is exceedingly rare, and even very difficult to ob-
tain*. The young are fometimes hatched as
early as the 15th of June; the cock and hen
may be feen toying with each other on the
brink of the half-formed fabric, and billing
with a fhrill expreffive chirp †; but they are
never obferved to copulate, which makes it
probable that this is done in the neft, fince this
chirping is heard early in the morning, and
fometimes during the whole night. Their
firft hatch confifts of five white eggs, with a
dufky ring near the large end; the fecond
hatch confifts of three or four, and the third,
when it does take place, of two or three. The
male feldom or never removes from his mate

---

* " It rarely builds in houfes," fays Ariftotle, which is confirmed by daily obfervation. The late M. Rouffeau, of Geneva,
after infinite pains, fucceeded to make them neftle in his chamber.
M. Hebert faw them build on the fpring of a bell; the bottom of
the neft refted on this fpring, the upper brim, which was femicircular, leaned againft the wall by its two ends, three or four inches
below the eave: the cock and hen, during the time they were employed in the conftruction, paffed the nights on the iron fpike to
which the fpring was faftened. The frequent concuffion given by
this fpring could not fail to difturb the action of nature in the developement of the little embryons; the hatch accordingly did not
fucceed: yet the pair would not forfake their tottering manfion,
but continued to inhabit it the reft of the feafon. The femicircular form which, on this occafion, they gave their neft, proves that
they can fometimes change their order of architecture.

† Frifch pretends that the males of this fpecies fing better than
thofe of the domeftic fwallow; but in my opinion it is quite the re-
verfe.

during

during incubation; he watches for her fafety
and that of the young brood, and darts impetuoufly on the birds that chance to approach too
near. After the eggs are hatched, both parents frequently carry food, and feem to beftow the moft affectionate care*. In fome
cafes, however, this paternal attachment appears to be forgotten: a young one which
was already fledged, having fallen out of the
neft upon the fole of the window, the parents
took no heed of it; but, finding itfelf thus abandoned, it ftrove to efcape, flapped its wings,
and, after three or four hours exertions, it
launched at laft into the air. I broke off, from
another window, a neft containing four young
ones juft hatched, and fet it in the fole of the
window; and yet the parents paffed and repaffed inceffantly and fluttered about the fpot,
without regarding the imploring cries of their
progeny †: a hen fparrow would, in fuch circumftances, have fed and tended her offspring
a fortnight. It would feem, therefore, that the
affection of the Martins for their young depends

---

* When the young are hatched, their excrements are faid to be
enveloped in a fort of pellicle; which enables the parents to roll
them eafily out of the neft. *Frifch.*

† A whole hatch having been put in the fame cage with the parents, thefe paffed the night fometimes on the bar of the cage,
fometimes on the brim of the neft, almoft always the one after the
other, and at laft one upon the other, without beftowing the fmalleft attention to their young: but it may be faid, that in this cafe
the paternal love was fwallowed up by the regret for the lofs of
liberty.

on

on the local fituation; however they continue to fetch them provifions for a long time, and even after they have begun to fly; thefe confift in winged infects, fnapped in the air, which is fo peculiarly their mode of catching *, that if they fee one fitting on a wall, they will fweep paft it to ftart the prey.

It has been faid that the fparrows often occupy the Martins' nefts, which is true. It has been added that the Martins thus thruft out return fometimes efcorted by auxiliaries and, in an inftant clofing up the aperture with the ufual mortar, take vengeance on the ufurpers † : whether this ever happened I cannot decide; but the inftances which have come under my obfervation do not countenance the opinions. The Martins returned frequently in the courfe of the fummer, quarrelled with the fparrows, and fometimes circled about for a day or two, but never attempted to enter the neft or to fhut it up. Nor can we fuppofe any antipathy to fubfift between thefe birds ‡ ; the fparrows will lay wherever they find it convenient.

Though the Martins are fhyer than the chimney fwallows, and though philofophers have

* This is the general opinion, and the moft confonant to daily obfervation; yet M. Guys affures me that thefe birds feek pinewood, in which they find caterpillars.

† Albertus firft broached this error; Rzaczynfki repeated it ; the Jefuit Batgowfki afferts his being a witnefs of the fact; and Linnæus gives it as a truth afcertained.

‡ Albertus *apud Gefnerum.*

believed

believed that they were incapable of being tamed *, yet is it very eafy to fucceed. They muft be fupplied with the proper food, which confifts of flies and butterflies †, and muft receive it often; above all, they muft be foothed into the lofs of liberty, a fentiment common to all animals, but in none fo lively or fo acute as in the winged tribes ‡. A tame Martin § was known to grow extremely fond of its miftrefs; it fat whole days upon her knees, and, when fhe appeared after fome hours abfence, it uttered joyous accents, clapped its wings, and fhewed every fign of lively feeling : it began to feed out of the hand, and its education would probably have com-

* M. Rouffeau, of Geneva.

† Some authors pretend that they cannot exift on vegetable fub-ftances ; yet we cannot fuppofe that thefe prove a poifon to them: bread was part of the food of the tame fwallow which I fhall pre-fently mention. But what is more fingular, children have been feen to feed young fwallows with dung that has dropt from the neft of another fwallow of the fame fpecies. The brood lived very well for ten days on this diet, and in all probability they would have fubfifted longer, had not the experiment been interrupted by a mother, who was fonder of cleanlinefs than of gaining knowledge.

‡ " I have often," fays M. Rouffeau, " had the pleafure of fee-ing them kept in my chamber while the windows were fhut, and fo tranquil as to chirp, frolic, and toy at their eafe, waiting till I fhould open for them, confident that I would not delay ; in fact I rofe, for that purpofe, every morning at four o'clock."

The voyager Leguat fpeaks of a tame fwallow that he had brought from the Canaries to the ifland of Sal; he let it out every morning, and it faithfully returned in the evening. *Voyage aux In-des Orientales, p.* 13. Leguat does not fay what fpecies it was. Other perfons have raifed fwallows. *See* Volfgang Franzius, *Hift. Anim. p.* 456 ; and the *Journal de Paris for* 1778.

§ In the noble Chapter of Leigneux in Fores.

pletely

pletely fucceeded, had it not efcaped. It did not fly far; it alighted on a young child, and foon fell a prey to a cat. The Vifcount Querhoent affures me, that he alfo trained, for feveral months, fome young Martins taken out of the neft; but he could never bring them to eat by themfelves, and that they always died when he gave over feeding them. When the one I have juft mentioned attempted to walk, it moved ungracefully, on account of its fhort legs; and, for this reafon, the Martins feldom alight but upon their nefts, and only in cafes where neceffity obliges them: for inftance, when they gather mud for building with, or when they fpend the night among the reeds towards the end of fummer, at which time they are become fo numerous as not to be all contained in their former lodgments*; or, laftly, when they affemble upon the ridges and corners of houfes previous to their migration. Hebert had a houfe in Brie, which was every year their general rendezvous; the number congregated was great, not only on account of their own multiplication, but becaufe many others of their kindred fpecies, the fand-martins and chim-

---

* About the end of fummer, they are obferved in the evenings circling in great numbers on the furface of water, almoft till dark: it is probably in order to repair to fuch fituations that every day they affemble an hour or two before fun-fet. Add, that they are much lefs frequent in towns about the evening than during the courfe of the day.

ney

ney fwallows, joined them : they had a peculiar
cry, which feemed to call them together. It
was remarked that fhortly before they began
their voyage they exercifed themfelves in foar-
ing to the clouds, thus preparing to wing their
courfe through the lofty regions*; a fact
which agrees with other obfervations related in
the preceding article, and which explains why
the fwallows are fo feldom feen in the air during their paffage from one country to another.

The Martins are widely diffufed through the
ancient continent; yet Aldrovandus afferts that
they are never feen in Italy, particularly in the
neighbourhood of Bologna. M. Hermann †
tells me, that in Alface they are caught with
the ftares, by fpreading a net about the clofe of
the evening over a marfh full of rufhes, and by
drowning next morning the birds that are entangled under it. Some of thefe drowned Martins may be reftored to life, and a fimple fact of
that kind might have given rife to the fable of
their annual immerfion and emerfion.

This fpecies appears to hold a middle rank
between the chimney fwallow and the black
Martin. It has little of the chirping and fami-
liarity of the former; but it builds its neft fi-

* Note communicated by M. Lottinger.
† This profeffor affures me that the White-rumps or Martins
grow fat in autumn, and are then very good to eat. Franzius fays
nearly as much: yet I publifh it with regret, as it tends to the deftruction of an ufeful fpecies.

milarly,

milarly, and its toes confift of the fame number of *phalanges*. It has the rough feet of the
black Martin, and its hind toe alfo turns forward; like that bird, too, it flies through heavy
rains, and then in larger flocks than ufual; it
clings alfo to the walls, and feldom alights on
the ground, and, when it does fo, it rather
creeps than walks. Its bill is wider than that
of the chimney-fwallow, at leaft apparently fo,
becaufe the mandibles open fuddenly as high as
the ears, and the edges form on each fide a pro-
jection: laftly, though it is fomewhat larger,
it feems rather fmaller, its feathers, and efpecially the inferior coverts of its tail, being not
fo fully webbed. The average weight of all
that I have weighed was conftantly from three
to four gros.

The rump, the throat, and all the under furface of the body, are of a fine white; the fide
of the coverts of the tail is brown; the upper
furface of the head and neck, the back, thofe
of the feathers and of the primary coverts of the
tail, are of a gloffy black, with blue reflections;
the feathers of the head and back cinereous at
the bafe, white in the middle; the quills of the
wings brown, with greenifh reflections on the
borders; the three laft of thofe next the body
are terminated with white; the legs clothed as
far as the nails with a white down; the bill
black, and the legs brown gray; the black of
the female is lefs diftinct, and its white not fo
                                        pure,

pure, even variegated with brown on the rump.
In the young ones the head is brown, and there
is a fhade of the fame colour under the neck :
the reflections from the upper furface of the
body are of a lighter blue, which has a greenifh
caft in certain pofitions, and, what is remarkable, the quills of the wings are of a deeper
tint. They frequently wag their tail upwards
and downwards, and the origin of the neck is
bare.

Total length five inches and a half; the bill fix
lines ; the infide pale red at the bottom, blackifh
near the point ; the noftrils round and open; the
tongue forked, a little blackifh near the end;
the tarfus five lines and a half, covered with
down rather on the fides than before or behind;
the middle toe fix lines and a half; the alar extent ten inches and a half; the tail two inches,
forked as far as fix, feven, or even nine lines;
in fome fubjects this forking reaches only five
lines, but in others it does not occur at all.

The inteftinal tube fix or feven inches ; the
cæca very fmall, and filled with a matter different from that contained in the true inteftines ;
it has a gall bladder; the gizzard is mufcular;
the *æfophagus* twenty lines, it dilates into a little glandular bag before its infertion ; the tefticles are of an oval fhape, and unequal ; the
greater diameter of the biggeft ones four or five
lines, the fmaller diameter three ; their furface
was marked with many circumvolutions, like
a fmall

a fmall veffel twifted and rolled in all directions.

What is fingular, the young Martins are
heavier than the adults : five that were taken
from the neft while they were fcarcely covered
with down, weighed together three ounces,
which give three hundred and forty-five grains
to each ; whereas both the parents weighed exactly an ounce, or each was two hundred and
eighty-eight grains. The gizzards of the young
birds were diftended with food, and weighed in
all one hundred and eighty grains, which was
equal to thirty-fix each ; but both the gizzards
of the parents, which contained hardly any
thing, weighed only eighteen, or they were
four times lighter than thofe of their brood.
This fact clearly proves that the parents neglect
their requifite fubfiftence in order to fupply their
young, and that, during infancy, the organs
concerned in nourifhment predominate*, as in
the adult period thofe fubfervient to generation.

Some individuals of this fpecies have their
whole plumage white ; and of this I can produce
two refpectable vouchers, Hebert and Hermann.
The white Martin of the laft had red eyes, as
is the cafe with fo many animals whofe hair or
feathers are white ; its legs were not covered
with down, like the reft of the fame hatch.

We may regard the fulvous bellied fwallow

* I have obferved the fame difproportion both in the gizzards
and in the inteftines of young fparrows, nightingales, fauvettes, &c.

of

of Barrere as a variety of this fpecies; and the whitifh breafted brown fparrow of Brown *, as occafioned by the influence of climate.

* This author calls it a *houfe-fwallow*, but it is more analogous to the white-rumped fwallow.

[A] Specific character of the Martin, *Hirundo Urbica*: " Its tail-quills are not fpotted, its back is bluifh black, and the whole of its under fide is white." The reader will find an excellent account of this bird by Mr. White in the Philofophical Tranfactions for 1774, or in his Natural Hiftory of Selborne, pp. 157, 162. We fhall extract the following paffage, as it further confirms the migration of the fwallows.

" As the fummer declines, the congregating flocks increafe in numbers daily by the conftant acceffions of the fecond broods; till at laft they fwarm in myriads upon myriads round the villages on the Thames, darkening the face of the fky as they frequent the aits of that river, where they rooft. They retire, the bulk of them I mean, in vaft flocks together about the beginning of October; but have appeared of late years in a confiderable flight in this neighbourhood, for one day or two, as late as November the 3d and 6th, after they were fuppofed to have been gone for more than a fortnight. They therefore withdraw with us the lateft of any fpe-cies. Unlefs thefe birds are very fhort lived indeed, or unlefs they do not return to the diftrict where they have been bred, they muft undergo vaft devaftations fomehow and fomewhere; for the birds that return yearly bear no manner of proportion to the birds that retire."

The SAND-MARTIN*.

Hirondelle de Rivage, Buff.
Hirundo Riparia, Linn. Gmel. Kram, Frif. Klein, &c.
Dardanelli, Aldrov.
The *Sand-weſtern* or *Bank-weſtern*, Charleton.
The *Sand-martin, Bank-martin,* or *Shore-bird*, Will.

WE have ſeen that the two preceding ſpecies beſtow much induſtry and labour
in conſtructing their little manſion; the two
following ſpecies, we ſhall find, breed in holes
in the ground, in walls or in trees, and are at
little pains to form theſe, ſtrewing coarſely ſome
litter.

The Sand-martins arrive in our climates, and
retire, nearly at the ſame time with the common
martins. Towards the end of Auguſt they gradually come nearer thoſe ſpots where they aſ-
ſemble, and about the end of September Hebert tells us that he ſaw a great number of both
ſpecies collected together on the houſe which

---

* Ariſtotle, *Hiſt. Anim.* Lib. I. 1, calls it Δρεπανις, from δρεπανον,
a hook; probably becauſe of its forked tail: In Greek, it had alſo
the name of Χελιδων Οαλαττια, or ſea-ſwallow.
Pliny terms it *Hirundo Riparia*, Nat. Hiſt. *Lib. XXX.* 4.
In Italian it has the names *Dardanelli, Rondoni, Tartari:* In Ger-
man, *Rhyn-vogel* (Rhine-bird), *Rhyn-Schwalme* (Rhine-ſwallow),
Waſſer-ſchwalme (water-ſwallow), *Erd-ſchwalme* (earth-ſwallow),
Ufer Schwalbe (ſhore-ſwallow): In Daniſh, *Dig-ſvale, Jord-ſvale,*
Blint-ſvale, Sol-bakke: In Norwegian, *Sand-ronne, Strand-ſvale,*
Dig-ſulu, Sand-ſulu: In Swediſh, *Strand ſwala, Back-ſwala:* In
Poliſh, *Jaſkotka:* In Siberian, *Streſchis.*

he

he poffeffed in Brie*, and particularly on the
fide of the roof that faced the fouth; and when
the flock was formed, it enrirely covered the
building. But all the Sand-martins do not
migrate. The commander Defmazys writes
me that they are always to be found in Malta
during the winter, and efpecially when the
weather is inclement †: and, as that fmall
rocky ifland has no lake or pool, we cannot
fuppofe that in the interval of ftorms they
plunge under water. Hebert has feen them
as often as fifteen or fixteen times in the
mountains of Bugey ‡, in the different winter
months: it was near Nantua, in a pretty high
fituation, in a glen of a quarter of a league
in length and three or four hundred paces in
width; the fpot was delicious, with a fouthern
afpect, and fheltered from the north weft by
vaft lofty rocks; it was clothed in perpetual

* This houfe was fituated in the fkirt of a fmall town, its principal afpect was towards a river, and it communicated with the country on feveral fides.

† " In St. Domingo," fays the Chevalier Lefebvre Defhayes,
" the fwallows are feen to arrive on the approach of a ftorm; if the
clouds difperfe, they alfo retire, and probably follow the fhower."
They are, in fact, very common in that ifland during the rainy feafon. Ariftotle afferted, two thoufand years ago, that the fhore-
fwallow appeared not in Greece but when it rained. Laftly, on all
feas birds of every kind repair in ftorms to the iflands, and fometimes feek fhelter aboard veffels, and their appearance is almoft
always the portent of fome furious guft.

‡ According to the fame obferver, it is much more unufual to fee
them during winter in the plains. Thefe birds feem to be the fpecies to which Ariftotle alludes when he fays, " Many fwallows
are feen in the narrow paffes of mountains." Lib. VIII. 16.

4   verdure,

verdure, its violets flowered in February, and, in that lovely recefs, winter wore the fmiles of fpring. There thefe fwallows might play, and circle, and catch their infect food; and if the cold becomes exceffively fevere, they could retire into their holes, where the froft can never penetrate, and where they may find earth-infects and cry-falids to fupport them during their fhort con-finement; or perhaps they pafs into a torpid ftate, to which Gmelin and many others affert they are liable, though that they are not always fo is proved by the experiments of Collinfon *. The country people told Hebert, that they ap-peared after the fnows of Advent were melted, if the weather was mild.

Thefe birds are found in every part of Europe: Belon obferved them in Romagna, where they breed with the king-fifhers and bee-eaters in the brinks of the Mariffa, anciently the *Hebrus*. Koenigsfeld found in his travels through the north, that the left bank of a brook which runs befide the village of Kakui in Siberia, was bored into a great number of holes, which ferved as retreats to fmall gray birds called *Strefchis*, which muft be Sand-martins: five or fix hundred may be feen flying confufedly about thefe holes, en-tering them or coming out, but conftantly in

* Klein. *Ordo Avium*, p. 202, 204. Philof. Tranf. *Vol. LIII*, p. 101. Gazette Litteraire, *T. V. p.* 364. Magafin de Stral-fund, &c.

motion

motion like flies *. The fwallows of this fpecies are very rare in Greece, according to Ariftotle, but they are pretty common in fome diftricts of Italy, Spain, France, England, Holland, and Germany †. They prefer fteep banks,
as affording the fafeft lodgment ; the margin of
ftagnant water, which abounds moft with infects ; and a fandy foil ‡, where they can more
eafily form their little excavations, and fettle themfelves in them. Salerne tells us that on the fides of
the Loire they breed in the quarries, others fay in
grottos ; and both accounts may in part be true.
The neft is only a heap of ftraw and dry grafs,
lined with feathers, on which the eggs are
dropt ||. Sometimes they make their own holes,
and at other times they take poffeffion of thofe
of the bee-eaters and king-fifhers ; the entrance
of the cavity is eighteen inches in length. It
has been alleged, that they can forefee inunda-
tions, and make a timely efcape § : but the fact
is, that they always dig their hole a little above
the higheft mark of the ftream.

The Sand-martins only hatch once a year,

---

* Delifle's travels into Siberia.

† In the banks of the Rhine, of the Loire, of the Saone, &c.

‡ Lottinger and Hebert.

|| Schwenckfeld fays that this neft is of a fpherical form ; but
this feems to be true rather of the holes than of the neft built in
them. "They make no nefts," fays Pliny. Aldrovandus is of the
fame opinion: Edwards fays, that thofe which Collinfon caufed to
be dug out were complete, but he does not fpecify their form.
Laftly, Belon doubts whether they excavate the holes themfelves.

§ Plin. *Lib. X.* 33.

according to Frifch; they lay five or fix eggs,
femi tranfparent and without fpots, fays Klein;
the young ones grow very fat, and may be compared for delicacy to the ortolans. The reafon is, becaufe they are able to procure a rich
fupply of food, fince, befides the numerous tribe
of winged infects, they find reptiles and chryfalids in the ground. In fome countries, as in
Valencia in Spain, there is a great confumption
of Sand-martins *; which would induce me,
notwithftanding the affertion of Frifch, to fuppofe that in thofe parts they hatch oftener than
once a year.

The adults hunt their prey on the furface of
the water with fuch activity that we might
imagine them to be fighting; they often run
upon each other in the purfuit of the fame flies,
and ftruggle with fhrill cries † to obtain the
plunder: but this conduct arifes entirely from
emulation.

Were we to judge from its manner of breeding, we fhould conclude, that this bird is the
wildeft of the European fwallows; yet is it
tamer than the black martin, which lives indeed
in towns, but never mingles with any kindred
fpecies, whereas the Sand-martin affociates with
the common martin, and even the chimney
fwallow: this happens particularly about the
time of migration, when the utility of uniting
is moft fenfibly felt. It differs from thefe two

---

* See Willughby. The young birds are however fubject to wood-
lice, which infinuate under the fkin, but never to bugs.

† Gefner.

fpecies in its plumage, in its voice, and, as we have already feen, in fome of its natural habits. It never perches, and it arrives much earlier in the fpring than the black martin. I know not on what ground Gefner pretends that it clings and hangs by the feet when it fleeps.

All the upper furface is of a moufe gray; there is a fort of collar of the fame colour at the lower part of the neck; all the reft of the under furface is white; the quills of the tail and of the wings are brown; the inferior coverts of the wings, gray; the bill blackifh, and the legs brown, clothed behind as far as the toes with a down of the fame colour.

The male is, according to Schwenckfeld, of a darker gray, and there is a yellowifh tint at the rife of the throat.

It is the fmalleft of the European fwallows. Total length four inches and nine lines; the bill a little more than five lines; the tail forked; the tarfus five lines; the hind toe the fhorteft; the alar extent eleven inches; the tail two inches and a quarter, forked eight lines, and confifting of twelve quills; the wings contain eighteen, of which the nine inner ones are equal; they proje 五 five lines beyond the tail [A].

[A] Specific charaéter of the Sand-martin, *Hirundo Riparia*: " It is cinereous, its throat and belly white." Thefe birds are not frequent in England. They are much fmaller than thofe of their kindred fpecies, and are moufe-coloured. They have a pe-culiar manner of flying; reeling and wavering, with odd jerks: Hence the peafants in Spain term them *Papiliones de Montagna*, or Mountain butterflies.

The CRAG-SWALLOW.

L'Hirondelle Grife de Rochers *, Buff.
Hirundo Montana. Gmel †.

THESE Swallows conftantly neftle in the
rocks, and never defcend into the plains,
but in purfuit of their prey. It commonly rains
in a day or two after their appearance; becaufe,
no doubt, the ftate of the air then drives the
infects from the mountains. The Crag-fwallows affociate with the common martins, but are
not fo numerous. Both fpecies are often feen
in the morning, wheeling about the caftle of
Epine in Savoy; the Crag-fwallows appear the
firft, and are alfo the firft to retire to the
heights; after half paft nine o'clock none is
found in the vale.

The Crag-fwallow arrives in Savoy about
the middle of April, and departs by the fifteenth
of Auguft; but fome loiter till the tenth of October. The fame may be faid of thofe which
inhabit the mountains of Auvergne and of
Dauphiny.

This fpecies feems to be intermediate to the
common martin, whofe cry and geftures it has,
and the fand-martin, which it refembles in its

---

* i. e. The Gray Rock-fwallow.

† My information with refpect to this fpecies was received from
the Marquis de Piolenc, who fent me two birds.

colours:

colours: all the feathers on the upper furface of the head and body, the quills and coverts of the tail, the quills and fuperior coverts of the wings, are of a dun gray, edged with rufous; the middle pair of the tail is lighter; the four lateral pairs, included between this middle and the outermoft one, are marked on the infide with a white fpot, which is not vifible unlefs the tail be fpread; the under furface of the body is rufous; the flanks rufous, tinged with brown; the inferior coverts of the wings brown; the legs clothed with a gray down, variegated with brown; the bill and nails black.

Total length five inches ten lines; the alar extent ten inches and two-thirds; the tail twenty-one lines, a little forked, confifting of twelve quills, and exceeding the wings feven lines.

The only thing which appeared to me worth noticing in its internal ftructure is, that inftead of a *cæcum* there was a fingle *appendix* of a line in diameter, and a line and a quarter in length. I have obferved the fame in the night-heron.

The S W I F T*.

Le Martinet Noir †, Buff.
Hirundo-Apus, Linn. Gmel. &c.
The *House-martin*, Charleton.
The *Black-martin*, or *Swift*, Will. and Penn.

THE Swifts are real fwallows, and poffefs
the characteriftic qualities even in a higher
degree : their neck, their bill, and their legs are
fhorter ; their head and throat larger ; their
wings longer ; their flight more lofty and rapid ‡.
They are continually on the wing, and when
they happen to fall by accident, they can
hardly rife if the ground be flat; they muft
clamber up fome clod or ftone, that they may
have room to wield their long pinions §, and

---

* Ariftotle, *Hift. Anim.* Lib. I. 1. applies to it the general name of
Απυς, or footlefs, meaning only that its feet are fhort and feldom
ufed : It was alfo called Κυψελος, from κυψελις, a bee's cell, on account of its mode of neftling ; for which reafon it had likewife the
appellation of Πετροχελιδων, or rock fwallow.   The two firft names
have been adopted by Pliny, *Apodes*, *Cypfelus* : In Arabic, *Abafic* :
In Spanifh, *Venceio*, *Arrexaquo* : In German, *Geyr-Schwalb* (vulture-
fwallow) : In Danifh, *Steen*, *Soe*, *Kirke-Muur-Svale* : In Norwegian,
Ring-Svale Swart-Sulu, *Field-Sulu* : In Swedifh, *Ring-Svala* : In
Dutch, *Steen-Swalemen*.

† i. e. The Black-martin.

‡ Ariftotle fays, that the Swifts may be diftinguifhed from the
fwallows by their rough feet : he was therefore unacquainted with
the fingular difpofition of their feet and toes, and with their habits
and economy, ftill more fingular.

§ A fowler affured me that they fometimes alight on heaps of
horfe-dung, where they find infects, and can eafily take wing.

7   commence

commence their motion. This is owing to their ftructure; for the tarfus is fo fhort, that they fit almoft on their belly *, and totter from fide to fide +. The Swifts have only two modes of life, that of violent exertion, or that of per- fect inaction; they muft either fhoot through the air, or remain fquat in their holes. The only intermediate ftate which they know, is that of clambering up walls and trunks of trees quite near their lodgment, and, by means of their bill, dragging themfelves into the cavity. Commonly they enter it full fpeed, after having paffed it and repaffed it above an hundred times; they dart in in an inftant, and with fuch celerity that we totally lofe fight of them.

Thefe birds are very focial with each other, but never mingle with the other kinds of fwal- lows; and we fhall find, in the fequel, that their difpofitions and inftincts are different. It has been faid that they have little fagacity; yet they can breed in our houfes without depending on our indulgence, and without regarding our controul. Their lodgment is a hole in the wall,

---

* Belon.

+ Two of thefe birds obferved by M. Hebert, when fet on a table or on the pavement, had only this motion: their feathers fwelled if a perfon approached his hand; a young one found at the foot of a wall in which was the neft, had already this habit of briftling up its feathers, which were not yet half grown. I have lately feen two that took their flight, the one from the pavement, and the other from a gravel walk: they did not walk at all, and never changed their place but by flapping their wings.

which

which widens into a larger cavity, and is preferred in proportion to its height from the
ground, as affording the fafeft retreat. They
neftle even in belfries and the talleft towers,
fometimes under the arches of bridges, where,
though the elevation is not fo great, they are
better concealed. Sometimes they fettle in hollow trees, or in fteep banks befide the king-
fifhers; the bee-eaters, and fand-martins. After
they have once occupied a hole, they return
every year to it*, and eafily diftinguifh it,
though hardly perceptible. It is fufpected, with
much probability, that they fometimes take poffeffion of the fparrows' nefts, and when, on
their return, they find the property reclaimed,
they, with little ceremony, expel the owners.

The Swifts are, of all the birds of paffage,
thofe which arrive the lateft in our climates and
retire the earlieft : in general they begin to appear about the end of April, or the beginning of
May, and they leave us before the end of
July †. Their progrefs is more regular than

* I know a church-porch and a belfry of which the Swifts
have kept poffeffion for time immemorial; M. Hebert, to whom I
owe many good obfervations on this fpecies, fees from his windows
a hole of the wall above a high cope, to which they have regularly
returned for thirteen years : the parents feem to tranfmit their manfion to their offspring.

† I am affured, that on the lake of Geneva they arrive not till
May, and retire about the end of July or the beginning of Auguft; and when the weather is fine and warm, as early as the fifteenth of July.

that

that of the other fwallows, and appears to be more affected by the variations of temperature. They are fometimes feen in Burgundy as early as the twentieth of April, but thefe flocks pufh farther; the fettlers feldom return to occupy their neft before the firft days of May *. They are noify on their appearance; rarely do two enter at once the fame hole, and never without fluttering much about its mouth; ftill more uncommon it is for a third to follow them, nor does it ever fettle.

I have in different times, and in different places, opened ten or twelve fwifts' nefts: in all of them I found the fame materials, and thefe confifting of a great variety of fubftances; ftalks of corn, dry grafs, mofs, hemp, bits of cord, threads of filk and linen, the tip of an ermine's tail, fmall fhreds of gauze, of muflin, and other light ftuffs, the feathers of domeftic birds, thofe of the partridge, and of the parrot, charcoal, in fhort, whatever they can find in the fweepings of towns. But how can birds which never alight on the ground gather thefe materials? A celebrated obferver fuppofes that they raife

---

* This year, 1779, though the fpring was uncommonly fine, they appeared not in the diftrict where I live till the firft of May, and returned before the ninth to the holes from which I had caufed their nefts to be taken. At Dijon, they were feen on the nineteenth of April, but thofe *domiciliated* did not take poffeffion of their holes till between the firft and fourth of May.

them

them by glancing along the furface of the ground,
as they drink by fkimming clofe on the water.
Frifch imagines that they catch the fubftances
in the air as they are carried up by the wind.
But it is evident that little could be collected in
the latter way, and, if the former were true, it
would not fail to have been obferved in towns.
I am inclined to think the account more probable which feveral plain people have told me;
that they have often feen the Swifts coming out
of fwallows or fparrows' nefts, and carrying materials in their claws. This obfervation is cor-
roborated by feveral circumftances; firft, the
Swifts' nefts confift of nearly the fame fubftances
with thofe of fparrows; fecondly, we know that
the Swifts enter fometimes into the nefts of
fmall birds to fuck their eggs, which we may
fuppofe they do for the fake of pillaging the
materials. With regard to the mofs which they
employ, it is in very fmall quantity, and they
may gather it with their little claws, which are
very ftrong, from trees, on which they can
clamber, and fometimes even they breed in their
hollow trunks.

Of feven nefts found under the head of a
church porch fifteen feet from the ground, there
were only three which had a regular cup-fhape,
and of which the materials were more or lefs
interwoven, and with greater order than ufual
in fparrows' nefts; they had alfo more mofs,
and

and fewer feathers, and were in general lefs bulky.*.

Soon after the Swifts have taken poffeffion of the nefts, fome plaintive cries iffue continually from it for feveral days, and fometimes during the night; at certain times, two voices may be diftinguifhed. Is it the expreffion of pleafure common both to the male and female? or is it the love fong by which the female invites the male to accomplifh the views of nature? The latter feems to be the moft probable conjecture, efpecially as the ardent cry of the male, when he purfues the female through the air, is fofter and lefs drawling. We are uncertain whether the female admits one or feveral males: we often fee three or four Swifts fluttering about the hole, and even ftretching out their claws to clamber on the wall; but thefe may be fuch as were hatched the preceding year, which ftill remember the place of their nativity. It is the more difficult to anfwer thefe queftions, fince the females have nearly the fame plumage with the males, and fince we can feldom have an opportunity of viewing their manœuvres.

Thefe birds, during their fhort ftay in our climates, have time only to make a fingle hatch;

---

* The beft formed of all weighed two ounces and one gros and a half, the feven together thirteen ounces and a half, and the largeft five or fix times more than the fmalleft: fome of them had a coat of dung, which could fcarcely be otherwife, confidering the fituation of thefe nefts, in holes of various depths.

this

this confifts of five white eggs, pointed, and of
a fpindle fhape : I have feen fome not yet hatched on the twenty-eighth of May. When the
young ones have pierced through the fhell, very
different from thofe of the other fwallows, they
are almoft filent, and crave no food: happily
the parents obey the voice of nature, and fupply
them with what is proper. They carry provifion only twice or thrice a-day ; but each time
that they return to the neft they bring ample
ftore, their wide throat being filled with flies,
butterflies, and beetles *. They alfo eat fpiders
which they find near their holes ; yet their bill
is fo weak that it cannot even bruife or hold that
feeble prey.

About the middle of June the young Swifts
begin to fly, and fhortly abandon their nefts,
after which the parents feem no more to regard
them. At every period of their lives they are
fubject to vermin, but which appear little to in-
commode them.

This bird, like all the reft of the kind, is
excellent for the table when fat; the young
ones, efpecially thofe taken out of the neft,
are reckoned, in Savoy and Piedmont, delicate
morfels. The adults are difficult to fhoot, becaufe they fly both high and rapidly ; but as,

* The only Swift that M. Hebert could kill, had a quantity of
winged infects in its throat. This bird catches thefe, according to
Frifch, by darting impetuoufly above them, with its bill wide
ftretched.

on

on account of this very rapidity, they cannot readily alter their courfe, they may, from this circumſtance, be hit not only by a fowling-piece, but alſo by the ſtroke of a ſwitch. The only attention required, is to place one's ſelf in their way by mounting to a belfry, a baſtion, &c. and to meet them with the blow as they dart directly on, or as they come out of their hole *. In the iſland of Zant the boys catch them with a hook and line; they place themſelves in the windows of ſome high tower, and uſe a feather for bait, which theſe birds try to ſnatch and carry to their neſt †. A ſingle perſon can catch in this way five or ſix dozen in a day ‡. Many of them appear at the ſea-ports, and, as a perſon can there more eaſily chooſe his ſtation, he is ſure of killing ſome.

The Swifts avoid heat, and, for this reaſon, they paſs the middle of the day in their neſts in the crevices of walls or rocks and in the loweſt row of tiles of tall buildings. In the morning and evening they go in queſt of proviſion, or flutter without any particular object, but for exerciſe. They return at ten o'clock in the forenoon when the ſun ſhines, and again at half

---

* Many are killed in this way in the little town which I inhabit, eſpecially thoſe which breed under the church-porch that I have mentioned.

† Perhaps alſo they miſtake the feather for an inſect; they have an acute ſight, but the rapidity of their motion muſt render objects leſs diſtinguiſhable.

‡ Belon.

an hour after it fets in the evening. They rove
in numerous flocks, defcribing an endlefs feries
of circles upon circles, fometimes in clofe ranks
purfuing the direction of a ftreet, and fometimes
whirling round a large edifice all fcreaming together, and with their whole might; often
they glide along without ftirring their wings,
and, on a fudden, they flap with frequent and
hafty ftrokes. We behold their motions, but
we cannot judge of their intentions.

A commotion may be perceived among thefe
birds as early as the firft of July, which announces their departure; their numbers increafe
confiderably, and, in the fultry evenings between the tenth and twentieth, their large af-
femblies are held. At Dijon, they conftantly
gather round the fame belfries*; and, though
thefe meetings are numerous, the Swifts appear
as frequent as ufual about the other edifices:
they are probably foreign birds, therefore on
their paffage to more fouthern climates. After
fun-fet they divide into fmall bodies, foar into
the air with loud fcreams, and fly quite differently from ordinary. They may be heard long
after they are gone out of fight, and they feem
to bend their courfe to the country; they no
doubt retire at night to the woods, for there it
is known they breed and catch infects, and that
thofe which haunt the plains during the day,

* Thofe of St. Philibert and of St. Benigne.

and

and even some of those which live in towns,
repair to the trees in the evening, where they
continue till dark. The city Swifts assemble
soon after, and all prepare to migrate into colder
countries. M. Hebert scarce ever saw them
later than the 27th of July; he supposes that
they travel during the night, and proceed to
no great distance, and cross not the sea. Indeed their aversion to heat is such that they
would shun the scorching air of Senegal *.
Many naturalists † pretend that they lie torpid
in their holes during winter, and even before
the end of the dog-days. But, in our climates,
they are undoubtedly migratory, and in the
nests which I searched, about the middle of
April, twelve or fifteen days before their first
appearance, I could not find a single bird.

Besides the regular periodical migrations, we
sometimes see in autumn numerous flocks,
which have by some accidents been separated
from the main body. Such was the one that
appeared to Hebert suddenly in Brie about the
beginning of November : it circled long round
a poplar, and then began to scatter, rose to a
great height, and vanished with the close of the
day. Hebert saw another flock about the end
of September in the vicinity of Nantua, where

* What Aristotle says of his Απυς, which lived in Greece the
whole year, would imply that it does not so much dread heat. But
may not the Απυς be our sand-martin ?

† Klein, Heerkens, Herman, &c.

they

they are not common. In both thefe ftray-
flocks there were many birds that had a cry different from that ufual to the Swifts; whether
that their voice alters in winter, that they were
young ones, or that they belong to a different
branch of the fame family.

In general, the Swift has no warble, but
only a fhrill whiftle, which varies little in its
inflections, and which is fcarce ever heard except when on the wing. In its hole it remains
ftill and filent, afraid, it would feem, of difclof-
ing its retreat: love alone roufes it from le-
thargy. At other times it is very unlike thofe
prattlers defcribed by the poet *.

Birds which fhoot through the air with fuch
rapidity muft have a quick eye, and, in the
prefent cafe, the fact corroborates the general
principle advanced in the " Difcourfe on the
Nature of Birds." But every thing has its limits, and I cannot believe that they will defcry
a fly at the diftance of half a quarter of a league,
as Belon afferts; that is, at twenty-eight thoufand times the fly's diameter, fuppofing that nine
lines, or nine times farther than a man could
fee. The Swifts are not only fpread through

* " Nigra velut magnas domini cum divitis ædes
" Pervolat, et pennis alta atria luftrat hirundo,
" Pabula parva legens, nidifque loquacibus efcas.
" Et nunc porticibus vacuis, nunc humida circum
" Stagna fonat."		Virg. Æneid. XII. 473.
Virgil feems to refer, in this paffage, to the houfe-fwallow.

all

all Europe; the Vifcount Querhoent faw them at the Cape of Good Hope, and I doubt not that they may be found alfo in Afia, and even in the new continent.

A moment's reflection will exhibit the fingularity of this bird: its life is divided between the extremes of motion and reft; it never receives the impreffions of touch, but during its fhort ftay in its hole; its joys are either exquifite or totally fufpended, nor can it have any idea of that languor which other beings feel from the dull continuance of even pleafurable fenfations; and, laftly, its character is a compound of temerity and fufpicion. It creeps by ftealth into its hole like a reptile, and obferves profound filence; but when it circles in its proper element, it feels its fuperiority, and, trufting to its powers, it overlooks or defpifes danger.

The Swift is larger than our other fwallows, and weighs ten or twelve gros; the eye is hollow, the throat afh-white, the reft of the plumage blackifh, with green reflections; the back and the inferior coverts of the tail are of a deeper caft; thefe coverts reach to the end of the two middle quills; the bill is black; the legs of a brown fieth colour; the fore part and the infide of the *tarfus* are covered with fmall blackifh feathers.

Total length feven inches and three quarters; the bill eight or nine lines; the tongue three

lines and a half, forked; the noſtrils like a long
ſhaped human ear, the convexity being turned
inwards, and their axis being inclined to the
ridge of the upper; the two eye-lids naked,
moveable, and ſhut near the middle of the ball
of the eye; the tarſus is near five lines, the four
toes turned forward *, and conſiſting each of
two *phalanges* only (a ſingular conformation,
peculiar to the Swifts); the alar extent about
fourteen inches; the tail near three inches,
compoſed of twelve unequal quills †, and forked more than an inch; it is exceeded eight or
ten lines by the wings, which contain eighteen
quills, that when cloſed reſemble the blade of
a ſcythe.

Œſophagus two inches and a half, and forms
near its bottom a ſmall glandulous bag; the
gizzard is muſcular in its circumference, lined
with a wrinkled looſe membrane, and contains
portions of inſects, but no pebbles; it has a gall
bladder, no *cæcum*; the inteſtinal tube from the
gizzard to the anus ſeven inches and a half;
the *ovarium* cluſtered with eggs of unequal ſizes
(this was on the 20th of May).

Having lately compared ſeveral Swifts of both
ſexes, I found that the males weighed more
than the females, that their feet were ſtronger,

* How can the genus in which it is ranged be deſcribed to have
three toes before and one behind?

† I know that Willughby reckons only ten; but perhaps he
confounded this ſpecies with the following.

that

that the white fpot on the throat is broader, and
that almoft all the white feathers which form
it have black fhafts.

The infect which infefts thefe birds is a
kind of loufe, of an oblong fhape, and orange
colour, but of different tints; having two
thread-like *antennæ*, its head flat and almoft
triangular, and its body confifting of nine rings,
befet with a few ftraggling hairs.

It is fomewhat remarkable that during their
ftay with us their plumage lofes its black glofs,
and bleaches by continual expofure to the fun
and the air. They arrive about the end of
April, and retire before the end of Auguft.

[A] Specific character of the Swift, *Hirundo-Apus:* " It is
blackifh, its throat white, all its four toes placed before." Mr.
White avers from many years' obfervation, that the Swifts even
copulate on the wing. In England they fly each day, in mid fummer, at leaft fixteen hours. Nor do they feem to be at all incom-
moded by the heat of our meridian fun; nay, they are never fo
lively as in fultry thundery weather: fo different is our climate to
that of the fouth of France, where they are confined to their holes
for fome hours at noon.

The WHITE-BELLIED SWIFT.

Le Grand Martinet a Ventre Blanc *, Buff.
Hirundo-Melba, Linn. and Gmel.
Hirundo Major Hispanica, Briff.
Hirundo Maxima Freti Herculei, Klein.
The *Greatest Martin* or *Swift*, Edw.

I FIND, in this bird, both the general qualities of the fwallow, and the peculiar characters of the Swift. Its legs are extremely
fhort; its four toes are turned forward, and
confift only of two *phalanges*; it never alights
on the ground, and never perches on trees :—
in thefe properties it agrees with the Swift :
but there are confiderable difparities that feparate it; for, befides the differences in the plumage, it is twice as large, its wings are longer,
and there are only ten quills in the tail.

Thefe birds delight in mountains, and breed
in the holes of crags. They appear annually
among the cliffs which border the Rhone in
Savoy, in thofe of the ifland of Malta, in the
Swifs Alps, &c. The one defcribed by Edwards was killed on the rocks of Gibraltar; but
it is uncertain whether it refides there, or was
only on its paffage. And though it were a fettler,
this would not be s fufficient reafon to call it
Spanifh Swallow, as Briffon has done : for, 1. it
is found in many other countries, and probably

* *i. e.* The Great White-bellied Martin.

in

in all thofe which abound with mountains and rocks. 2. It is rather a Swift than a fwallow. One was killed, in 1775, in our diftricts on a pool, at the foot of a high mountain.

The Marquis de Piolenc (to whom I am indebted for my acquaintance with thefe birds, and who has fent me feveral) writes me that they arrive in Savoy about the beginning of April, and that they fly at firft over the pools and marfhes, and in a fortnight or three weeks they reach the high mountains; that they do not fly fo lofty as the common Swifts, and that the time when they retire is not fo precife or fixed as that of their appearance, and depends much on the ftate of the weather, and on the warmth or chillnefs of the air *: laftly, he fubjoins that they live on beetles, fpiders, &c. that they are difficult to fhoot, that the flefh of the old ones is unpleafant †, and that the fpecies is not numerous.

It is probable that thefe White-bellied Swifts breed alfo among the fteep rocks on the fea-fide, and that we may apply to them, as well as to the common Swifts, what Pliny fays of certain birds without feet that fly in the open fea at all diftances from the fhore, circling round the veffels. Their cry is nearly the fame with that of the common Swift.

* In the country of Geneva they remain a fhorter time than the Swift or black Martin.

† Sportfmen ufually fay that thefe birds are hard both to kill and to eat.

The

The whole of the upper furface is brown-gray, but deeper on the tail and wings, with reddifh and greenifh reflections; the throat, the breaft, and the belly, white; on the neck there is a brown-gray collar, variegated with black-ifh; the fides are variegated alfo with blackifh and with white; the lower belly and the inferior coverts of the tail are of the fame brown with the back; the bill black; the legs flefh coloured, covered with down before and on the infide; the ground of the feathers brown beneath the body, and light gray above; almoft all the white feathers have a black fhaft, and the brown ones are edged delicately with whitifh at the tip. In one male which I obferved, the feathers on the head were deeper coloured than in two others with which I compared it: it weighed two ounces five gros.

Total length eight inches; the bill an inch, flightly hooked; the tongue four lines, of a triangular fhape; the iris brown; the eye-lids naked; the tarfus five lines and a half; the nails ftrong, the inner one the fhorteft; the alar extent above twenty inches; the wings compofed of eighteen quills; the tail three inches and a half, confifting of ten unequal quills, forked eight or nine lines, and exceeded by the lines two inches at leaft.

The gizzard flightly mufcular, very thick, lined with a loofe membrane, containing fragments of infects, and fome whole ones, and

4   among

among others was one whose skinny wings
reached more than two inches; the intestinal
tube nine or ten inches; the *œsophagus* dilating
below into a glandulous bag; no *cæcum*, nor could
I perceive any gall bladder; the testicles very
long and small (this was on the 18th of June).
It appeared to me that the mesentery was stronger, the skin thicker, the muscles more elastic,
and the brain firmer, than in other birds: every
thing denoted strength, and indeed the swiftness
of its motion necessarily implies that.

We may remark that the subject described by
Edwards was smaller than ours. He asserts that
it resembles the sand-martin so exactly that the
same description will serve both. It is true that
their plumage is nearly alike, and that all the
swallow tribe are similar; but that naturalist
should have noticed that the toes are differently
disposed.

[A] Specific character of the White-bellied Swift, *Hirundo-Melba*: " It is brown; its throat and belly white; all its toes placed
before."

FOREIGN BIRDS,

WHICH ARE RELATED TO THE SWALLOWS, THE MARTINS, AND THE SWIFTS.

THOUGH the Swallows of the two continents form only one tribe, and are analogous in their shape and principal properties *, they have not all the same instincts and natural habits. In Europe, and on the nearest borders of Africa and of Asia, they are almost wholly birds of passage. At the Cape of Good Hope a part only migrate, and the rest are stationary. In Guiana, where the temperature is pretty uniform, they remain the whole year, without shifting their abodes; nor is the manner of life the same in them all; some prefer the settled and cultivated spots; others indifferently frequent inhabited places, or the wildest solitude; some inhabit the uplands, others the fens; some appear to be attached to particular districts; but none of them build their nest with earth like ours, though some breed in hollow trees, like the Swifts, and others in banks, like the Sand-martins.

It is remarkable, that almost all the late observers agree, that in this part of America and

* Perhaps we should except the bill, which is stronger in some American Swallows.

5

in

in the adjacent iflands, fuch as Cayenne, St.
Domingo, &c. the fpecies of Swallows are more
numerous and various than in Europe, and that
they relide there the whole year; while, on
the contrary, Father Dutertre, who travelled
through the Caribbees when the colonies were
juft planted, affures us that Swallows are very
rare in thefe iflands, and that they are migratory as in Europe *. If both thefe obfervations
be regarded as well afcertained, they will fhew
the influence of civilized man on nature, fince
his prefence is alone fufficient to invite whole
fpecies to fettle and multiply. There is a cu-
rious remark made by Hagftroem in his *Swedifh
Lapland*, that corroborates this remark: he relates that many birds and other animals, whether from a predilection to human fociety or
from views of intereft, gather near the new fettlements; he excepts, however, the geefe and
ducks, who obferve a different conduct, and,
both in the mountains and in the vallies, lead
their migrations in a direction oppofite to thofe
of the Laplanders.

I fhall conclude by remarking with Bajon and
many other obfervers, that, in the iflands and continent of America, there is often a great difference
between the plumage of the male and female of
the fame fpecies, and often a ftill greater in the

---

* " During the feven or eight years that I lived there, I never
faw more than a dozen : they appear not (he fubjoins) except in the
five or fix months that they are feen in France."

fame

fame individual at different ages. And this fact will juftify the liberty which I fhall take in ranging them.

I.

The BLACK SWALLOW.

Le Petit Martinet Noir *, Buff.
Hirundo Nigra, Gmel.
Hirundo Dominicenfis, Briff.

THIS bird, which inhabits St. Domingo, is fomewhat differently fhaped from the Swift: the bill is rather fhorter; the legs rather longer; fo is the tail, and alfo lefs forked; the wings much longer; laftly, the feet do not feem in the figure to have their four toes turned forward, nor does Briffon tell how many *phalanges* the toes have.

This doubtlefs is the fame with the fpecies defcribed by Bajon as almoft quite black, which frequents the dry favannas, and breeds in holes in the ground, and perches often on withered trees. It is fmaller than the Swift, and of a more uniform blackifh, moft of the individuals not having a fingle fpot of any other colour in their whole plumage.

Total length five inches and ten lines; the

---

* *i. e.* The Little Black *Martinet* (Swift).

bill

bill fix lines; the tarfus five lines; the alar extent fifteen inches and a half; the tail two inches
and a half, forked fix lines, and exceeded by the
wings fourteen lines, and in fome eighteen lines.
In one fpecimen there was a fmall very narrow
white bar on the front. I faw another in Mauduit's excellent cabinet, that had been brought
from Louifiana; it was of the fame fize and
nearly of the fame plumage, being of a blackifh
gray without any glofs, and its legs not clothed
with feathers [A].

II.

The WHITE-BELLIED SWIFT *.

Le Grand Martinet Noir à Ventre Blanc, Buff.
Hirundo Dominicenfis, Gmel.
Hirundo cantu Alaudam referens, Klein.
The *St. Domingo Swallow,* Lath.

I CONCEIVE this bird to be a Swift from the
account given by Father Feuillée, who faw
it in St. Domingo. He calls it indeed a *Swallow,* but then he compares it to the Swifts, with
regard to fize, fhape, and even colours. He
faw it in the month of May fitting on a rock,

---

* i. e. The White-bellied great black *Martinet* (Swift).
[A] Specific chara&ter of the *Hirundo Nigra:* " It is entirely
black."

and took its fong for that of a lark, till the opening dawn enabled him to diftinguifh it. He affures us that numbers of thefe birds are feen in the American iflands in the months of May, June, and July.

The predominant colour of the plumage is fine black, with the glofs of burnifhed fteel ; it fpreads not only on the head, and all the upper fide of the body, including the fuperior coverts of the tail, but alfo on the throat, the neck, the breaft, the flanks, the thighs, the fmall coverts of the wings ; the quills, the great fuperior and inferior coverts of the wings, and the quills of the tail, are blackifh, the inferior coverts of the tail and of the belly are white; the bill and legs brown.

Total length feven inches ; the bill eight lines ; the tarfus fix ; the alar extent fourteen inches and two lines ; the tail two inches and three quarters, forked nine lines, compofed of twelve quills; it does not project beyond the wings.

Commerfon brought from America three birds much like the one defcribed by Briffon, and which feem to belong to the fame fpecies [A].

[A] Specific character of the *Hirundo Dominicenfis:* " It is black, with a fteel glofs ; its belly white."

III.

The PERUVIAN SWALLOW.

Le Martinet Noir & Blanc à Ceinture Grise *, Buff.
Hirundo Peruviana, Gmel.
Hirundo Peruviana Major, Briff.

THE plumage of this bird confifts of three principal colours : black is fpread over the back, and as far as the fuperior coverts of the tail inclufively ; fnowy white on the under fide of the body ; light cinereous on the head, the throat, the neck, the fuperior coverts of the wings, their quills, and thofe of the tail : all thefe quills are edged with yellowifh gray, and on the belly there is a light afh-coloured girdle.

This bird is found in Peru, where it was defcribed by Father Feuillée. Like all the Swifts, it has fhort legs ; the bill fhort and broad at its bafe ; the nails hooked and ftrong, black like the bill ; and the tail forked.

* i. e. The black and white Swift with a gray girdle.

IV.

The WHITE-COLLARED SWIFT.

Le Martinet a Collier Blanc, Buff.
Hirundo Cayanenfis, Gmel.

THIS is a new fpecies which we received from the ifland of Cayenne. We have ranged it with the Swifts, becaufe it has the four toes turned forward.

The collar which diftinguifhes it is of a pure white, and very confpicuous on the bluifh black, which is the prevailing colour of its plumage. The part of this collar which paffes under the neck forms a narrow band, and terminates on each fide in a large white fpot, which occupies the throat and all the upper fide of the neck; from the corners of the bill rife two fmall diverging white bands, the one ftretching above the eye to form an eye-lid, the other paffes under the eye to fome diftance; laftly, on each fide of the lower belly there is a white fpot placed in fuch a manner, that it appears below and above; the reft of the upper and under fides, including the fmall and middle coverts of the wings, is of a velvet black with violet reflections; what appears of the great coverts of the wings next the body is brown edged with white; the great quills and thofe of the tail are black; the former bordered interiorly with rufty brown; the bill and

legs

legs black; the feet feathered to the nails. Bajon fays that this Swift breeds in houfes. I
have feen its neft at Mauduit's ; it was very large,
well ftuffed, and conftructed with the cotton
of dog's-bane * ; it had the fhape of a truncated
cone, of which one of the bafes was five inches
in diameter, and the other three inches ; its
length was nine inches; it appeared to have ad-
hered by its large bafe, compofed of a fort of
pafteboard made of the fame fubftance: the cavity of this neft was divided obliquely near its
middle by a partition, which extended near the
bafe where the eggs were lodged, and near that
part there was a fmall heap of very foft dog's
bane, which formed a kind of valve, and feemed
intended to fcreen the young from the external
air.   Such precautions, in fo warm a climate,
fhews that thefe Swifts feel acutely the fenfa-
tion of cold.   They are as large as the common
Martins.

Total length, being the average of feveral fpe-
cimens, five inches, and from three to eight
lines ; the bill fix or feven ; the tarfus three or
five ; the hind nail flender ; the tail from two
inches to two inches and two lines, forked eight
lines, and exceeds the wings from feven to
twelve lines.

* _Apocynum Cannabinum,_ Linn.

V.

The ASH-BELLIED SWALLOW.

Le Petite Hirondelle Noire a Ventre Cendre *, Buff.
Hirundo Cinerea, Gmel.
Hirundo Peruviana, Briff.

THIS Peruvian Swallow, according to Father
Feuillée, is much fmaller than the European Swallows; its tail is forked, its bill very
fhort and almoft ftraight; the head and all the
upper fide of the body, including the fuperior
coverts of the wings and tail, are of a fhining
black; all the under fide of the body cinereous;
laftly, the quills of the wings and of the tail
are of a dull afh-colour, edged with yellowifh
gray [A].

VI.

THE

BLUE-SWALLOW of LOUISIANA.
Buff.

Hirundo Violacea, Gmel.

THE whole plumage is of a deep blue, yet
not uniform, but gloffed with different
tints of violet; the great quills of the wings are

* The Afh-bellied little black Swallow.
[A] Specific chara&ter of the *Hirundo Cinerea*: " It is black,
below cinereous; the quills of its wings and tail gray; its orbits
brown."

alfo

alſo black, though only on the inſide, and not
ſeen unleſs the wings are ſpread ; the bill and
legs are black ; the bill a little hooked.

Total length ſix inches and ſix lines ; the bill
ſeven lines and a half; the tail very forked, and
exceeded five lines by the wings, which are
very long.

M. Lebeaux has brought from the ſame country another ſpecimen, which is evidently of the
ſame kind, though larger, and the quills of the
tail and wings, and the primary coverts of the
wings, are blackiſh, without any ſteel gloſs.

Total length eight inches and a half; the bill
nine lines, pretty ſtrong and ſomewhat hooked;
the tail three inches, forked an inch, and falls
a little ſhort of the wings [A].

VARIETIES.

THE BLUE SWALLOW of Louiſiana ſeems
to be the principal ſtem which has given origin
to four varieties, two of which are ſpread through
the north, and the other two through the ſouth.

I. The Cayenne Swallow * of the *Planches*

[A] Specific character of the *Hirundo Violacea:* " It is dark blue,
tinged with violet; the inſide of its greater wing-quills, its bill, and
its legs, are black."

* *Hirundo Chalybea,* Gmel.
  Hirundo Cayanenſis, Briſſ.
  The *Chalybeate Swallow,* Lath.

Enlu-

Enluminées. It is the moſt common ſpecies in
the iſland of Cayenne, where it remains the
whole year. It is ſaid to ſit frequently among
bruſh wood, and on half-burnt trunks that bear
no leaves. It does not build a neſt, but breeds
in hollow trees. The upper ſurface of the
head and of the body is blackiſh, gloſſed with
violet; the wings and the tail the ſame, but
edged with a lighter colour : all the under ſurface of the body is ruſty gray veined with brown,
and growing more dilute on the lower belly and
the inferior coverts of the tail.

Total length ſix inches; the bill nine lines
and a half, ſtronger than that of the Swallows';
the tarſus five or ſix lines; the hind toe and nail
are the ſhorteſt ; the alar extent fourteen inches;
the tail two inches and a half; forked ſix or
ſeven lines, and exceeded by the wings about
three lines.

II. I have ſeen four blue Swallows brought
from South America by Commerſon, which
were of a middle ſize between thoſe of Cayenne
and Louiſiana, and which differed only in the
colours of the lower ſurface of the body. In
three of them the throat was brown gray, and
the under ſide of the body white. The fourth,
which came from Buenos Ayres, had its throat
and all the under ſide of its body white, ſprinkled
with brown ſpots, which are more frequent on
the fore parts, and wider ſcattered on the lower
belly.

III. The

III. The Carolina bird * which Catefby calls
the *Purple Martin*. It belongs to the fame eli-
mate, and is of the fame bulk with the one
from Buenos Ayres, juft mentioned. A fine
deep purple is fpread over all its plumage, and
ftill more intenfe on the quills of the tail and of
the wings ; the bill and legs are rather longer
than thofe of the preceding varieties, and its tail,
though fhort, projects fomewhat beyond the
wings. It neftles in holes made on purpofe for
it around the houfes, and in gourds which are
fet on poles to invite it. It is efteemed ufeful
becaufe it fcares away the birds of prey, and
the ravenous beafts, or rather gives notice of
their appearance. It leaves Virginia and Carolina on the approach of winter, and returns in
the fpring.

Total length feven inches and eight lines;
the bill ten lines; the tarfus eight lines ; the tail
two inches and eight lines, and forked fourteen
lines, projecting a little beyond the wings.

IV. The Swallow from Hudfon's Bay, termed
by Edwards *The Great American Martin* †. Like

* *Hirundo Purpurea,* Gmel.
  The *Purple Swift,* Penn.
  The *Purple-Swallow,* Lath.
Specific character : " It is entirely purple, its tail forked."

† *Hirundo-Subis,* Linn. and Gmel.
  Hirundo Freti Hudfonis, Briff.
  The *Canada Swallow,* Lath.
Specific character: " It is bluifh black; its mouth and its
under fide whitifh cinereous."

o o 2   the

the preceding varieties, it has a stronger bill than usual; the upper surface of the head and body is of a shining purplish black, and there is a little white at the base of the bill; the great quills of the wings, and all those of the tail, are black without any gloss, and edged with lighter colour; the upper edge of the wing whitish; the throat and breast deep gray; the sides brown; the under surface of the body white, shaded with a brown cast; the bill and legs blackish.

Total length near eight inches; the bill eight lines; the edges of the upper mandible scalloped near the point; the tarsus seven lines; the tail near three inches, forked seven or eight lines, and exceeding the wings three lines.

VII.

The BRASILIAN SWALLOW.

La Tapere, Buff.
Hirundo Tapera, Linn. Gmel. Ray, Sloane.
Hirundo Americana, Briff. and Klein.
Tapera, Marcgr.

THIS Brasilian Swallow, Marcgrave tells us, resembles much the European; its size the same; its manner of circling also the same; and its legs as short, and feet of a similar shape. The
upper

upper fide of the head and body, including the wings and the tail, are brown gray, but the quills of the wings and the extremity of the tail browner than the reft; the throat and breaft gray mixed with white; the belly white, and alfo the inferior coverts of the tail; the bill and the eyes are black; the legs brown.

Total length five inches and three quarters; the bill eight lines, and its opening extends beyond the eyes; the tarfus fix lines; the alar extent twelve inches and a half; the tail two inches and a quarter, compofed of twelve quills, forked three or four lines, and a little exceeded by the wings.

According to Sloane, this bird belongs to the Swifts, only its plumage is darker. It frequents moftly the meadows and favannas; and is faid to perch, from time to time, on the bufhes: and fince none of our Swallows, Martins, or Swifts, have that habit, I am inclined, notwithftanding the opinion of Sloane and Oviedo *, to think that the *tapera* is a fpecies peculiar to America; at leaft, it is diftinct from thofe of Europe.

Edwards fufpects it to be the fame with his *great American Martin*; but on comparing the defcriptions, I find differences in the plumage, in the fize, and in the proportions [A].

---

* Oviedo reckons the *tapera* among the birds that are common to both continents.

[A] Specific character of the *Hirundo-Tapera:* " Its tail-quills are equal; its body blackifh, and white below."

VIII.

THE
BROWN-COLLARED SWALLOW.

Hirondelle Brune & Blanche a Ceinture Brune*, Buff.
Hirundo Torquata, Gmel.

IN general, all the upper furface is brown, and all the under white or whitifh, except a broad brown girdle on the breaft and thighs: there is a flight exception, however; it is a fmall white fpot on each fide of the head between the bill and the eye. This bird was brought from the Cape of Good Hope.

Total length fix inches; the bill eight lines, ftronger than ufual in Swallows, the upper mandible a little hooked, its edges fcalloped near the point; the tail twenty-feven lines and fquare, and falling eight lines fhort of the wings, which grow very narrow near the extremities, for the fpace of about two inches.

---

* *i, e.* The Brown and White Swallow with a brown cinĉture.

IX.

The WHITE BELLIED CAYENNE SWALLOW, *Buff.*

Hirundo Leucoptera, Gmel.
The *White-winged Swallow,* Lath.

A SILVERY white spreads not only over all the under surface of the body, including the inferior coverts of the tail, but also on the rump, and it borders the great coverts of the wings ; and this edging extends more or less in different individuals ; the upper side of the head, neck, and body, and the small superior coverts of the wings, are cinereous, with reflections which are more or less conspicuous, and fluctuate between green and blue, and of which there are also some traces on the quills of the wings and of the tail, whose ground colour is brown.

This handsome Swallow skims along the ground like ours, circles in the overflowed savannas of Guiana, and perches on the lowest branches of leaflefs trees.

Total length from four and a quarter to five inches ; the bill six or eight lines ; the tarsus five or six ; the hind nail strongest after the middle one ; the tail an inch and a half, forked two or three lines, and exceeded, from three to six lines, by the wings.

We may regard the Spotted-bellied Cayenne

Swallow

Swallow as a variety of this species, differing
only in its plumage, and the ground colours
being still nearly the same, always brownish-
gray and white ; but the upper side of the body,
and of the quills of the wings and of the tail, is
of an uniform brown, without any reflection or
mixture of white ; the under side, on the contrary, which in the other is of an uniform
white, is in this bird white sprinkled with
oval brown spots, thicker on the fore part of
the neck and of the breast, and thinner towards
the tail.—In some of the White-bellied Swallows there is a mixture of white on the upper
coverts of the wings, and the gray or brown of
the upper surface of the body is less glossy.

X.

The ESCULENT SWALLOW.

La Salangane, Buff.
Hirundo Esculenta, Linn. and Gmel.
Apus Marina, Rumphius and Olearius.
Hirundo Riparia Cochinchinensis, Briss.
Hirundo nido eduli, Bontius.
Layong-Layong, Marsd. Sumatra.

SALANGANE is the name which the inhabitants of the Philippines bestow on a small
Sand-martin, celebrated for the singular quality
of its nest, which is eaten and esteemed a great
delicacy

delicacy in China, and in many of the other adjacent countries *. The high price which it bears tempts frequently to adulterate it, which, together with the fables that have been propagated on the fubject, occafions much obfcurity and contradiction.

Thefe nefts have been compared to thofe of the *Halcyons*, and many have haftily concluded that they were the fame. The ancients conceived the latter to be real birds' nefts, compofed of flime, froth, and other impurities that float on the furface of the fea, and they diftinguifhed them into feveral kinds. The one mentioned by Ariftotle was of a fpherical form, its mouth narrow, of a rufty colour, and of a fpongy cellular fubftance, confifting chiefly of fifh-bones †. But a flight comparifon with the defcription which Doctor Vitaliano Donati gives of the *Alcyonium* of the Gulf of Venice‡ will convince

* In China thefe nefts are called *Saroi-Bouras*; in Japan, *Jenwa, Jinku*; in India, *Patong*.

† Arift. *Hift. Anim.* Lib. IX. 14. Plin. *Lib XXXII.* 8. There are always many of thefe bones and fcales of fifhes in the neft of our halcyon or king-fifher, but they are chiefly fcattered among the duft on which this bird lays its eggs, and do not enter into the compofition of the neft; for our king-fifher never builds one.

‡ The *Alcyonium* is a marine body . . . approaches the round or convex figure above . . . its furface tuberous . . . completely invefted with very thick fpines . . . of an earthy colour, but free from filth, of a wax-colour . . . the heart much fofter . . fpongy and cavernous . . . with many fpines much entangled and cloathed with flefh, &c." *Storia Naturale marina dell Adriatico*, p. 58.

us

us that they are exactly the same, and only
the *nidi* of fea infects. The only difference is
that Donati fays its entrance is large, and Ariftotle that it is fmall; but thefe terms are evidently vague: the Italian finds the mouth to be
one fixth of the width of the whole.

But the *patong* of the Eaft-Indies is the real
neft of a fpecies of Swallow. Writers are not
agreed either with regard to its materials, its
form, or the places where it is found: fome
affert that it is attached to rocks, clofe on the
furface of the water*; others, that it is lodged
in the hollows of thefe rocks †; and others, that
it is concealed in holes made in the ground ‡.
And Gemelli Carreri adds, "that the failors are
always in fearch along the beach, and when
they find earth thrown up, they open the fpot
with a ftick, and take the eggs and the young,
which are reckoned equally delicate §."

With regard to the form of thefe nefts, fome
affirm that they are hemifpherical ‖, while others
fay, " that they have many cells, which are
like large conglutinated fhells, and marked as
fuch with *ftriæ* and rugofities ╪."

---

* *Curiofites de la Nature & de l'Art*, *p.* 170.
† John de Laet, Van Neck, Kircher, &c.
‡ Gemelli Carreri, *Voyage round the World*, *t. V. p.* 268.
§ The fame thing has been faid of our Sand-martin. *Salerne,*
and Willughby.
‖ *Mufæum Worm.*
╪ Father Philip Marin. *Hift. de la Chine*, p. 42.

Of

Of its fubftance, the accounts are ftill more various. Some affert that it is ftill unknown *; others, that it is the froth of the fea, or fifh-fpawn; others, that it is ftrongly aromatic; others, that it is infipid; others, that it is a juice gathered by the *Salanganes* from the tree called *calambouc*; others, that it is compofed of a vifcous fubftance difcharged from the bill in the love feafon; others, that it is formed of the fifh-plants found on the fea: but the greater number agree, that the fubftance of thefe nefts is tranfparent and like ifinglafs, which is the fact. The Chinefe fifhers affured Kæmpfer that thofe ufually fold were nothing but a preparation of the marine polypi, and he adds, that by his receipt the colour may be imitated. All thefe difcordant relations prove that various fubftances, natural or artificial †, have at different times, and in different countries, been regarded as the nefts of the *Salangane*. In this ftate of uncertainty, I could not do better than apply for in-

* Kircher, Du Halde, &c.

† Here is Kæmpfer's recipe: Firft fkin the polypes, and fteep the flefh in a folution of alum for three days; then rub, wafh, and clean it till it become tranfparent, and afterwards pickle it. *Hiſt. du Japan, t. I. p.* 120. In thofe countries many other preparations are made of the fame kind; in China, with the tendons of ftags, and with the fins of fharks. See Olof Toren, *Voy. aux Indes Orient. p.* 76; *Etabliſſ. Europ. dans les Indes, t. I. l.* 2. (N. B. Ifinglafs is made of the fwimming bladders of a fifh common in the Ruffian feas.) In Tonquin fowls' eggs are feafoned in fuch manner as to preferve them, and to fit them for feafoning to other difhes. *Hiſtory of Tonquin*, in Churchill's Collection, *Vol. VI. p.* 6.

formation

formation to that philofophical traveller, M. Poivre, formerly *Intendant* of the Iflands of France
and Bourbon. That gentleman was fo obliging
as to fend me the following account.

" In 1741 I embarked in the fhip Mars,
bound for China, and in the month of July, the
fame year, we reached the ftraits of Sunda, very
near Java, and between two fmall iflets, called
the *Great and Little Tocque*. We were .there
becalmed, and went afhore on Little Tocque
to hunt green pigeons. While the reft of the
party were clambering among the precipices, I
walked along the beach to gather fhells and
jointed corals, which are found here in great
abundance. After having made almoft an entire circuit of the iflet, it was growing late,
when a failor who accompanied me, difcovering
a deep cavern in the rocks on the brink of the
fea, went into it, and fcarce advanced two or
three fteps when he called aloud to me. I
haftened to the mouth of the cavern, and found
it darkened by an immenfe cloud of fmall birds,
which poured out like fwarms. I entered it,
and knocked down with my cane many of thefe
poor little birds, with which I was then unacquainted ; as I penetrated farther, I perceived
the roof of the cavern to be covered entirely with
fmall nefts fhaped like holy-water-pots*. The
failor

---

* Each of thefe nefts contained two or three eggs or young ones,
which lay foftly on feathers, like thofe which the parents had on
their

failor had already broken off feveral, and had filled his frock with them and with birds. I alfo detached fome of the nefts, and found them glued firmly to the rock. Night now came on, and we returned to the fhip with the fruits of our excurfion.

" The nefts which we brought were known by many of our people on board, who had made feveral voyages to China, to be the fame with thofe fo highly valued in that country. The failor kept feveral pounds, which he fold to good account at Canton. For my part, I delineated and coloured thefe birds with their nefts and their young; and I difcovered them to be real Swallows: they were about the fize of the larger kind of humming-birds (colibris).

" Since that time I have obferved, in feveral voyages, that, in the months of March and April, the feas which extend from Java to Cochin-China, and from the promontory of Sumatra to New Guinea, are covered with fifh-fpawn, which floats on the water like ftrong glue half-melted. I have learnt from the Malays, the Cochin-Chinefe, and from the natives of the Philippines and Moluccas, that this is the fubftance of which the *Salangane* conftructs its neft *. They all agree in this

their breaft. As thefe nefts foften in water, they could not withftand rain, or bear an expofure near the furface of the fea.

* It gathers the fpawn either by razing the furface of the fea,

this account. On paffing the Moluccas in April and the ftraits of Sunda in March, I fifhed up fome of this fpawn with a bucket, and after having drained off the water and dried it, I found it refembled exactly the fubftance of thofe nefts.

" About the end of July and the beginning of Auguft, it is cuftomary with the people of Cochin-China to rove the iflets, which fkirt their coaft to the diftance of twenty leagues, in fearch of the nefts of thefe little Swallows . . .

" The Salanganes are feldom ever found, but in that immenfe Archipelago which encircles the eaftern extremity of Afia . . . . All that Archipelago, where the iflets may be faid almoft to touch each other, is extremely favourable to the breeding of fifh ; their fpawn is very abundant ; the water is there warmer than in the ocean."

I have obferved feveral of thefe nefts ; they refembled the half of an oblong hollow ellipfoid, made by cutting it at right angles through the middle of the larger axis ; and the plane of this fection had ftuck to the rock : they confifted of a yellowifh white fubftance, femi-tranfparent ; compofed externally of exceeding thin and nearly concentric layers ; the infide was formed of irregular net-work, the mefhes very unequal,

or by alighting on the rocks on which it is caft and coagulated. Sometimes threads of this vifcous fubftance are feen hanging at the bills of thefe birds, and which have been fuppofed, but without foundation, to be extracted from their ftomach in the love-feafon.

and

and placed one above another, the threads being drawn from the fame fubftance with the outer layers, and much interwoven.

In the nefts which were entire, no feather could be perceived; but on cutting carefully into their fubftance, we found fome entangled, which diminifhed the tranfparency of the part. Sometimes, though much more rarely, we dif-covered fragments of egg-fhells; and almoft all of them had veftiges of the birds' excre-ments *.

I held in my mouth, a whole hour, a fcale de-tached from one of thefe nefts; it had at firft a flight faline flavour, afterwards it was infipid as pafte; it did not diffolve, but foftened and fwelled. M. Poivre alfo found it had only the tafte of ifinglafs, and he affures us that the Chinefe value it folely for its nutritious invigor-ating quality; he adds, that he never ate any thing fo rich and ftrengthening as the foup made with it and meat +. If the *Salanganes* feed on the fame fubftance with which they build their neft, and which is fo plentiful in thofe feas, and if it has the prolific property, which the Chinefe afcribe to it, no wonder that the fpe-cies is very numerous. It is faid that a thou-

* Moft of thefe obfervations were firft made by Daubenton the younger, who communicated them to me with feveral nefts of Sa-langanes, where I obferved the fame things.

+ Might not this foup owe part of its qualities to the flefh ufed in making it?

fand

fand cafks of thefe nefts are annually exported
from Batavia, having been procured from the
iflets of Cochin-China and the Eaft. Each cafk
weighs one hundred and t'wenty-five pounds,
and each neft half an ounce * ; hence the whole
muft amount to one hundred and twenty-five
thoufand pounds, and contain four millions of
nefts.

I muft confefs, however, that the philofopher
Redi†, judging from experiments made by others,
and perhaps incomplete, entertains great doubts
with regard to the reftorative virtue of thefe
nefts, which is attefted by many other writers,
who on that point agree with Poivre ‡.

Nothing better fhews that the Salangane
has remained long unknown, than the different
names beftowed on it. It is called the *Sea-
Swallow* and the *Halcyon:* its wings have been
fuppofed to be blue, and it has been reprefented
as fometimes equal in fize to the ordinary Swallows, fometimes as larger, fometimes as fmaller.
In fhort, nothing accurate was known before
M. Poivre.

---

* *Etabliffemens Europeens dans les Indes Orientales, t. I. liv. 2.*

† *See* the obfervations of Redi in the *Coll. Acad. part. etram.
t. IV. p.* 567. If it be true, as alleged, that the Hollanders begin to import thefe nefts into Europe, the fact will be foon afcertained.

‡ *Comedunt in primis ii qui in caftris venereis ftrenue fe exercere volunt.* Mufæum Wormianum, *Lib. III.* 21. " It is a great reftorative to Nature, and the luxurious Chinefe make much ufe of it."
Sprat's Hiftory of the Royal Society of London, p. 206.

Kircher

Kircher had afferted that thefe birds appear on the coaft only in the breeding feafon, and that it is uncertain where they live during the reft of the year; but M. Poivre informs us, that they remain conftantly in thofe iflets and rocks where they were hatched, and that they live like the European Swallows, only they circle rather lefs; in fact, their wings are fomewhat fhorter.

They have only two colours, blackifh on the upper furface, and whitifh on all the under fur-face, and alfo on the tips of the tail-quills; the iris is yellow, the bill black, and the. legs brown.

It is rather fmaller than the. wren; its total length two inches three lines; the tarfus as many; the hind toe the fhorteft; the tail ten lines, forked three, compofed of twelve quills, and projects three fourths of its length beyond the wings.

[A] Specific character of the *Hirundo Efculenta*: " All its tail-quills are marked with a white fpot."

XI.

The WHEAT SWALLOW.

La Grande Hirondelle Brune a Ventre Tacheté, ou Hirondelle
des Bles,* Buff.
Hirunɕo Borbonica, Gmel.

THIS bird is ufually called the Corn Swallow
in the Ifle of France; it haunts the wheat
fields, the glades of the woods, and prefers the
uplands; it frequently fits on trees and on
ftones; it follows the herds, or rather the infects which infeft them: it is alfo feen, from
time to time, flying in large troops behind the
veffels lying in the roads, and conftantly in the
purfuit of winged infects; its cry is much like
that of the common Swallow.

The Vifcount de Querhoent obferved that
the Wheat Swallows circled frequently in the
evening near a cut which had been made in a
mountain; and he thence conjectures that, like
our Sand-martins and Swifts, they pafs the night
in holes under ground, or in crevices of the
rocks. They, no doubt, breed in the fame retreats; which may be the reafon that their nefts
are unknown in the Ifle of France. The Vifcount de Querhoent could procure no account
of their incubation but from an old perfon of the
Ifle of Bourbon, and born in the country, who
told him that they fat in September and Octo-

* *i. e.* The Great Brown Swallow with a fpotted tail.

5

ber,

ber, that he had caught many on their nefts in caverns and the holes of rocks, &c. that thefe are compofed of ftraw and a few feathers, and that he never faw more than two eggs, which were gray and dotted with brown.

This Swallow is as large as our Swift; the upper furface of its body blackifh brown; the under furface gray, fprinkled with long brown fpots; the tail fquare; the bill and the legs black.

VARIETY.

THE Little Brown Swallow with a fpotted tail, from the Ifle of Bourbon, muft be regarded as a variety of fize from the preceding fpecies. It has alfo fome flight differences of colours; the upper furface of its head, of its wings, and tail, is blackifh brown; the three laft quills of the wings are terminated with dirty white, and edged with greenifh brown; this laft colour fpreads over all the reft of the upper furface; the throat and all the under fide of the body, including the inferior coverts of the tail, are marked with longitudinal brown fpots, on a gray ground.

Total length four inches nine lines; the bill feven or eight lines; the tarfus fix lines; all the nails fhort and flightly hooked; the tail near two inches, fquare, and falls about feven lines fhort of the wings.

XII.
The GRAY-RUMPED SWALLOW.

La Petite Firondelle Noire a Croupion Gris *, Buff.
Hirundo 'rancica, Gmel.

COMMERSON brought this new species from
the Isle of France; it is scarce, though
there are abundance of insects in that island; it
is even lean, and its flesh unpalatable; it lives
indifferently in the town or the country, but always near springs; it is never observed to sit;
it flies with great celerity; it is as large as a tit-
mouse, and weighs two gros and a half. The
Viscount de Querhoent saw it frequently towards the evening in the skirts of the woods,
and he thence infers that it chooses the sylvan
shades for its nocturnal retreats.

All the upper surface is of an uniform blackish, except the rump, which is whitish, and this
is likewise the colour of the under surface.

Total length four inches two lines; the bill
five lines; the tarsus four lines; the alar extent
nine inches; the tail near two inches, and, in
the subject described by Commerson, it had only
ten quills, which were nearly equal; it is ten
lines shorter than the wings, which consist of
seventeen or eighteen quills.

A specimen brought from the East Indies by

---

* *i. e.* The Little Black Swallow with a gray rump.

Sonnerat,

Sonnerat, appears to me to belong to this fpecies, or rather to form the fhade between this
fpecies and the preceding variety; for the under
fide of the body was fpotted like the latter, and
it refembled the former in the colour of the
upper fide of the body, and in the dimenfions;
only the wings projected feventeen lines beyond
the tail, and the nails were flender and hooked.

XIII.

THE

R UFOUS-RUMPED SWALLOW.

L'Hirondelle a Croupion Roux et Queue Carrée, Buff *.
Hirundo Americana, Gmel.

ALL the upper furface except the rump is
blackifh brown, with reflections fluctuat-
ing between brown green and deep blue; the
rufous colour of the rump is a little mingled,
each feather being edged with whitifh; the
quills of the tail brown; thofe of the wings alfo
brown, with fome greenifh reflections; the primaries edged interiorly with whitifh, and the
fecondaries edged with the fame colour, which
rifes a little on the outfide; all the under furface
of the body is dirty white; and the inferior coverts of the tail rufty.

* i. e. The Swallows with a rufous rump and fquare tail.

Total

Total length fix inches and a half; the bill
nine or ten lines; the tarfus five or fix; the
toes difpofed three one way and one the other ;
the alar extent about ten inches; the tail two
inches, almoft fquare at the end, and a little
fhorter than the wings.

Commerfon faw this Swallow on the banks
of the De la Plata in May 1765. He brought
from the fame country another bird which may
be regarded as a variety of this fpecies; it differed in having its throat rufty, more white than
rufous on the rump and the lower coverts of the
tail; all the quills of the tail of the wings were
deeper, with more diftinct reflections; no white
on the great quills of the wings, which projected ten lines beyond the tail, which was a
little forked; the alar extent eleven inches.

XIV.

The SHARP-TAILED BROWN SWALLOW of LOUISIANA.

Hirundo Pelafgia, Var. Gmel.

THERE are fome Swallows in America whofe
tail-quills are entirely deftitute of webs at
the ends, and terminate in a point.

The prefent was brought from Louifiana by
Lebeau ; its throat and the fore fide of the neck
were

were dirty white, fpotted with greenifh brown; all the reft of the plumage appeared of a pretty uniform brown, efpecially at firft fight; for, on a nearer infpection, we perceive that the head and the upper furface of the body, including the fuperior coverts of the wings, are of a deeper caft; the rump and the under furface of the body are lighter, the wings blackifh, edged interiorly with black; the legs are brown.

Total length four inches and three lines; the bill feven lines; the tarfus fix lines; the middle toe fix lines; the hind toe the fhorteft; the tail feventeen or eighteen lines, including the pointed fhafts, a little rounded at the end; thefe fhafts black, and four or five lines long; thofe of the middle quills of the wings largeft, and twenty-two lines fhorter than the wings.

The American Swallow of Carefby, and the Carolina Swallow of Briffon *, is much fhorter winged than that from Louifiana, but refembles it exactly in fize, in its general proportions, in its plumage, and in the fharp fhafts; and as the climate is nearly the fame, if the great difference in the length of the wings were not conftant, we might regard it as a variety of the fame fpecies. The times of its arrival in Carolina and Virginia, and of its departure from thofe countries, correfpond, fays Catefby, to the ap-

* *Hirundo Pelafgia,* Linn. and Gmel.
The *Aculeated Swallow,* Penn. and Lath.

pearance

pearance and retreat of the Swallows in England.
He fufpects that they winter in Brazil, and he
tells us that they breed in the chimneys in
Carolina.

Total length four inches and three lines; the
bill five lines, the tarfus the fame, the mid toe
fix; the tail eighteen lines, and three lines
fhorter than the wings.

The Sharp-tailed Swallow of Cayenne, called
camaria, refembles more that of Louifiana, in its
fize, than that of Carolina; its wings being
longer than the latter and fhorter than the former. On the other hand, it differs rather more
in its colours, for the upper fide of the body is
deeper brown, and verging on blue; the rump
gray; the throat and the fore part of the neck
gray, with a rufty caft; the under fide of the
body grayifh fhaded with brown; in general,
the colour of the higher parts is rather brighter
and more diftinct than that of the lower. Perhaps it is a fexual variety, efpecially as the
Cayenne Swallow has been reckoned a male.

It is faid in Guiana never to approach the
fettlements; and certainly it does not breed in
the chimneys, for there are no chimneys in that
country.

Total length four inches and feven lines; the
bill four lines, the tarfus five; the tail twenty
lines including the points, which are two or
three lines; the wings extend about an inch
beyond it.

XV.
The SHARP-TAILED BLACK SWALLOW of MARTINICO.

Hirundo Acuta, Gmel.
Hirundo Martinicana, Briff.
The *Sharp-tailed Swallow,* Lath.

IT is the fmalleft of all the Sharp-tailed Swallows; not larger than a gold crefted wren; the points which terminate the quills of the tail very fine.

All the upper fide of the head and body black without any exception, the throat gray brown, and the reft of the under fide of the body dull brown; the bill black and the legs brown.

Total length three inches and eight lines; the bill four lines, the tarfus the fame; the mid toe four lines and a half; the alar extent eight inches and eight lines, and eight lines fhorter than the wings.

END OF THE SIXTH VOLUME.

Printed in the United States
By Bookmasters